Lecture Notes in Economics and Mathematical Systems

Volume 692

This series reports on new developments in mathematical economics, economic theory, econometrics, operations research and mathematical systems.

The series welcomes proposals for:

1. Research monographs
2. Lectures on a new field or presentations of a new angle in a classical field
3. Seminars on topics of current research
4. Reports of meetings provided they are of exceptional interest and devoted to a single topic.

In the case of a research monograph, or of seminar notes, the timeliness of a manuscript may be more important than its form, which may be preliminary or tentative.

The series and the volumes published in it are indexed by Scopus and ISI (selected volumes).

For further information on the series and to submit a proposal for consideration, please contact Johannes Glaeser (Senior Editor Economics) johannes.glaeser@springer.com.

Pedro Macedo · Victor Moutinho · Mara Madaleno
Editors

Advanced Mathematical Methods for Economic Efficiency Analysis

Theory and Empirical Applications

Springer

Editors
Pedro Macedo
Department of Mathematics
University of Aveiro
Aveiro, Portugal

Victor Moutinho
Management and Economics Department
University of Beira Interior
Covilhã, Portugal

Mara Madaleno
Department of Economics, Management, Industrial Engineering and Tourism
University of Aveiro
Aveiro, Portugal

ISSN 0075-8442 ISSN 2196-9957 (electronic)
Lecture Notes in Economics and Mathematical Systems
ISBN 978-3-031-29582-9 ISBN 978-3-031-29583-6 (eBook)
https://doi.org/10.1007/978-3-031-29583-6

This Springer imprint is published by the registered company Springer Nature Switzerland AG
The registered company address is: Gewerbestrasse 11, 6330 Cham, Switzerland

Preface

The analysis of economic efficiency, since the formal works of Koopmans, Debreu, Shephard, and Farrell in the 1950s, has assumed a very important role in human activities. We would say that it is almost impossible not to discuss economic efficiency nowadays. Knowing how much we intend to produce or consume in an economy is easy. However, doing the best with the limited resources available is harder and economies need to weigh both sides, i.e., inputs and outputs, at the moment decisions are to be taken. Thus, the best combinations need to be done to ensure that economic efficiency is to be reached. Although the literature is massive in this area, two methodologies for the analysis of economic efficiency have emerged and proved to be fundamental over the years: the Data Envelopment Analysis and the Stochastic Frontier Analysis. It is on their formalisms, applications, and recent developments that this book intends to make an important contribution.

From the contact with these methodologies in our professional activities, whether in the industry or academia, was born the idea to create a book that would allow an overview of the concept of economic efficiency and both methodologies, and simultaneously reveal some of the promising recent research. It was intentional that the book is broad enough for readers who are practitioners, or just curious about the area. To guarantee the success of such endeavor, we would have to invite some of the most renowned authors in these areas. It is a great honor to have in this book some of the authors whose work has always been a scientific reference for us and which can be found on the shelves of our offices.

We are deeply grateful to all the authors who participated in the book. Without their valuable contributions and expertise, this book simply would not exist. Thank you so much! We also would like to thank Johannes Glaeser, Vijay Selvaraj and Sudhany Karthick for their patience with deadlines, guidance, and strong support in all stages of book production.

Aveiro, Portugal — Pedro Macedo
Covilhã, Portugal — Victor Moutinho
Aveiro, Portugal — Mara Madaleno

Contents

Contributors

Christine Amsler Michigan State University, East Lansing, USA

Ana S. Camanho Faculdade de Engenharia, Universidade do Porto, Porto, Portugal

Yi Yi Chen Tamkang University, New Taipei, Taiwan

Cinzia Daraio DIAG, Sapienza University of Rome, Rome, Italy

Giovanna D'Inverno Department of Economics and Management, University of Pisa, Pisa, Italy;
Faculty of Economics and Business, KU Leuven, Leuven, Belgium

William H. Greene Department of Economics, Stern School of Business, New York University, New York, USA

Samah Jradi EM Normandie University, Le Havre, France

Magdalena Kapelko Department of Logistics, Wroclaw University of Economics and Business, Wroclaw, Poland

Subal C. Kumbhakar University of Miami, Miami, USA;
Department of Economics, Binghamton University, Binghamton, USA

Daniel Pacheco Lacerda Production Engineering and Systems - PPGEPS/UNISINOS, Research Group on Modeling for Learning - GMAP | UNISINOS, São Leopoldo, Brazil

Pedro Macedo CIDMA – Center for Research and Development in Mathematics and Applications, Department of Mathematics, University of Aveiro, Aveiro, Portugal

Mara Madaleno GOVCOPP—Research Unit in Governance, Competitiveness and Public Policy, DEGEIT—Department of Economics, Management, Industrial Engineering and Tourism, University of Aveiro, Aveiro, Portugal

Mónica Meireles IBS—Iscte Business School, Business Research Unit (bru_iscte), Iscte—Instituto Universitário de Lisboa, Lisbon, Portugal

Victor Moutinho NECE—Research Unit in Business Sciences, Department of Management and Economics, University of Beira Interior, Covilhã, Portugal

Thyago C. C. Nepomuceno DIAG, Sapienza University of Rome, Rome, Italy; Campus Agreste, Federal University of Pernambuco, Caruaru, Brazil

Alfons Oude Lansink Business Economics, Wageningen University, Wageningen, The Netherlands

Christopher F. Parmeter Department of Economics, University of Miami, Coral Gables, USA

Fabio Sartori Piran Production Engineering and Systems - PPGEPS/UNISINOS, Research Group on Modeling for Learning - GMAP | UNISINOS, São Leopoldo, Brazil

John Ruggiero University of Dayton, Dayton, OH, USA

Peter Schmidt Michigan State University, East Lansing, USA

Maria Conceição Silva Católica Porto Business School, CEGE - Centro de Estudos em Gestão e Economia, Lisbon, Portugal

Alexander D. Stead Institute for Transport Studies, University of Leeds, Leeds, UK

Spiro E. Stefanou Economic Research Service, United States Department of Agriculture, Washington D.C., USA

Hung Jen Wang National Taiwan University, New Taipei, Taiwan

Phill Wheat Institute for Transport Studies, University of Leeds, Leeds, UK

Andreia Zanella Universidade Federal de Santa Catarina, Florianópolis, Brazil

Introduction

Mara Madaleno, Pedro Macedo, and Victor Moutinho

1 General Insights and Concepts

Readers will easily understand that the book **Advanced Mathematical Methods for Economic Efficiency Analysis** is all about economic efficiency measurement. The spectrum of the intended material is very broad coming across economic theory, econometrics, mathematics, statistics, and applications, among others, revealing the special importance of efficiency within the context. But what is efficiency, what is economic efficiency, and why should we care about it? **Efficiency** is usually associated with the highest level of performance, i.e., using the least amount of inputs to achieve the greatest amount of output. Therefore, efficiency requires lowering the number of needless resources used to produce a given output, including physical materials, personal time, and energy (Cambridge Advanced Learner's Dictionary & Thesaurus, 2023). It is a measurable concept typically defined using the ratio of useful output to total input.

In reality, there are daily life concepts that are used with no difference, although they mean different things. For example, the helplessness in the everyday language

M. Madaleno (✉)
GOVCOPP—Research Unit in Governance, Competitiveness and Public Policy, Department of Economics, Management, Industrial Engineering and Tourism, University of Aveiro, 3810-193 Aveiro, Portugal
e-mail: maramadaleno@ua.pt

P. Macedo
CIDMA—Center for Research and Development in Mathematics and Applications, Department of Mathematics, University of Aveiro, 3810-193 Aveiro, Portugal
e-mail: pmacedo@ua.pt

V. Moutinho
NECE—Research Unit in Business Sciences, Department of Management and Economics, University of Beira Interior, 6200-209 Covilhã, Portugal
e-mail: ferreira.moutinho@ubi.pt

P. Macedo et al. (eds.), *Advanced Mathematical Methods for Economic Efficiency Analysis*, Lecture Notes in Economics and Mathematical Systems 692,
https://doi.org/10.1007/978-3-031-29583-6_1

with the use of "efficient" is striking. Often one does not know whether one should say effectively or efficiently and uses simply the phrase "effectively and efficiently". Also, the comparative is used more efficiently, although efficient is already a superlative, provided "most efficient" is comparable to "most optimal". **Effectively** means "in a way that is successful and achieves what you want". The term **Efficiently** is used when "working or operating in an organized, quick, and effective way" (Cambridge Advanced Learner's Dictionary & Thesaurus, 2023). Another important distinction to be made as well is between the concepts of efficiency and effectiveness, which are often used interchangeably but they are not the same thing. **Efficiency** is defined as the ability to accomplish something with the least amount of wasted time, money, and effort or competency in performance. However, **effectiveness** is defined as the degree to which something is successful in producing a desired result, or successful.

When we are evaluating complex activities like the public ones such as schools, hospitals, airports, water utilities, energy, environment, and many more, efficiency and effectiveness become complementary concepts. Usually, efficiency analysis evaluates the performance of complex decision-making units (DMUs), by transforming inputs into outputs (Farrell, 1957). **Inputs** are any resources used to create goods and services. Examples of inputs include labor (workers' time), fuel, materials, buildings, financial capital, research and development, innovation, and equipment. When used properly, through a production function, inputs generate outputs. **Output** is the number of goods or services produced in a specific period. For a business producing one good, output could simply be the number of units of that good produced in each period, such as a month or a year. The **production function**, in economics, is usually an equation that expresses the relationship between the quantities of productive factors or inputs (like labor and capital) used and the amount of product (output) obtained. Therefore, reflects the amount of product that can be achieved from every combination of inputs, assuming that the most efficient available methods of production are used. Through a production function, we can measure the marginal productivity (the change in output obtained from one additional unit of input) of a particular input, but we can also determine the cheapest combination of productive factors that can be used to produce a given output.

If efficiency analysis evaluated the performance of DMUs accounting for the ability to transform inputs into outputs, effectiveness studies measure the performance of these DMUs considering predefined goals (Cherchye et al., 2019; Golany, 1988). If these goals are specified through policies, effectiveness assessment relates to the analysis of the impact of the intervention on some variables of interest, revealing the ability of the policy to influence them (Abadie & Cattaneo, 2018). In simple words, effectiveness analysis indicates if we are doing the correct thing, whereas efficiency dictates if we are doing it correctly (Asmild et al., 2007; Drucker, 1977; Førsund, 2017). Combining both perspectives in a given policy intervention is essential to detect inefficiencies at the policy level (Kornbluth, 1991; Mergoni & De Witte, 2022).

Economic efficiency, in mathematical terms, is given by a function of the ratio of the actual value of an economic variable divided by the potential value of that same economic variable. It is just a measure of how good things are economically equated to how good they could potentially be. The formula for determining economic efficiency is represented by

$$\text{Economic Efficiency} = f\left(\frac{\text{Actual Value of Economic Variable}}{\text{Potential Value of Economic Variable}}\right). \quad (1)$$

In economics, the concept of efficiency most commonly used is that of **Pareto Efficiency**, named due the Italian economist and political scientist Vilfredo Pareto (1848–1923), which is a major pillar of welfare economics. An allocation is Pareto efficient if it is impossible to, from that point, make someone better off without making someone else worse off. An outcome is said to be Pareto inefficient if it is possible to make at least one agent better off without making any other agent worse off (Debreu, 1983a, b, 1951, 1954; Dasgupta & Heal, 1980).

But, let's go step by step on the definition and purpose of economic efficiency. Assessing the efficiency of entities is a powerful means of evaluating the performance of these entities, and the performance of markets and whole economies (Debreu, 1983a, b, 1951, 1954; Dasgupta & Heal, 1980). There are various types of efficiency, comprising allocative and productive efficiency, technical efficiency, 'X' efficiency, dynamic efficiency, and social efficiency, just to mention a few. **Allocative efficiency** arises when consumers pay a market price that reflects the private marginal cost of production (the change in the producer's total cost brought about by the production of an additional unit of a good or service). The condition for allocative efficiency for a firm, for example, is to produce an output where marginal cost just equals price. **Productive efficiency** appears when an entity is combining resources in such a way as to produce a given output at the lowest possible average total cost (total cost divided by the total quantity of output produced). Costs will be minimized at the lowest point on a firm's short-run average total cost curve. This also means that the average total cost equals the marginal cost because the marginal cost always cuts the average total cost at the lowest point on the average total cost curve (Debreu, 1983a, b). To identify which output a firm would produce, and how efficient it is, we need to combine data on both costs and revenue. **Technical efficiency** relates to how much output can be obtained from a given input, such as a worker or a machine, or a specific combination of inputs. Therefore, the maximum technical efficiency occurs when output is maximized from a given quantity of inputs. The simplest way to distinguish productive and technical efficiency is to think of productive efficiency in terms of cost minimization by adjusting the mix of inputs, whereas technical efficiency is output maximization from a given mix of inputs. The **'X' efficiency** concept was originally applied to management efficiencies by Leibenstein (1960, 1978), applied particularly to circumstances where there is more or less motivation of management to maximize output, or not. Thus, in opposition to the other concepts, 'X' efficiency occurs when the output of firms, derived from a given amount of input, is the greatest it can be. It is likely to arise when entities operate in highly competitive markets where managers are motivated to produce as much as they can with the resources they have. Thus, when markets lack competitive market characteristics, as in the case of oligopolies and monopolies, there is likely to be a loss of 'X' efficiency. In this case, the output is not being maximized due to a lack of managerial motivation.

Before moving to the definitions of the rest of the economic efficiency types, we should explain some more economic concepts in the meanwhile introduced. Namely, a competitive market in economics refers to a marketplace where there are a large number of buyers and sellers and no single buyer or seller can affect the market. Other characteristics of competitive markets rely on the fact that they have no barriers to entry, contain lots of buyers and sellers, and have homogeneous products. On the other extreme, with absolute power over the price settlement, we have the monopoly market structure, characterized by a single seller, selling a unique product in the market. In a monopoly market, the seller faces no competition, as he is the sole seller of goods with no close substitutes. In the middle, we may position the oligopoly, as a market characterized by a small number of firms (small enough to give each some market power), who realize they are interdependent in their pricing and output policies. Therefore, it is placed within the other market structures presented previously and we can distinguish it from the others. An oligopoly is distinguished from perfect competition due to each oligopoly needing to have into account their interdependence, the products being heterogeneous, and oligopolists having some control over price. It is distinguished from a monopoly provided the monopolist has no rivals. These market structures' presentation is important in the sense that the long-run equilibrium in perfectly competitive markets meets two important conditions which are allocative efficiency and productive efficiency. These two conditions have important implications. On one hand, resources are allocated to their best alternative use. On the other hand, they provide the maximum satisfaction attainable by society.

From the above-mentioned types of efficiency, we are still lacking two of them. **Dynamic efficiency** as a concept emerged with the Austrian Economist Schumpeter (1934) and means technological progressiveness and innovation. The neo-classical economic theory suggests that when existing firms are in an industry, the incumbents are highly protected by barriers to entry and they will tend to be inefficient. In his very known work of 1911, The Theory of Economic Development, he argued that this is not necessarily the case. It means, firms that are highly protected are more likely to undertake risky innovation and generate dynamic efficiency. In practice, entities can benefit from four types of innovation: process innovation, product innovation, organizational innovation, and marketing innovation. Lastly, **social efficiency** occurs when all the private and external costs and benefits are taken into account while producing an extra unit. Private firms only have the incentive to consider external costs into account if they are forced to internalize them through taxation or through the purchase of a permit to pollute.

A state of economic efficiency is essentially theoretical, meaning a limit that can be somehow approached but never reached. Instead, and in mathematical terms to compute economic efficiency, economists look at the amount of loss, referred to as waste, between pure efficiency and reality to see how efficiently an economy operates. Thus, deviations from the efficiency frontier are what we compute through models. Indeed, we cannot dissociate economic efficiency from the mathematical concepts of Stochastic Frontier Analysis (SFA) and Data Envelopment Analysis (DEA), which have been the main dominant approaches in the literature for its analysis, accounting, and assessment. It is around these frontier models able to compute efficiency that this book reflects and presents the chapters in their three different parts.

Before moving to the book's composition and contents, we need to mention that the principles of economic efficiency are based on the concept that resources are scarce. **Scarcity** means that there are not sufficient resources to ensure that all aspects of an economy function at their highest capacity at all times. Considering this scarcity, in economies products must be distributed to meet the needs of that same economy in an ideal way while also limiting the amount of waste produced. When we talk about waste in this context we are talking about the inputs used for the production of the outputs. The idyllic state is related to the welfare of the population. Here, peak efficiency results in the highest level of welfare possible based on the resources available.

The most efficient production entities are those that maximize their profits or outputs having simultaneously high revenues with the minimum costs. Thus, they are productive if they choose the most suitable combination of inputs that minimizes their costs while producing as much output as possible. By doing so, they operate efficiently. When all firms in the economy do that, economists refer to it as productive efficiency. That is also why competitive markets are the most efficient market form which is known, however, hard to reach in practice. In the same vein, consumers wish to maximize their well-being or welfare, or do efficient consumption. That is to say, they wish to consume final goods which ensure the maximum satisfaction to their needs and wants, but at the minimum or lowest cost. These consumer demands will guide productive firms to produce the right quantities of consumer goods in the economy which will provide the greatest consumer satisfaction under the input costs, in agreement with the supply and demand laws as argued by Smith (1776). Additionally, when economically scarce resources are allocated across different industries and firms, each being guided by the principles of productive efficiency, in a way that ensures the right quantity of final consumer goods to individuals, then we talk about allocative efficiency. Lastly, considering that each values goods in a different way and considering that we have diminishing marginal utility, the distribution of final goods in an economy may be either efficient or inefficient. Just as a note, the **law of diminishing marginal utility** holds that as we consume more of an item, the amount of satisfaction produced by each additional unit of that good declines. A good example of teaching is glasses of water. If I am thirsty, drinking the first glass is good, the second as well, and the third starts being at cost, whereas more than that starts to be unbearable. This happens since satisfaction has a limit that we know is reached due to diminishing **marginal utility**. Marginal utility represents the change in utility gained from utilizing an additional unit of a product. As such, another type of efficiency is **distributive efficiency** which occurs when the consumer goods in an economy are distributed so that each unit is consumed by the individual who values that unit most highly compared to all other individuals. However, this type of efficiency assumes that the amount of value that individuals place on economic goods can be quantified and compared across individuals, when in reality all individuals are heterogeneous, with different needs, different satisfaction levels, and different marginal utilities.

We have mentioned the concept of welfare and related it to economic efficiency previously. But, measuring economic efficiency is often subjective, relying on assumptions about the social good, or welfare, created and how well that serves consumers. **Welfare** relates to the standard of living and relative comfort experienced by people within the economy. When the economy reaches productive and allocative efficiency simultaneously, or else, at the peak economy, the welfare of one cannot be changed upwards without consequently decreasing the welfare of another individual (Pareto Efficiency). We should bear in mind that reaching the Pareto Efficiency state in economics has nothing to do with fairness or equality considering that at this point the standard of living of all individuals within the economy may not be equal. The focus is simply on reaching a point of optimal operation considering the use of limited and scarce resources. Thus, to finalize, Pareto Efficiency is a state that when reached means that a distribution was made where one party's situation cannot be improved without making another party's worst.

A very recent literature review by Mergoni and De Witte (2022), discusses policy evaluation and efficiency in several areas of economic activity. The article provides a systematic literature review of studies investigating the effect of an interference on the efficiency of a DMU when efficiency is computed using nonparametric frontier approaches (DEA). Their findings indicate that, despite the prominent role of frontier techniques in the analysis of public sector performances and the importance of the effectiveness and the policy perspective, these two approaches have long been kept separate. They recommend the combination of efficiency and effectiveness as key elements to evaluate public interventions and detect inefficiencies at the policy level, particularly in fundamental sectors such as education, health, and the environment. Other important literature surveys of economic efficiency measurement include Murillo-Zamorano (2004), Kalirajan and Shand (1999), Battese (1992), and Førsund et al. (1980), among many others.

As previously mentioned, efficiency is not dissociated from **frontier evaluation approaches**. These offer us a mathematical formulation of the concept of technical efficiency (Førsund et al., 1980). Koopmans (1951) and Farrell (1957) refer that a combination of input and output is efficient if it is not possible to increase the level of any output without increasing also the level of at least one input, or decreasing the level of any input without decreasing the current level of at least one output. As will be explained in the chapters, two main approaches for frontier estimation can be distinguished, parametric and nonparametric approaches, according to the criterion used to specify the (functional) form. Both approaches are characterized by a number of models, the most representative models that of SFA (Aigner & Chu, 1968; Aigner et al., 1977) and the DEA (Banker et al., 1984; Charnes et al., 1978, 1981) model, respectively for the parametric and nonparametric frameworks. Provided the increased number of deterministic models already include stochastic components and vice versa, the boundary between these strands is becoming blurred (Daraio & Simar, 2007). Some of these new specifications are to be discussed in the present book chapters. More details of frontier models and recent approaches to measuring efficiency within economics are to be presented in the chapters.

Efficiency measurement deals with scarce resources within economics and the minimization of waste to maximize output. The climate change fighting and sustainability efforts and concerns created the need for agents to care as well with the concept of **eco-efficiency**. This was first introduced by the World Business Council for Sustainable Development (WBCSD) in the early 1990s. It is similar to economic efficiency but based on the concept of using fewer resources to generate more goods and services (products) while decreasing the levels of waste and environmental pollution. A popularly used definition of eco-efficiency by WBCSD is 'being achieved by the delivery of competitively priced goods and services that satisfy human needs and bring the quality of life, while progressively reducing ecological impacts and resource intensity throughout the life cycle, to a level at least in line with the Earth's estimated carrying capacity'. The WBCSD considers seven aspects of eco-efficiency, namely, (1) reducing the material intensity of goods and services; (2) reducing the energy intensity of goods and services; (3) reducing the dispersion of any toxic materials; (4) enhancing the recyclability of materials; (5) making the maximum possible utilization of renewable resources; (6) enhancing the durability (shelf time) of products; and (7) improving service intensity of goods and services. Thus, eco-efficiency is all about reducing ecological damage to a minimum while at the same time maximizing efficiency, namely, maximizing the efficiency of a company's production process. Being an eco-efficient entity means that it uses less water, material, and energy while recycling more. Those that embrace the concept also seek to eliminate hazardous emissions or by-products. Putting it simpler, they aim to reduce their ecological impact, trying to reduce the ecological load. So, eco-efficiency means being an efficient business while at the same time protecting our environment. Thus, a new paradigm in economic efficiency assessment deals with the measurement of eco-efficiency considering that awareness about the need of reducing environmental burdens emerged. As a concept, eco-efficiency means doing 'more with less'—using environmental resources more efficiently in economic processes, or else, providing a way of thinking about breaking the nexus between economic activity and environmental impacts, and therefore achieving sustainable development (Caiado et al., 2017; Lueddeckens, 2023).

2 The Book Chapters and Contents

The book is divided into three parts, the first of which is devoted to basic concepts to make the book self-contained. The second is devoted to DEA and the last to SFA. In Part 2 the topics range from stochastic DEA to multidirectional dynamic inefficiency analysis, including directional distance functions, the elimination and choice translating algorithm, benefit-of-the-doubt composite indicators, and internal benchmarking for efficiency evaluations. Part 3 includes also exciting and cutting-edge theoretical research, such as robustness, nonparametric stochastic frontier models, hierarchical panel data models, and estimation methods like corrected ordinary least squares and maximum entropy. In the following subsections, we detail further the contents of each chapter.

2.1 Part 1

Part 1 comprises three chapters, one of which develops further the concepts with which we are dealing through the book, namely, *Production Economics and Economic Efficiency*, by Mónica Meireles, the second refers to *Data Envelopment Analysis: A Review and Synthesis* written by Ana S. Camanho and Giovanna D'Inverno, being the third devoted to *Stochastic Frontier Analysis: A Review and Synthesis* embracing the eco-efficiency concept by Mara Madaleno and Victor Moutinho.

This Chap. 3 in Part 1 focuses on the different efficiency concepts, highlighting their special importance and continuing this introductory chapter. Here the author details the main differences between the meanings of efficiency and effectiveness often confused and misunderstood. A literature review of efficiency is provided based on the seminal works of Debreu (1951, 1954) and Dasgupta and Heal (1979) by addressing the significance and meaning of efficiency in their works. Also, the productivity concept is distinguished from that of efficiency. A review of efficiency and productive measures is to be presented. Thus, the Malmquist and the Luenberger Productivity Indexes, as well as the DEA and SFA methodologies, as productive measures, based on the huge importance of assessing and improving the efficiency of producers, as well as comparing it between different producers, are to be presented and discussed. To highlight to the reader the importance of efficiency in economics, and to explain the importance and applicability of the efficiency and productivity concepts and methodologies, this chapter also presents some real-world applications of efficiency measures, from different perspectives such as those of energy, environment, health, banking, education, and tourism, the areas where efficiency has been highly explored.

As mentioned previously DEA and SFA are the two approaches that are more commonly applied in empirical research to assess efficiency. Thus, Chap. 3 in this Part 1 introduces the main concepts and models underlying the evaluation of efficiency using the Data Envelopment Analysis (DEA) technique. By performing initially an historical overview of the origin of DEA models, the authors present the theory underlying the representation of the technology of production and that of the efficient frontier. Before presenting recent DEA developments, the main models for evaluating efficiency are reviewed, also including a discussion of well-established and emerging areas of analysis. Applications of successful management strategies using DEA and policy measures implemented are also examined. While including a literature review the authors end up pointing directions for future research.

Finally, the last chapter of Part 1 intends to define SFA models and besides their usefulness in efficiency assessment, the authors present a literature review of articles using SFA in Eco-Efficiency (EE) assessment within environmental and economic fields. A comparison between SFA and DEA is presented, discussing its advantages and disadvantages comparatively. Being as well a literature review of SFA applications in EE, future research ideas were identified and shared with readers. Results allowed the authors to favor the SFA methodology and evidence the existence of more room to implement SFA in EE assessment.

2.2 Part 2

Part 2 is composed of five chapters, one from Thyago Nepomuceno and Cinzia Daraio about *Combining Directional Distances and ELECTRE Multicriteria Decision Analysis for Preferable Assessments of Efficiency*, a second which explores *Benefit-of-the-Doubt Composite Indicators and use of Weight Restrictions* written by Ana S. Camanho, Andreia Zanella, and Victor Moutinho, a third concerning *Multidirectional Dynamic Inefficiency Analysis: An Extension to Include Corporate Social Responsibility* written by Magdalena Kapelko, Alfons Oude Lansink, and Spiro E. Stefanou, a fourth which was named simply *Stochastic DEA*, written by Samah Jradi and John Ruggiero, and finally a fifth of *Internal Benchmarking for Efficiency Evaluations using Data Envelopment Analysis: A Review of Applications and Directions for Future Research*, from the authors Fabio Sartori Piran, Ana S. Camanho, Maria Conceição Silva, and Daniel Pacheco Lacerda.

Chapter 5 in Part 2 starts by evidencing that traditional nonparametric frontier models used to asses technical, allocative, cost, and scale efficiencies, based on DEA, reflect not only the most favorable way of weighing outputs over inputs but also tradeoffs of compensations among the many production possibilities. These tradeoffs may impede the correct estimation of efficiency, implying an incorrect way of doing an evaluation. This is particularly evident when managers or policymakers have an explicit preference for some production resources or products. Moreover, some DMUs' good performance on some production variables may offset the bad performance on others, which could result in a bad qualification of DMUs as efficient (or less inefficient) in most DEA rankings, but not under the subjective perspective of the decision-maker, as with non-discretionary inputs, bad outputs, or less desirable production configurations. With these conflicts in mind, the authors discuss this issue offering a perspective on how we can advance in this avenue by developing multicriteria non-compensatory directions for the expansion of outputs or contraction of inputs. In the end, a numerical example was provided.

In Chap. 6 the construction of Benefit-of-the-Doubt Composite Indicators (BoD CI) is discussed. These allow the aggregation of individual indicators to obtain an overall measure of performance, forcing frontier methods to reflect the relative performance of multidimensional concepts beyond the traditional production setting involving the transformation of inputs into outputs. Reviewing alternative formulations of CI, it is included the Directional BoD CI based on a Directional Distance

Function model, which allows the aggregation of desirable and undesirable indicators. As pointed out by the authors, CI models often require the specification of weight restrictions to reflect the relative importance of indicators and so alternative formulations for indicator-level and category-level restrictions are discussed. The use of virtual weight restrictions advantages and limitations, expressing the importance of indicators in percentage terms, are also explored. This chapter ends with an empirical application of assessments involving Directional Composite Indicators with weight restrictions.

The Chap. 7 contributes to research on inefficiency measurement by developing a method for evaluating inefficiency considering firms' corporate social responsibility (CSR) engagement. The applied method is based on the dynamic multidirectional inefficiency analysis (dynamic MEA). Here, inefficiency is reached through adjustments in inputs, outputs, and investments in proportion to the improvement potential defined by an ideal input–output-investment point. Authors recognize that including CSR in the production, function is not new, but the novelty of their work relies on doing that through the dynamic MEA method provides a novel contribution to this field of research. The authors do an empirical exercise using data from European firms in three different industries during 2010–2017. Results demonstrate that the highest inefficiency source is related to investments, independently of the industry. Additionally, it is argued that the lowest dynamic inefficiencies occur for other industries, followed by consumption, and finally, capital, relating these results to different pressures put by the firms' stakeholders on CSR engagement within specific industries.

Chapter 8 is devoted to stochastic DEA (SDEA) presenting approaches for handling both the normal/half-normal and normal/exponential models using SDEA. They start by presenting the SDEA methodology. Afterward, they consider the additional structure allowing them to estimate the most likely quantile consistent with the production frontier under both distributional assumptions for technical efficiency, providing an alternative measure of firm-level technical efficiency. Moreover, the authors introduce a measure of individual firm efficiency relative to the median which, following the authors, provides consistent measures of technical efficiency across all estimators including DEA. Even so, the authors recognize that the measure will also be contaminated by statistical noise which can be a drawback.

Finally, in Part 2 we have Chap. 9 pointing out that the literature often neglects the possibility of using DEA within an organization when comparable units are not available. This is attributed to the fact that efficiency evaluations based on DEA are often associated with external benchmarking, requiring an expressive sample of comparable firms and access to sensitive information. In practice, organizations present unique characteristics that make it challenging to find appropriate comparators. Having this point in mind, internal benchmarking represents an alternative that enables conducting relative efficiency assessments by introducing the time dimension in the assessment of a single firm. The chapter provides a literature review of internal longitudinal benchmarking assessments conducted with DEA, exploring applications in different sectors, and analyzing the conditions under which the use of DEA for internal benchmarking is appropriate, presenting advantages and disadvantages.

2.3 Part 3

Part 3 is also composed of five chapters. It starts with *Recent Advances in the Construction of Nonparametric Stochastic Frontier Models* prepared by Christopher F. Parmeter and Subal C. Kumbhakar, a second about *A Hierarchical Panel Data Model for the Estimation of Stochastic Metafrontiers: Computational Issues and an Empirical Application* from Christine Amsler, Yi Yi Chen, Peter Schmidt, and Hung Jen Wang, the third about *Robustness in Stochastic Frontier Analysis* written by Alexander D. Stead, Phill Wheat, and William H. Greene, a fourth which explores *Is it MOLS or COLS?* by Christopher F. Parmeter, and finally a joint work of the current book editors *Stochastic Frontier Analysis with Maximum Entropy Estimation*, namely Pedro Macedo, Mara Madaleno, and Victor Moutinho.

The first chapter in Part 3 explores the literature growth of semi- and nonparametric methods to estimate the stochastic frontier model. This chapter provides a critical analysis of this burgeoning and important literature, highlighting the different approaches to achieving near-nonparametric identification. Thus, the chapter curates the large literature using a consistent notation and describes the pros and cons of the available estimators for various features of the stochastic frontier model. After, the importance of the relaxation of various modeling assumptions, issues of implementation, and interpretation are offered to ease access to these approaches. In the end, insights into what to date has seen limited focus, and inference, are provided along with avenues for future research.

For Chap. 11 the authors start highlighting that in the meta-frontier literature, firms are placed into groups, generally defined by technology or geography. Considering that each group has its technological frontier, the meta-frontier is the upper bound of these group frontiers. This literature aims to measure a firm's inefficiency and to decompose it into its inefficiency relative to its group's frontier and the inefficiency of its group's frontier relative to the meta-frontier. Besides presenting the hierarchical stochastic frontier model, where the hierarchy is firms in groups in the overall set of groups, the chapter presents an empirical implementation of this model, emphasizing computational issues.

Robustness in the context of stochastic frontier analysis, and alternative models and estimation methods, that appear more robust to outliers, is the focus of Chap. 12 in Part 3. Putting forward that several models assuming heavy-tailed noise distributions appeared in the literature, including the logistic, Laplace, and Student's t distributions, the authors based their exploration around the fact that even so there has been little explicit discussion of what is meant by 'robustness' and how models might be compared in terms of robustness to outliers. Thus, this chapter discusses two different aspects of robustness in stochastic frontier analysis. Initially, they explore the robustness of parameter estimates, by comparing the influence of outlying observations across different specifications—a familiar approach in the wider literature on robust estimation. Finally, the robustness of efficiency predictions to outliers across different specifications was also explored.

Chapter 13 of Part 3 assesses the terminology of modified and corrected ordinary least squares (MOLS/COLS) in efficiency analysis. These two approaches, while different, are often conflated. Beyond this several remarks on the practicality and utility of the methods are provided. The author concludes that given that the type of frontier model (stochastic or deterministic) being deployed is more important than the amount of adjustment (in conditional mean), this should be the dominant force driving the information conveyed to a reader/listener when using either of the terms.

Finally, the last chapter of Part 3, Chap. 14, concentrates on maximum entropy estimation in SFA. Considering that maximum entropy in the estimation of parameters in stochastic production frontier models could be an attractive procedure in economic efficiency analysis, this chapter reviews the generalized maximum entropy and the generalized cross entropy estimators. Their implementation in the stochastic frontier analysis is discussed, including advantages and possible concerns, and the chapter ends with an application to eco-efficiency analysis of European countries to illustrate the procedures of maximum entropy estimation.

3 Concluding Remarks

The goal of this introductory chapter was to reveal to readers the use of frontier methods to assess economic efficiency. It starts by describing economic concepts of efficiency with in this chapter of Part 1 being its extension. Detailed elements based on authors' abstracts were as well presented and from the reading of the following chapters, it will be clear the usefulness and broad applications of efficiency measurements in different economic activity sectors. Different methodologies to explore efficiency were presented, discussed in comparative terms, and applied empirically in different contexts evidencing concrete results, whereas other chapters explore different specifications of efficiency assessment, presenting empirical results as well or by simply presenting literature reviews and opening room for more future research possibilities.

Acknowledgements This work is supported by the Center for Research and Development in Mathematics and Applications (CIDMA), the Research Unit on Governance, Competitiveness and Public Policies (GOVCOPP), and the Research Unit in Business Science and Economics (NECE-UBI) through the Portuguese Foundation for Science and Technology (FCT—Fundação para a Ciência e a Tecnologia), references UIDB/04106/2020, UIDB/04058/2020 and UID/GES/04630/2021, respectively.

References

Abadie, A., & Cattaneo, M. (2018). Econometric methods for program evaluation. *Annual Review of Economics, 10*, 465–503.

Aigner, D., & Chu, S. (1968). On estimating the industry production function. *The American Economic Review, 58*(4), 826–839.

Aigner, D., Lovell, C., & Schmidt, P. (1977). Formulation and estimation of stochastic frontier production function models. *Journal of Econometrics, 6*(1), 21–37.

Asmild, M., Paradi, J., Reese, D., & Tam, F. (2007). Measuring overall efficiency and effectiveness using DEA. *European Journal of Operational Research, 178*(1), 305–321.

Banker, R., Charnes, A., & Cooper, W. (1984). Some models for estimating technical and scale inefficiencies in data envelopment analysis. *Management Science, 30*(9), 1078–1092.

Battese, G. E. (1992). Frontier production functions and technical efficiency: A survey of empirical applications in agricultural economics. *Agricultural Economics, 7*(3–4), 185–208.

Caiado, R. G. G., Dias, R. de F., Mattos, L. V., Quelhas, O. L. G., & Filho, W. L. (2017). Towards sustainable development through the perspective of eco-efficiency—A systematic literature review. *Journal of Cleaner Production, 165*, 890–904. https://doi.org/10.1016/j.jclepro.2017.07.166.

Cambridge Advanced Learner's Dictionary & Thesaurus. (2023). Definition of efficiency from the Cambridge Advanced Learner's Dictionary & Thesaurus © Cambridge University Press. Retrieved from https://dictionary.cambridge.org/dictionary/english/efficiency.

Charnes, A., Cooper, W., & Rhodes, E. (1978). Measuring the efficiency of decision making units. *European Journal of Operational Research, 2*(6), 429–444.

Charnes, A., Cooper, W., & Rhodes, E. (1981). Evaluating program and managerial efficiency: An application of data envelopment analysis to program follow through. *Management Science, 27*(6), 668–697.

Cherchye, L., De Witte, K., & Perelman, S. (2019). A unified productivity-performance approach applied to secondary schools. *Journal of the Operational Research Society, 70*(9), 1522–1537.

Daraio, C., & Simar, L. (2007). *Advanced robust and nonparametric methods in efficiency analysis: Methodology and applications*. Springer Science & Business Media.

Dasgupta, P. & Heal, G. M. (1979). *Economic Theory and Exhaustible Resources*. Cambridge economic handbooks. Publisher J. Nisbet, ISBN: 0720203120, 9780720203127.

Dasgupta, P., & Heal, G. (1980). *Economic theory and exhaustible resources (Cambridge Economic Handbooks)*. Cambridge University Press. https://doi.org/10.1017/CBO9780511628375

Debreu, G. (1951). The coefficient of resource utilization. *Econometrica, 19*, 273–292.

Debreu, G. (1954). A classical tax-subsidy problem. *Econometrica, 22*, 14–22.

Debreu, G. (1983a). The coefficient of resource utilization. In G. Debreu & W. Hildenbrand (Authors), *Mathematical economics: Twenty papers of gerard debreu* (Econometric Society Monographs, pp. 30–49). Cambridge: Cambridge University Press. https://doi.org/10.1017/CCOL052123736X.002

Debreu, G. (1983b). A classical tax-subsidy problem. In G. Debreu & W. Hildenbrand (Authors), *Mathematical economics: Twenty papers of gerard debreu* (Econometric Society Monographs, pp. 59–67). Cambridge: Cambridge University Press. https://doi.org/10.1017/CCOL052123736X.004

Drucker, P. (1977). *An introductory view of management*. Harper & Row.

Farrell, M. (1957). The measurement of productive efficiency. *Journal of the Royal Statistical Society: Series A (general), 120*(3), 253–281.

Førsund, F. (2017). Measuring effectiveness of production in the public sector. *Omega, 73*, 93–103.

Førsund, F., Lovell, C., & Schmidt, P. (1980). A survey of frontier production functions and of their relationship to efficiency measurement. *Journal of Econometrics, 13*(1), 5–25.

Golany, B. (1988). An interactive MOLP procedure for the extension of DEA to effectiveness analysis. *Journal of the Operational Research Society, 39*(8), 725–734.

Kalirajan, K. P., & Shand, R. T. (1999). Frontier production functions and technical efficiency measures. *Journal of Economic Surveys, 13*(2), 149–172.

Koopmans, T. (1951). Efficient allocation of resources. *Econometrica: Journal of the Econometric Society, 19*, 4, 455–465.

Kornbluth, J. (1991). Analysing policy effectiveness using cone restricted data envelopment analysis. *Journal of the Operational Research Society, 42*(12), 1097–1104.

Leibenstein, H. (1960). *Economic backwardness and economic growth: Studies in the theory of economic development*. Wiley.

Leibenstein, H. (1978). *General X-efficiency theory and economic development*. Oxford University Press. ISBN 978–0–19–502380–0.

Lueddeckens, S. (2023). A review on the handling of discounting in eco-efficiency analysis. *Clean Technologies and Environmental Policy, 25*, 3–20. https://doi.org/10.1007/s10098-022-02397-9

Mergoni, A., & De Witte, K. (2022). Policy evaluation and efficiency: A systematic literature review. *International Transactions in Operational Research, 29*, 1337–1359. https://doi.org/10.1111/itor.13012

Murillo-Zamorano, L. R. (2004). Economic efficiency and frontier techniques. *Journal of Economic Surveys, 18*(1), 33–77. https://doi.org/10.1111/j.1467-6419.2004.00215.x

Schumpeter, J. A. (1911). *The theory of economic development*. Harvard University Press.

Schumpeter, J. A. (1934). *The theory of economic development: An inquiry into profits, capital, credits, interest, and the business cycle*. Transaction Publishers.

Smith, A. (1776). *The Wealth of Nations*. W. Strahan and T. Cadell, London. Smith, Adam (1776). *An Inquiry into the Nature and Causes of the Wealth of Nations* (Vol. 1, 1 ed.). London: W. Strahan. ISBN 978–1537480787.

Part 1

Production Economics and Economic Efficiency

Mónica Meireles

This introductory chapter focuses on the different efficiency concepts, which are the base for the following chapters, highlighting their special importance. A particular emphasis is devoted to clarifying the main differences between the meanings of efficiency and effectiveness often confused and misunderstood. Exploring the main literature review on efficiency, this chapter stresses the studies of Debreu (1951, 1954) and Dasgupta and Heal (1979) by addressing the significance and meaning of efficiency in their works. Another concept often confused with efficiency is productivity, which is also clarified and analyzed. Based on the huge importance of assessing and improving the efficiency of producers, as well as comparing it between different producers, it also provides a review of efficiency and productive measures, such as the Malmquist and the Luenberger Productivity Indexes, as well as the DEA and SFA methodologies. In order to better explain the importance and applicability of the efficiency and productivity concepts and methodologies, this chapter also mentions some real-world applications of efficiency measures, for instance on energy, environment, health, banking, education and tourism.

Efficiency and **effectiveness** are fundamental concepts in assessing a firm's performance. Although being different concepts, they are often confused and misunderstood. Efficiency focuses on the firm's ability to improve its competitive advantages, through earnings appropriation. Effectiveness is the firm's ability to create new growth opportunities in the market, through differentiation and innovation. Therefore, efficiency is a measure of operational excellence or productivity, and thus it is concerned with minimizing costs and improving operational margins, whereas effectiveness is related to the firm's own strategy to generate sustainable production

M. Meireles (✉)
IBS—Iscte Business School, Business Research Unit (bru_iscte), Iscte—Instituto Universitário de Lisboa, Av. Das Forças Armadas, 1649-026 Lisbon, Portugal
e-mail: monica.meireles@iscte-iul.pt

P. Macedo et al. (eds.), *Advanced Mathematical Methods for Economic Efficiency Analysis*, Lecture Notes in Economics and Mathematical Systems 692,
https://doi.org/10.1007/978-3-031-29583-6_2

growth and its capacity of achieving the defined goals (Mouzas, 2006). In the literature, there is a consensus that efficiency is obtained by minimizing the inputs or maximizing the outputs, where both inputs and outputs should be correlated (Rebelo et al., 2013).

The concept of efficiency is thus related to the notion of performance and productivity, which are relative concepts that convert inputs into outputs. The performance is obtained through the ratio of outputs to inputs, the so-called productivity ratio, whose values show a better performance the larger they are (Coelli et al., 2005). Productivity refers to the total factor productivity that involves all production factors. Therefore, any measure of productivity related to just one factor is a partial productivity measure that can mislead and misrepresent the overall productivity or performance of a firm when considered in isolation (Coelli et al., 2005).

Productivity is a static or level concept that can be measured to compare a firm's performance at a given point in time, allowing measuring productivity level differences across firms. Conversely, productivity change is a dynamic concept that assesses the evolution of the productivity performance of a firm or industry over time (Coelli et al., 2005).

Productivity growth has been of interest to researchers and policymakers, as it is the engine that drives economic prosperity, the standards of living and the competitiveness of a country (Lin et al., 2013). Two main theories have been proposed to explain productivity growth: the convergence theory and the endogenous growth theory. The former states that there is a general tendency for the *per capita* income or total factor productivity (TFP) in low-income countries to converge towards those of high-income countries. The rationale behind this theory lies on the concept of diminishing returns to scale, as demonstrated in the work of Solow (1956). The latter claims that *per capita* income or productivity of low- and high-income countries stays constant or diverges over time. The rationale behind this theory is based on the concept of increasing returns to scale advocated by Arrow (1962) and further developed by Romer (1986) and Lucas (1988). According to this theory, the increasing returns to scale result from the externalities associated with the acquisition of technical knowledge. Therefore, even if a firm faces diminishing returns, the spillover effect allows technical knowledge to diffuse, resulting in the exhibition of increasing returns to scale at the aggregate level (Lin et al., 2013).

Although efficiency and productivity have been often used as synonymous, they are not exactly the same thing (Coelli et al., 2005). In the literature, several studies can be found that try to define and measure efficiency and productivity. Initially, economists attributed productivity changes to **technological changes**, that is, to shifts in the production frontier. In the 1980s, it became accepted that productivity change could also be caused by **efficiency change**, that is, by shifts over time of firms relative to their frontier (Hollingsworth, 2008). Consequently, productivity is a broad concept that embraces technological changes and different efficiency concepts.

Efficiency concepts

In the literature, several studies can be found using different efficiency concepts. Sometimes, those concepts, used in different contexts, might result in some confusion and misperception. Therefore, it is important to clarify and elucidate these concepts.

Based on the seminal definitions of efficiency by Farrell (1957), let's assume that in a production process, a single input is used to produce a single output. The production function (frontier) reflects the state of the art of the technology in that industry. This frontier concept is especially important for the analysis of efficiency since efficiency is measured as the relative distance to the frontier (Ji & Lee, 2010). Firms that operate along the technology defined by that frontier are **technically or technologically efficient**, while those beneath it are technically inefficient since technically they could increase output, until reaching the production frontier, without requiring more input. Therefore, the production frontier represents the maximum output that is produced from all input combinations (output-oriented perspective) or a given output that is produced with the minimum quantity of inputs (input-oriented perspective). If technical efficiency is equal to 1, then the firm is technically efficient. If it is inferior to 1, then the firm is technically inefficient. Thus, the smaller the technical efficiency is, the more inefficient the firm will be. Nevertheless, a technically efficient firm does not mean that it is maximizing its productivity. Indeed, productivity is measured by the slope of the ray y/x (y stands for outputs and x for inputs), where the greater the slope the higher the productivity. If the ray from the origin is tangential to the production function, then that point is the feasible production point that maximizes productivity, representing the technically optimal productive scale (TOPS) or **scale efficiency**. Any firm on any other point in the production frontier, although also technically efficient, is less productive, and thus is not maximizing productivity, due to the scale effects. Indeed, any firm located at any point to the left of this TOPS point is operating in the increasing returns to scale portion of the production frontier, which can improve its productivity by increasing its economies of scale towards point TOPS. Those firms on the right side of that point are operating in the decreasing returns to scale portion, which means they can increase their productivity by decreasing their scale of operation towards point TOPS.

If time is considered, productivity can change with advances in technology. This **technical or technological change** implies an upward shift in the production frontier. In this case, all firms can technically produce more output for each input level, than previously.

Besides these physical and technical relationship concepts, there are others related to costs and profits (Coelli et al., 2005). If the information on prices is available, in addition to technical efficiency, it is possible to consider **allocative efficiency.** This efficiency occurs when, given the input prices, the input combination minimizes cost, or when, given the output prices, the output combination maximizes revenue.

Both technical and allocative efficiency concepts comprise total **economic efficiency** (Hollingsworth, 2008). The former expresses the firm's ability to reach the maximum output from a given set of inputs, whereas the last express the firm's ability to use the inputs at optimal proportions given their prices and production technology.

Therefore, a firm is economically efficient if it knows both how to obtain maximum outputs from given inputs and how to choose input and output mixes to maximize revenues and minimize costs, thus maximizing profit.

In summary, technological change means all the feasible combinations of inputs and outputs quantities that can expand or contract depending on the exogenous environment. Technical efficiency change means that the firm moves closer to or further away from this boundary. These two concepts of productivity change are independent of each other. An increase in the first means that the technological frontier has moved (measures the frontier-shift effect or innovation), whereas the second means that the firm's position relative to the frontier has changed (measures the catch-up effect). Both components allow testing the two previously referred opposing theories of productivity growth: the endogenous growth theory and the convergence growth theory. Therefore, the increase in productivity can result from *technical change* (TC), which results from a shift in the production technology; from *technical efficiency change* (TEC), which results from the firm's ability to operate closer to the technology frontier given the available technology; from *scale efficiency change* (SEC), which results from improvements in the scale of operations or the scale efficiency towards the technologically optimum scale (TOPS) point; from *allocative efficiency change* (AEC), which results from the firm's ability to make an optimal proportion use of its inputs given their prices and the available technology or, in case of multi-output firms, from the ability of the firm to exploit economies of scope with changes in the output mix, the *output mix effect*. In case the technology only exhibits constant returns to scale (CRS), then the only sources of productivity change are the technical change and the technical and allocative efficiency.

Therefore, when a productivity increase is observed, it might have resulted either from an increase in technical efficiency, an increase in allocative efficiency, technical change, the exploitation of economies of scale, or a combination of these four factors. If there are no reliable price data, the allocative efficiency measures are not possible and the efficiency and productivity measures become restricted to technical change, technical efficiency and scale efficiency (Coelli et al., 2005).

Earlier, Debreu (1951) had intended to measure the efficiency of the economy by introducing the concept of the coefficient of resource utilization, ρ. This coefficient is measured through the distance between the actual physical quantity resource and the optimal quantity resource, which represents the non-utilized resources. To evaluate such a distance, this difference is multiplied by the price, for each commodity. Therefore, this distance function reaches its minimum for an optimum quantity by reducing all non-optimal quantities by the ratio ρ. This coefficient of resource utilization of the economy, ρ, is the smallest fraction of all available physical resources that would allow each level of consumption to benefit from at least the same satisfaction as before. For a Pareto optimal situation, ρ equals one, whereas for a non-optimal situation, it is smaller than one. Therefore, ρ measures the efficiency of the economy by measuring the underemployment of physical resources, the technical inefficiency and the inefficiency of the economic organization. Hence, economic efficiency refers to a situation where, multiplying the quantities of all available resources by a fraction

ρ and maintaining the same technical/technological knowledge as before, at least the same individual satisfaction level is obtained.

Following these works, Dasgupta and Heal (1979), in their economic theory and exhaustible resources study, analyze the allocative efficiency concept in the use of exhaustible natural resources, particularly energy resources. According to the authors, to solve the exhaustible resources problem the market system needs to produce an optimal allocation of these resources over time. It is this efficiency allocation that will allow the depletion of the resource at an optimal rate.

Productivity measures

To measure productivity, four methods are usually used: the least-squares econometric production models, the total factor productivity (TFP) indices, the data envelopment analysis (DEA) and the stochastic frontier analysis (SFA). The first two measure the technical change and/or the TFP and assume that all firms are technically efficient. The last two provide relative efficiency measures among firms. The DEA is a non-parametric method, while the SFA is a parametric method. In the former, the deterministic specifications can be solved by using mathematical programming or econometric techniques. In the latter, the stochastic specifications can only be estimated by econometric techniques (Murillo-Zamorano, 2004). These two methods can also be distinguished by other features namely their data requirements, their behavioral assumptions and whether they recognize random errors in the data or not (Coelli et al., 2005). These characteristics will be further discussed in this and later chapters.

Regarding total factor productivity (TFP), it can be measured through different approaches, namely the Hicks-Moorsteen approach, the Malmquist approach, the profitability ratios or the source-based approach. The type of approach to be selected depends on the purpose of the measure. If we intend to measure productivity changes without needing to specify their sources, then both the Hicks-Moorsteen approach and the Malmquist productivity index are the most appropriate methods. However, if our aim is at measuring productivity change in a more business-oriented way, then the profitability ratios are the most adequate. Nevertheless, if we choose to measure the TFP change using the Malmquist index, we need a large number of firm-level data on input and output quantities. In case there is only data for a single firm over time, then only the Hicks-Moorsteen approach is feasible (Coelli et al., 2005).

When there is more than one input, the productivity calculation is not so straightforward. In these situations, an index of inputs that aggregates all the inputs is needed to obtain a ratio measure of productivity (Coelli et al., 2005). Therefore, productivity growth has been measured using index numbers. The index numbers are the most widely used instruments to measure changes in levels of several economic variables from a reference period, the so-called "base period", relative to the "current period". Some examples are the consumer price index (CPI), price deflators, indices of import and export prices and financial indices like the Dow Jones Index. The index numbers are used in DEA and SFA to evaluate efficiency and productivity change. Indeed, applying the DEA and SFA typically requires the use of a large number of input and output variables that usually results in the loss of degrees of freedom. To avoid this estimation problem it is important to use index numbers that aggregate the data into a smaller number of inputs and outputs. Some examples of index numbers are the Laspeyres, Paasche, Fisher and Törnqvist, which focus on price or quantities. Laspeyres uses the quantities of the base period as weights, while Paasch uses the quantities of the current period as weights. The Fisher index is a geometric mean of the Laspeyres and Paasche indices. The Törnqvist is a weighted geometric average of the price relatives (for example, a convenient computational form is a logarithmic change in the price), with the weights given by the average of the value shares in both base and current periods. It is one of the most used indices in TFP studies to measure productivity (Coelli et al., 2005).

The use of index numbers has a limitation, though. It requires data on prices for all inputs and outputs and price information does not always exist, in particular, if we are considering undesirable outputs. Furthermore, the productivity index ignores the contribution of scale economies and the differences in technology by considering all the Decision-Making Units (DMUs) as homogeneous, thus using the same production technology (Lin et al., 2013). To overcome these problems productivity can be measured using a distance function since it only requires quantities for all inputs and outputs.

The distance function concept is related to efficiency and the production function concept, as it measures deviations from the boundary of technology. It was first introduced by Malmquist (1953) and Shephard (1953) and allows for measuring efficiency and productivity. It has been widely employed to estimate energy and environmental efficiency and the shadow prices of pollutants because it can provide a total-factor efficiency indicator and can include undesirable outputs in the model (Choi et al., 2012). The distance function can be estimated using the non-parametric DEA approach and the parametric SFA approach. Zhou et al. (2012), for instance, use the Shephard energy distance function to estimate the economy-wide energy efficiency performance from a production efficiency perspective. These types of Shephard distance functions are based on the concept of radial efficiency measure that assumes a proportional adjustment for all inputs or outputs. The big advantage of the distance functions is that they allow describing a production technology without the need to specify a behavioral objective, such as cost minimization or profit maximization. They can be estimated through econometric or mathematical programming methods. Chung et al. (1997) extended the classical Shephard's output distance function to the directional output distance function. The directional distance function measures the smallest changes in inputs and outputs in a given direction, which are necessary for a producer to reach the production frontier (Barros et al., 2008). The merit of this function is that it can be used to measure the DMU efficiency and productivity in increasing one output and contracting another output simultaneously. Its use is flexible due to the variety of direction vectors it allows for (Briec & Kerstens, 2009). However, this conventional directional distance function is a radial efficiency measure that may overestimate efficiency when there is some slack. To overcome this limitation, non-radial efficiency measures are often encouraged.

In the literature, the most popular approach in estimating productivity has been the non-parametric methods—the DEA and the Malmquist productivity index. The non-parametric methods have the advantage of not imposing an a priori functional form to the technology, nor any restrictive assumptions regarding input remuneration. Furthermore, the frontier nature of these technologies allows for capturing any productive inefficiency and offers a benchmarking perspective. Nevertheless, there are distinctions between productivity measures based on ratios (indices) and differences (indicators). The Luenberger productivity indicator is an example of a difference-based productivity indicator. It has the advantage of allowing the evaluation of profit-maximizing organizations. The Luenberger productivity indicators embrace the Malmquist productivity approach and, if necessary, can also specialize in an input- or output-oriented perspective according to the cost minimization or revenue maximization cases (Barros et al., 2008). These approaches will be further developed later in this chapter.

In summary, to measure productivity changes, index numbers can be used to measure the changes in the levels of outputs produced and inputs used in the production process over two periods of time or across two firms. The calculation of the TFP index, as well as of the DEA and SFA, requires the use of index numbers (Coelli et al., 2005). For productivity measurement, production technology plays a crucial role as it allows transforming a vector of inputs into a vector of outputs. For allocative efficiency, the prices are essential for determining the composition of the inputs and outputs that maximize profits or revenues and minimize costs (Coelli et al., 2005).

The Malmquist TFP Index

In the literature, there are two conventional indices used to investigate efficiency change: the Malmquist index and the Luenberger index.

The Malmquist TFP Index was first proposed by Caves et al. (1982) as a ratio of two distance functions to measure productivity, and was further developed by Färe et al. (1994). It results from defining the TFP Index by using the Malmquist input and output distance functions. It measures the radial distance of the observed output and input vectors in two different periods with respect to a reference technology. Since the distances can be either output- or input-oriented, the Malmquist indices differ according to the orientation adopted. The productivity, measured through the output, focuses on the maximum output level that could be produced with a given input vector and a given production technology. The productivity, measured through the input, focuses on the minimum level of inputs needed to produce the observed output vectors under a reference technology. This Index has become a popular approach to computing the TFP index to measure productivity change, though it is rather incomplete as it only captures technological change and technical efficiency change (Coelli et al., 2005). If the production technology has constant returns to scale, the Malmquist productivity index can be interpreted as a TFP index (Zhou et al., 2010).

In 1994, Färe et al. extended the original Malmquist productivity index by calculating it within a non-parametric framework for 17 OECD (Organization for Economic Co-operation and Development) countries. This method has several advantages: it is more flexible than other techniques in estimating productivity because no a priori technology function is needed nor any limitation on input remuneration is imposed, it can capture productive inefficiency and it can provide a standard baseline for comparison (Lin et al., 2013).

An interesting study was conducted by Zhang and Choi (2013) to overcome the already-mentioned problem associated to the conventional directional distance function. As a radial efficiency measure, it can overestimate efficiency when there is some slack. Therefore, they suggested a non-radial Malmquist performance index to measure dynamic changes in total-factor CO2 emission performance over time by solving several non-radial DEA models. This new performance index compounds an efficiency change index, a best-practice gap change index and a technology gap change index.

In summary, the Malmquist TFP index is appropriate to capture the efficiency and the technical change in case the technology exhibits constant returns to scale (CRS). Conversely, if the technology exhibits variable returns to scale (VRS) then the Malmquist TFP index fails to capture the different sources of productivity change, though its decomposition into technical and efficiency change components remains valid. It is worth noting that when panel data is available, under CRS, the Malmquist TFP index is considered to be the most appropriate approach, whereas if only limited data is available, the Hicks-Moorsteen or the index number approaches are usually the best choices. The Hicks-Moorsteen approach uses a measure of output growth as a percentage of the growth in the input used. If the output growth is attained by less than 100% growth in input use, then productivity growth is achieved (Coelli et al., 2005).

Luenberger Productivity Indicator

The Luenberger productivity indicator was introduced into production theory by Chambers et al., (1996, 1998), by using the directional distance function. They have transposed the benefit function developed by Luenberger (1992) for the consumer theory into the production theory. The directional distance function is a generalization of the traditional Shephard distance function, and thus it embraces the Malmquist productivity Index. The main advantage of this approach is that it can simultaneously contract inputs and expand outputs or it can simultaneously maximize desirable outputs while reducing undesirable ones at the same rate. The directional distance function projects the input and/or the output vector from itself to the technology frontier in a preassigned direction. Its calculation is based on the arithmetic mean of the productivity change between periods t and $t + 1$.

The Luenberger productivity indicator comprehends two components as in the Malmquist productivity index: the efficiency change between periods t and $t + 1$, representing enhancement in management (marketing initiatives, quality improvements) and the technological change captured by the arithmetic mean of the last two differences, representing the shift of technology associated to innovation (investment in methodologies, procedures, techniques) between these two periods.

In an attempt to develop an embracing measure of productivity change, Balk (2001) extended the Malmquist-Luenberger measure of productivity index by adding the scale efficiency change to the other two traditional factors: the technological change and the technical efficiency change. The scale efficiency change means that the firm has moved along the frontier to the point where the input–output ratio is better. Later, Peypoch and Solonandrasana (2008) extended the Luenberger (1996) directional technology distance function to an aggregate measure, which is just the sum of the individual directional distance functions. Barros et al. (2008) employ the directional distance function and the Luenberger productivity indicator to analyze hospital efficiency and productivity growth. A Malmquist productivity index is also applied for comparison. The productivity indicator is decomposed into the usual components of productivity growth: technological change and efficiency change.

Data Envelopment Analysis (DEA)

Data Envelopment Analysis (DEA) is a mathematical programming technique. This non-parametric statistical approach has been commonly used in assessing the relative efficiency and productivity of Decision-Making Units (DMUs) that transform multiple inputs into multiple outputs (Wang & He, 2017). It was first developed by Charnes et al. (1978), based on the work of Farrell (1957) who uses a radial distance function, as a managerial and organizational performance measurement tool. In its early stage, DEA was mainly applied to evaluating public sectors' relative operating efficiency, for example, banks, hospitals, schools airlines, railways, utilities, age-care facilities and police stations among others (Emrouznejad et al., 2008). Over the last decades, considerable interest has also arisen in measuring the efficiency and productivity of the production units focusing on DEA applications, with more variables and complicated models.

Compared to the traditional econometric methods such as regression analysis and simple ratio analysis, the DEA approach is a numerical method that uses linear programming to convert inputs into outputs to measure the performance of comparable products or organizations (Iram et al., 2020). In DEA, each DMU is free to choose any combination of inputs and outputs to maximize its efficiency score, which is the ratio of the total weighted results to the total weighted inputs (Iram et al., 2020).

This multi-factorial approach emerged as an important and essential tool in a large number of management areas. It has been widely applied to sectors such as banking, education (including higher education), hospitals and health care (Hollingsworth, 2008), manufacturing, hotels and tourism (Barros & Santos, 2006), airlines, defense firms and other enterprises and institutions. DEA assesses how efficient is a production unit in transforming a set of inputs into a set of outputs. It allows multiple inputs and outputs to be considered simultaneously without any assumption on data distribution. Efficiency is measured in terms of a proportional change in inputs or outputs. It can be an input-oriented model, by minimizing inputs for a given output level, or an output-oriented model, by maximizing outputs without requiring more inputs.

The DEA approach uses two main models: the original formulation, known as the CCR model, developed by Charnes et al. (1978), who generalize Farrell's approach to the multiple output case (Murillo-Zamorano, 2004), and the BCC model, developed by Banker et al. (1984). The former analyses the productive efficiency of one unit, assuming constant returns to scale (CRS), ignoring the scale effects. The BCC model is an extension of the CCR model, by enabling the estimation of efficiency under variable returns to scale (VRS). The inefficiency level of each unit is determined by comparing it to a benchmarking decision-making unit (Murillo-Zamorano, 2004). The CRS assumption is more appropriate when all the DMUs are operating at their optimal scale and when there is a small variability among the inputs. The VRS assumption is more suitable when there are different scales.

In 1994, Färe et al. used the DEA technique to calculate and decompose the Malmquist productivity index. This index is decomposed into three components of productivity change: technical efficiency, scale efficiency and technical change. Each component is greater than, equal to, or less than one, depending, respectively, on their positive, null or negative contribution to productivity growth. These components are based on output-oriented measures of technical efficiency and the DEA approach provides a non-parametric method of calculating these measures. The advantage of using the DEA approach to calculate these measures is that, as a non-parametric method, it avoids the risk of confounding the effects of each component of productivity change with those of an inappropriate functional form (Lovell, 1996).

In 1995, it was created the website www.DEAzone to include a full set of DEA resources, beginning the maturity phase for DEA (Emrouznejad et al., 2008). This methodology provides an appropriate method to deal with multiple inputs and outputs in examining relative efficiency, allowing the measurement of efficiency and productivity of large organizations that involve a complex multi-input/output structure. DEA surveys both input and output data seeking the points where the largest output can be produced by the minimum input. By linking these points, through a linear programming model, we can get an efficient production frontier and the units being out of the frontier are inefficient. DEA allows confirming how much it should adjust the inputs and outputs to reach an efficient level of converting inputs into outputs. The efficient unities, with a coefficient of efficiency equal to 1, will serve as a benchmark to the inefficient ones whose coefficient of efficiency is below 1. DEA is thus, not only a methodology that allows assessing the performance of a DMU, but also a benchmarking technique that allows detecting management failures and helping in the improvement decisions. More recently, as energy and environmental problems have become more prominent, this approach has also been applied to estimating the circular economy development levels, as for instance in China (Fan & Fang, 2020; Wu et al., 2014), in Austria (Jacobi et al., 2018), in the OECD countries (Iram et al., 2020) and in the European Union and the World (Haas et al., 2015).

Fan and Fang (2020) applied the DEA to assess the energy and water-saving potential in different regions in China. The DEA method has also been used to assess environmental efficiency (Wang and He, 2017) and energy efficiency for evaluating energy or electricity consumption efficiency in residential buildings (Grösche, 2009), industries (Chen & Gong, 2017; Liu et al., 2017), energy-intensive firms (Moon & Min, 2017), regions (Hu et al., 2012), provinces (Wu et al., 2017), cities (Gonzalez-Garcia et al., 2018) and countries (Moutinho et al., 2017). Indeed, since the DEA approach is useful when the output measures are in the form of output indicators and when the prices are not available or not relevant, as it is the case for most of the non-market services (Coelli et al., 2005), it has become a mainstream method for studying energy efficiency and environmental efficiency world-wide.

Wang et al. (2016) analyze the determinants of efficiency change from the perspectives of technical efficiency and technical change, using an index decomposition analysis approach. Since directional distance function models may lead to biased estimations due to different directions given to different units, they propose a non-radial efficiency evaluation model. This methodology can increase desirable outputs while reducing undesirable outputs. They applied it to analyze the economic efficiency and CO2 emissions efficiency of the Asia Pacific Economic Cooperation (APEC) members.

The main advantages of the DEA techniques are their flexibility and adaptability. Indeed, DEA is easy to use, as it does not need the imposition of a functional form on the underlying technology (that is, on the production function) nor does it require price information. It is possible to use multiple inputs and outputs simultaneously, allowing them to have different unit measures (unlike ratio analysis, where the imposition of weights attached to inputs and outputs is important). It identifies the best practices, which serve as a basis for comparison with the DMU (efficiency is measured relative to the highest observed performance rather than against some average). Finally, it can decompose the efficiency into several components. DEA is a holistic technique that combines all the relevant indicators into an overall index for performance comparisons (Zhou et al., 2010). The main drawback of DEA is its deterministic nature, which does not allow the statistical treatment of the noise. Bootstrapping techniques can avoid this disadvantage by providing a suitable way to analyze the sensitivity of efficiency scores relative to the sampling variations and statistical inferences on the index scores (Simar & Wilson, 2007).

SFA

The Stochastic Frontier Analysis (SFA), developed by Aigner et al. (1977) and by Meeusen and van den Broeck (1977) measures efficiency relative to a stochastic parametric frontier. The great virtue of SFA over DEA is that it is stochastic, allowing it to be capable of distinguishing the effects of statistical noise from those of inefficiency. However, its drawback is that it is parametric, requiring the adoption of a specific form for the production function (usually the translog function). Consequently, in case of adopting an inappropriate functional form, SFA is unable to distinguish inefficiency from other deviations resulting from the assumption of a misspecified production function (Lovell, 1996).

While DEA assumes that all deviations from the production frontier are the result of technical inefficiency, SFA is motivated by the idea that those deviations might not be entirely under the control of the firm. Therefore, SFA allows for such deviations to be also the result of shocks (weather, strikes, random equipment failures, etc.), measurement errors and other random factors. Hence, in the framework of SFA, firms may operate beneath the production frontier not only because they may be technically inefficient but also due to exogenous factors. Furthermore, since the random shocks may affect positively or negatively their operations, firms may even temporarily operate above the (non-stochastic) frontier.

The regression model underlying SFA has 3 components: the deterministic production function; a symmetrical disturbance representing random shocks; and an asymmetric, negatively skewed term representing firm-specific inefficiency. While the estimation of the parameters of the deterministic frontier does not require further assumptions besides the parametric specification of the production function, the most popular estimators of efficiency scores require distributional assumptions for both the components of the error term to separate the noise component from the inefficient factor. For two recent exceptions that circumvent the need for distributional assumptions on the noise and/or the inefficiency components see Kumbhakar and Barnstein (2019) and Belotti and Ferrara (2021).

Most regression models used in SFA assume a normal distribution for the random shock, differing in the choice of the distribution for the inefficient term. The simplest model assumes a half-normal distribution for the latter term (Aigner et al. (1977), while popular variants are based on truncated normal (Stevenson, 1980) and exponential and gamma distributions (Greene, 1990). Different distributional assumptions can naturally give rise to different predictions of technical efficiency, but when firms are ranked based on those predictions the rankings tend to be quite robust to distributional choice (Coelli et al., 2005). Estimation is performed in all cases using the maximum likelihood (ML) method.

Firm's operations are often influenced by exogenous variables that characterize the environment in which production takes place (Coelli et al., 2005). Examples of these so-called environmental, contextual or non-discretionary variables, which often are beyond the control of the firm's manager, are government regulations, firm's ownership and size, labor force age and geographical location. There are two main ways of accounting for environmental variables in SFA, depending on whether we assume that the environmental variables influence the production frontier itself or the inefficiency effects. The first method is to incorporate them directly in the deterministic part of the production frontier, treating the environmental variables in a similar way to the production factors (Coelli et al., 1999). This procedure implies that the estimated measures of technical efficiency have into account the effect of both traditional inputs and environmental variables. The second alternative is to allow the environmental variables to directly influence the stochastic component of the production frontier by specifying the mean of the inefficient term as a function of the environmental variables (Kumbhakar et al., 1991). In this case, although the parameters associated to the environmental variables are still estimated simultaneously with the parameters of the production function, environmental variables are interpreted as determinants of technical efficiency and not of the production frontier.

When panel data is available, it is possible to consider time-varying inefficiency. Kumbhakar (1990), Battese and Coelli (1992), Cornwell et al. (1990) and Cuesta (2000) propose specific functions that determine how technical efficiency varies over time. However, only in the two last cases is a change in the rank ordering of firms over time allowed. A more general approach was proposed by Battese and Coelli (1995), which allows for time-varying inefficiency by using the second method described above to incorporate environmental variables. Thus, assuming for the inefficiency term the truncated normal distribution, for example, the mean of that distribution is specified as a function of time dummies and trend variables. If environmental variables are also present in the specification of their mean, then the inefficiency effects are allowed to vary across firms and over time. SFA can also be used in conjunction with distance functions. For example, Lin et al. (2013), using the parametric directional output distance function, construct the group frontier by using the SFA. As opposed to conventional methods that measure productivity considering desirable and undesirable outputs symmetrically, they adopted a directional distance production function method that differentiates between desirable and undesirable outputs. The reasoning behind this is to allow considering nonproportional changes in output, as it is possible to expand desirable outputs while contracting undesirable outputs. The directional output distance function is parameterized using a quadratic flexible function form to calculate the three components of technical efficiency change, technological change and scale efficiency change. The parameters of this function can be estimated using either ordinary least square (OLS) or ML methods. After choosing the most suitable method, the estimated parameters are used to calculate the generalized metafrontier Malmquist productivity index.

References

Aigner, D. J., Lovell, C. A. K., & Schmidt, P. (1977). Formulation and estimation of Stochastic Frontier Production Function Models. *Journal of Econometrics, 6*(1), 21–37.

Arrow, K. (1962). The economic implications of learning by doing. *Review of Economic Studies, 29*, 155–173.

Balk, B. (2001). Scale efficiency and Productivity efficiency. *Journal of Productivity Analysis, 15*, 159–183.

Banker, R. D., Charnes, A., & Cooper, W. W. (1984). Some models for estimating technical and scale efficiencies in data envelopment analysis. *Management Science, 30*(9), 1078–1092.

Barros, C. P., & Santos, C. (2006). The measurement of efficiency in Portuguese hotels with DEA. *Journal of Hospitality and Tourism Research, 30*, 378–400.

Barros, C. P., Menezes, A. G., Peypoch, N., Solonandrasana, B., & Vieira, J. C. (2008). An analysis of hospital efficiency and productivity growth using the Luenberger indicator. *Health Care Management Science, 11*, 373–381.

Battese, G. E., & Coelli, T. J. (1992). Frontier production functions, technical efficiency and panel data: With application to paddy farmers in India. *Journal of Productivity Analysis, 3*, 153–169.

Battese, G. E., & Coelli, T. J. (1995). A model for technical inefficiency effects in a stochastic frontier production function for panel data. *Empirical Economics, 20*, 325–332.

Belotti, F., Ferrara, G. (2021), "Imposing monotonicity in stochastic frontier models: an iterative nonlinear least squares procedure", CEIS Tor Vergata Research Paper Series 17(5), No.462.

Briec, W., & Kerstens, K. (2009). Infeasibility and directional distance functions with application to the determinateness of the Luenberger productivity indicator. *Journal of Optimization Theory and Applications, 141*, 55–73.

Caves, D. W., Christensen, L. R., & Diewert, W. E. (1982). Multilateral comparisons of output, input and productivity using superlative index numbers. *The Economic Journal, 92*, 73–86.

Chambers, R. G., Chung, Y., & Färe, R. (1996). Benefit and distance functions. *Journal of Economic Theory, 70*, 407–419.

Chambers, R. G., Chung, Y., & Färe, R. (1998). Profit, directional distance functions, and Nerlovian efficiency. *Journal of Optimization Theory and Applications, 98*, 351–364.

Charnes, A., Cooper, W. W., & Rhodes, E. (1978). Measuring the efficiency of decision-making units. *European Journal of Operational Research, 2*(6), 429–444.

Chen, X., & Gong, Z. (2017). DEA efficiency of energy consumption in China's manufacturing sectors with environmental regulation policy constraints. *Sustainability, 9*(2), 210.

Choi, Y., Zhang, N., & Zhou, P. (2012). Efficiency and abatement costs of energy-related CO2 emissions in China: A slacks-based efficiency measure. *Applied Energy, 98*, 198–208.

Chung, Y.H., Färe, R., Grosskopf, S. (1997), "Productivity and undesirable outputs: a directional distance function approach", Journal of Environmental Management", Vol. 51, pp. 229–240.

Coelli, T. J., Perelman, S., & Romano, E. (1999). Accounting for environmental influences in stochastic frontier models: With application to international airlines. *Journal of Productivity Analysis, 11*, 25–273.

Coelli, T., Rao, D. S. P., O'Donnell, C., & Battese, G. (2005). *An introduction to efficiency and productivity analysis* (2nd ed.). Springer.

Cornwell, C., Schmidt, P., & Sickles, R. C. (1990). Production frontiers with cross-sectional and time-series variation in efficiency levels. *Journal of Econometrics, 46*, 185–200.

Cuesta, R. A. (2000). A production model with firm-specific temporal variation in technical inefficiency: With application to Spanish dairy farms. *Journal of Productivity Analysis, 13*, 139–158.

Dasgupta, P. S., & Heal, G. M. (1979). *Economic Theory and Exhaustible Resources.* Cambridge University Press.

Debreu, G. (1951). The coefficient of Resource Utilization. *Econometrica, 19*(3), 273–292.

Debreu, G. (1954). A Classical Tax-Subsidy Problem. *Econometrica, 22*(1), 14–22.

Emrouznejad, A., Parker, B., & Tavares, G. (2008). Evaluation of research in efficiency and productivity: A survey and analysis of the first 30 years of scholarly literature in DEA. *Socio-Economic Planning Sciences, 42*, 151–157.

Fan, Y., & Fang, C. (2020). Circular economy development in China-current situation, evaluation and policy implications. *Environmental Impact Assessment Review, 84*, 106441.

Färe, R., Grosskopf, S., Norris, M., & Zhang, Z. (1994). Productivity growth, technical progress and efficiency change in industrialized countries. *The American Economic Review, 84*, 66–83.

Farrell, M.J. (1957), "The measurement of Productive Efficiency", Journal of the Royal Statistical Society (Series A: General), 120 (2), pp. 253–281.

Gonzalez-Garcia, S., Manteiga, R., Moreira, M. T., & Feijoo, G. (2018). Assessing the sustainability of Spanish cities considering environmental and socio-economic indicators. *Journal of Cleaner Production, 178*, 599–610.

Greene, W. H. (1990). A Gamma-distributed stochastic frontier model. *Journal of Econometrics, 46*, 141–164.

Grösche, P. (2009). Measuring residential energy efficiency improvements with DEA. *Journal of Productivity Analysis, 31*(2), 87–94.

Haas, W., Krausmann, F., Wiedenhofer, D., & Heinz, M. (2015). How circular is the global economy?: An assessment of material flows, waste production, and recycling in the European Union and the World in 2005. *Journal of Industrial Ecology, 19*(5), 765–777.

Hollingsworth, B. (2008). The measurement of efficiency and productivity of health care delivery. *Health Economics, 17*, 1107–1128.

Hu, J. L., Lio, M. C., Kao, C. H., & Lin, Y. L. (2012). Total-factor energy efficiency for regions in Taiwan. *Energy Sources, Part b: Economics, Planning and Policy, 7*(3), 292–300.

Iram, R., Zhang, J., Erdogan, S., Abbas, Q., & Mohsin, M. (2020). Economics of energy and environmental efficiency: Evidence from OECD countries. *Environmental Science and Pollution Research, 27*, 3858–3870.

Jacobi, N., Haas, W., Wiedenhofer, D., & Mayer, A. (2018). Providing an economy-wide monitoring framework for the circular economy in Austria: Status quo and challenges. *Resources, Conservation and Recycling, 137*, 156–166.

Ji, Y.-B., & Lee, C. (2010). Data envelopment analysis. *The Stata Journal, 10*(2), 267–280.

Kumbhakar, S. C. (1990). Production frontiers, panel data and time-varying technical inefficiency. *Journal of Econometrics, 46*, 201–211.

Kumbhakar, S.C., Bernstein, D.H. (2019), "Does xistence of inefficiency matter to a neoclassical xorcist? Some econometric issues in panel stochastic frontier models", in Parmeter, C.F., Sickles, R.C. (eds.) *Advances in Efficiency and Productivity Analysis*, pp. 139–161.

Kumbhakar, S. C., Ghosh, S., & McGuckin, J. T. (1991). A generalized production frontier approach for estimating determinants of inefficiency in U.S. dairy farms". *Journal of Business and Economic Statistics, 9*, 279–286.

Lin, E., Chen, P.-Y., Chen, C.-C. (2013), "Measuring green productivity of country: a generalized metafrontier Malmquist productivity index approach", Energy, pp. 340–353.

Liu, J. P., Yang, Q. R., & He, L. (2017). Total-factor energy efficiency (TFEE) evaluation on thermal power industry with DEA, Malmquist and multiple regression techniques. *Energies, 10*(7), 1–14.

Lovell, C. A. K. (1996). Applying efficiency measurement techniques to the measurement of productivity change. *The Journal of Productivity Analysis, 7*, 329–340.

Lucas, R. (1988). On the mechanics of economic development. *Journal of Monetary Economics, 22*, 3–42.

Luenberger, D. C. (1992). Benefit function and duality. *Journal of Mathematical Economics, 21*, 461–481.

Luenberger, D. C. (1996). Welfare from a benefit viewpoint. *Economic Theory, 7*, 445–462.

Malmquist, S. (1953). Index Numbers and Indifference Surfaces. *Trabajos De Estadistica, 4*, 209–242.

Meeusen, W., & van den Broeck, J. (1977). Efficiency estimation from Cobb-Douglas Production functions with composed error. *International Economic Review, 18*(2), 425–444.

Moon, H., & Min, D. (2017). Assessing energy efficiency and the relatrd policy implications for energy-intensive firms in Korea: DEA approach. *Energy, 133*, 23–34.

Moutinho, V., Madaleno, M., & Robaina, M. (2017). The economic and environmental efficiency assessment in EU cross-country: Evidence from DEA and quantile regression approach. *Ecological Indicators, 78*, 85–97.

Mouzas, S. (2006). Efficiency versus effectiveness in business networks. *Journal of Business Research, 59*, 1124–1132.

Murillo-Zamorano, L. (2004). Economic efficiency and frontier techniques. *Journal of Economic Surveys, 18*(1), 33–77.

Peypoch, N., & Solonandrasana, B. (2008). Aggregate efficiency and productivity analysis in the tourism industry. *Tourism Economics, 14*(1), 45–56.

Rebelo, S., Matias, F., & Carrasco, P. (2013). Application of the DEA methodology in the analysis of efficiency of the Portuguese hotel industry: An analysis applied to the Portuguese geographical regions. *Tourism & Management Studies, 9*(2), 21–28.

Romer, P. M. (1986). Increasing returns and long run growth. *Journal of Political Economy, 94*, 1002–1057.

Shephard, R. W. (1953). *Cost and Production Function.* Princeton University Press.

Simar, L., & Wilson, P. (2007). Estimation and inference in two-stage, semi-parametric models of production processes. *Journal of Econometrics, 136*, 31–64.

Solow, R. M. (1956). A contribution to the theory of economic growth. *Quarterly Journal of Economics, 109*, 65–94.

Stevenson, R. E. (1980). Likelihood functions for generalised stochastic frontier estimation. *Journal of Econometrics, 13*, 49–78.

Wang, Z., & He, W. (2017). CO2 emissions efficiency and marginal abatement costs of the regional transportation sectors in China. *Transportation Research Part D, 50*, 83–97.

Wang, Z., He, W., & Che, K. (2016). The integrated efficiency of economic development and CO2 emissions among Asia Pacific Economic Cooperation members. *Journal of Cleaner Production, 11*, 765–772.

Wu, A. H., Cao, Y. Y., & Liu, B. (2014). Energy Efficiency evaluation for regions in China: An application of DEA and Malmquist indices. *Energy Efficiency, 7*, 429–439.

Wu, J., Xiong, B., An, Q., Sun, J., & Wu, H. (2017). Total-factor energy efficiency evaluation of Chinese industry by using two-stage DEA model with shared inputs. *Annual Operational Research, 255*(1–2), 257–276.

Zhang, N., & Choi, Y. (2013). Total-factor carbon emission performance of fossil fuel power plants in China: A metafrontier non-radial Malmquist index analysis. *Energy Economics, 40*, 549–559.

Zhou, P., Ang, B.W., Han, J.Y. (2010), "Total factor carbon emission performance: A Malmquist index analysis", Energy Economics, Vol. 32, pp. 194–201.1

Zhou, P., Ang, B. W., & Zhou, D. (2012). Measuring economy-wide energy efficiency performance: A parametric Frontier approach. *Applied Energy, 90*, 196–200.

Data Envelopment Analysis: A Review and Synthesis

Ana S. Camanho and Giovanna D'Inverno

1 Introduction

Performance evaluation and benchmarking are continuous improvement tools with a very relevant contribution to the prosperity of organisations. They contribute positively to help organisations to evolve in order to remain viable in highly competitive environments. Performance assessments enable (i) revealing the strengths and weaknesses of operations within organisations (internal benchmarking); (ii) identifying the units/companies/sectors/countries with the best performance and learn from their policies and best practices (external benchmarking); (iii) understanding the causes of differences among the performance levels of organisations.

The performance of an organisation can be defined in several ways. A natural measure of performance is a productivity index, calculated as the ratio between outputs (results) and inputs (resources), where higher values of this index are associated with better performance. But performance is also a relative concept. For example, the performance of a factory can be measured in relation to the previous year's levels, or it can be measured in relation to that observed in other factories. These three alternatives correspond to the three key concepts of performance measurement: 'productivity', 'productivity change', and 'efficiency'. The concept of 'effectiveness' corresponds to the alignment of results with strategy and implies 'doing the right things'. In the literature on quantitative methods of performance measurement, it is understood

A. S. Camanho (✉)
Faculdade de Engenharia, Universidade do Porto, Porto, Portugal
e-mail: acamanho@fe.up.pt

G. D'Inverno
Department of Economics and Management, University of Pisa, Pisa, Italy
e-mail: giovanna.dinverno@unipi.it

Faculty of Economics and Business, KU Leuven, Leuven, Belgium

P. Macedo et al. (eds.), *Advanced Mathematical Methods for Economic Efficiency Analysis*, Lecture Notes in Economics and Mathematical Systems 692,
https://doi.org/10.1007/978-3-031-29583-6_3

that the choice of outputs should be aligned with what is really important for the organisation, so efficiency ('doing things right') and effectiveness are expected to be aligned.

Assuming that high levels of efficiency and productivity, as well as high levels of productivity growth, are desirable objectives, then it is important to define and measure efficiency and productivity in a way that respects economic theory and provides useful information for managers and policy makers. Indeed, improving efficiency is one of the key components of productivity growth, which in turn is the main driver of economic welfare. The benefits of understanding the relationship between efficiency and productivity, as well as of quantifying these quantities in order to be able to act on their determinants, are the main contributions of the literature associated with quantitative methods of performance measurement.

The efficiency of a productive unit, referred to as a Decision Making Unit (DMU), is defined by comparing its inputs and outputs to those of the best performing peers. The inputs correspond to the resources used, whereas the outputs are the products or services obtained as a result of the production process. The level of outputs produced must be related in some way to the level of inputs used to secure them. This relationship is called the technology of production and defines the maximum possible output obtainable from given inputs. Exact knowledge of the technology of production is not usually available. Thus, for a long time, economists and management scientists have developed alternative methods for deriving empirically the the technology of production from a set of DMUs observed. Despite the differences in the methods available for the estimation of the technology of production, efficiency is always defined by comparing observed to optimal productive performance.

The chapter proceeds as follows. First, we start with a brief historical overview on the measurement of efficiency, focusing on the evolution of frontier analysis methods. As implied by their name, frontier methods estimate production technologies that go through the boundary of the production space. For this reason, they are deemed the most appropriate for the assessment of efficiency, as they are based on 'best practices' rather than 'average performance'. Then, we provide an introduction to the DEA method. It includes a description of the theory underlying the representation of the technology of production and the efficiency frontier in DEA, which is based on the Axiomatic Approach (Debreu, 1951; Koopmans, 1951; Shephard, 1970). The main DEA models for the evaluation of efficiency are reviewed along with recent developments in the DEA literature. To conclude, we discuss the DEA most recent applications, available softwares and opportunities for future studies.

2 The Origin of Frontier Methods

Traditional approaches to efficiency measurement consist of a comparison between observed and optimal values of the outputs, or the inputs, of a Decision Making Unit (DMU). The comparison can take the form of observed to maximum output obtainable from the given input, or the ratio of minimum input required for producing the given output to the observed input. In these two comparisons, the optimal is defined in terms of the physical production possibilities and efficiency is called technical.

It would also be possible to define the optimal incorporating the economic goal of the DMU. In this case, efficiency is called economic and is measured by comparing observed and optimum cost, revenue or profit, subject to appropriate constraints both on quantities (i.e., reflecting the technology of production) and prices (i.e., reflecting the market conditions).

Even at this conceptual stage of efficiency measurement two problems arise. How many and which inputs and outputs should be included in the analysis, and how should the optimal production levels of a DMU be determined?

In relation to the first problem, it is clear that the efficiency results obtained are highly dependent on the selection of variables to be included in the assessment, as well as how they are measured. These variables should be chosen to reflect the primary aims of the assessment. For example, when assessing the performance of schools, one can examine the ability of individual schools to utilise their resources in order to achieve high examination results. In this case, it would be appropriate to choose as inputs the resources available at a school (e.g., number of teachers, facilities and expenditure) and as outputs the examination achievements of pupils. Conversely, if the objective of the assessment concerned the value added at schools, the inputs should include information on the entry standards and socio-economic background of pupils (see Thanassoulis & Dunstan, 1994).

The second problem, concerning the determination of the optimal production level of a DMU, is the most difficult to answer. Traditional economic approaches theoretically define a production function, which is a mathematical representation of the relationship between inputs and outputs, and is defined as the maximal possible output obtainable from given inputs. The seminal work by Cobb and Douglas (1928) on the estimation of average production functions contributed substantially to the development of this field of economics. Since then, more flexible production function forms were developed and tested on empirical data. However, although the estimation of average production functions has become commonplace in economics, the estimation of frontier production functions has only attracted widespread attention recently. As Aigner et al. (1977, p. 21) mentioned: "*The theoretical definition of a production function expressing the maximum amount of output obtainable from given input bundles with fixed technology has been accepted for many decades. And for almost as long, econometricians have been estimating average production functions.*" Despite the key contributions from economic theory to frontier analysis, for many years the productivity literature ignored the efficiency component due to the difficulties in estimating optimal, as opposed to average, input-output relationships.

The underpinnings of efficiency measurement date back to the work of Debreu (1951) and Koopmans (1951). Debreu provided the first measure of efficiency, which was called the 'coefficient of resource utilisation' and Koopmans was the first to define the concept of technical efficiency. Farrell (1957) extended their work in a seminal paper whose key development was to show how to bring data to bear on Debreu's formulation of the 'coefficient of resource utilisation'. Farrell also remarked that the main reason why all attempts to solve the efficiency measurement problem had previously failed was because of the inability to combine multiple inputs and outputs, measured at different scales, into a satisfactory measure of efficiency. Farrell proposed an approach called 'activity analysis' that would deal more adequately with this problem from a multi-dimensional perspective while considering the existence of multiple inputs and outputs. His measures were intended to be applicable to any productive organisation, in his words 'from a workshop to a whole economy'.

Farrell (1957) work provided the foundations to the estimation of empirical frontier production functions. In most production processes, the conversion of inputs into outputs does not follow a known functional form. Therefore, the traditional economic method, based on theoretically defined production functions requiring *a-priori* specification of a functional form, is likely to identify as best performance some unattainable ideal. Farrell (1957) suggested changing the focus from absolute to relative efficiency by promoting the comparison of a DMU to the best actually achieved by peers performing a similar function. In these circumstances, the frontier should be an empirical representation of best practices. This allowed opening new perspectives of analysis, moving from the search for efficiency in the sense of the Pareto-Koopmans definition (or absolute efficiency) to the concept of relative efficiency, that was latter specified as follows:

> *"A DMU is to be rated as fully (100%) efficient on the basis of available evidence if and only if the performances of other DMUs does not show that some of its inputs or outputs can be improved without worsening some of its other inputs or outputs."* (Cooper et al., 2011, p. 3)

Although Farrell (1957) presented the basic concepts of frontier methods for measuring efficiency, he did not present a method that was able to quantify efficiency in a multidimensional context. Thus, his work ended up not receiving much attention from his peers for almost two decades. However, Farrell's ideas made it possible to lay the foundations for frontier methods, which consider that performance is a function of the state of technology and of the efficiency level. The state of technology depends on the frontier relationship between inputs and outputs, where the position of this frontier defines the limit of the Production Possibility Set (PPS) associated with a given technology. This limit should be estimated based on what was effectively observed in the organisations. The efficiency level indicates the positioning relative to this frontier, where a bad positioning reveals waste (or inefficiencies) that may have several origins. Improvements in the performance of an organisation can thus occur through two ways: technological evolution (change in the position of the frontier due to productivity gains) or efficiency gains (change in the distance to the frontier).

Farrell (1957) graphical illustration of the efficiency concepts has now become classical. In order to provide a pictorial representation of his ideas, consider a set of DMUs that produce a single output (Y) using two inputs (X_1 and X_2) in varying quantities, as shown in Fig. 1.

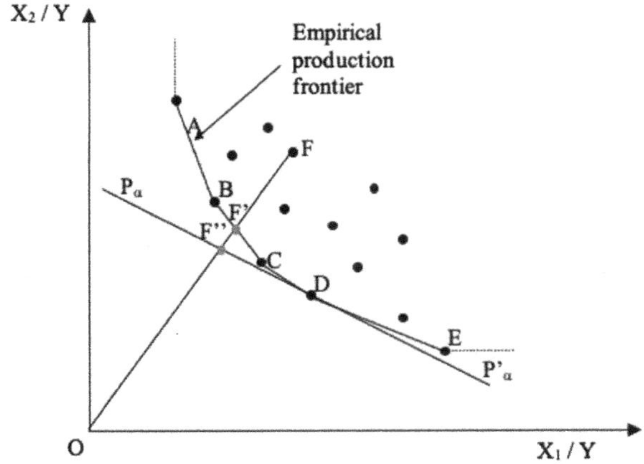

Fig. 1 Efficiency measurement

Farrell (1957) analysis had an input-reducing focus and assumed constant returns to scale (CRS). Under efficient input to output transformations, CRS means that scaling the input levels by a factor β leads to an equally proportionate scaling in the outputs by the same factor β. As the DMUs are each producing a normalised level of output, this allows their representation in a two-dimensional diagram. The segments linking DMUs A, B, C, D and E form the technical efficient production frontier. Note that the frontier has a piecewise linear shape.

DMU F will be used to illustrate the efficiency concepts. Its technical efficiency is given by the ratio OF'/OF. A ratio less than one indicates that it is possible to build a composite DMU that employs the same proportion of inputs (or input mix) and can produce the same output as the assessed DMU using only a fraction of its inputs.

Note that Farrell's measure of technical efficiency is the inverse of the distance function introduced by Shephard (1970). Looking beyond technical efficiency, Farrell (1957) also proposed a measure of economic efficiency based on a cost minimising behaviour. The measure of Farrell's cost efficiency is illustrated for DMU F. It requires the specification of an isocost line, whose slope is equal to the observed prices ratio at the DMU (i.e., $-P_1/P_2$, where P_1 is the price of input 1 and P_2 the price of input 2). This is represented in Fig. 1 by the line $P_\alpha P_\alpha^{'}$. Comparing points F' and D on the production frontier, although they both exhibit 100% technical efficiency, the costs of production at D will only be a fraction OF''/OF' of those at F'. This ratio is defined as the input allocative efficiency of DMU F. Input allocative efficiency attempts to capture the inefficiency arising solely from the wrong choice of technically efficient input combinations given input prices, that is, it measures the extent to which a DMU uses the various factors of production in the best proportions in the light of their prices. If DMU F were perfectly efficient, both technically and allocatively, its costs would be a fraction OF''/OF of their current level. This ratio gives a measure of cost efficiency. It indicates the extent to which the DMU is supporting its current level of outputs at minimum cost.

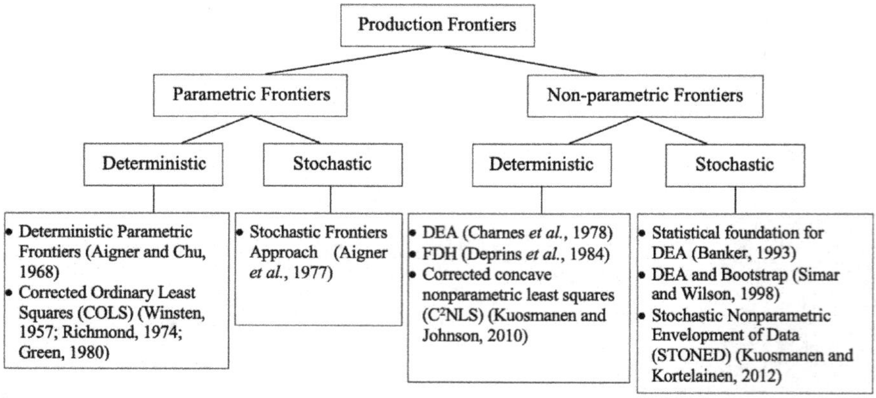

Fig. 2 Pioneering works on production frontier methods

In summary, the work of Farrell (1957) was innovative for a number of reasons. It relaxed the need for specifying a functional form prior to estimating efficiency from empirical data. It introduced the principle of constructing a hypothetical DMU (such as F') as a convex combination of observed DMUs. It recognised the existence of multi-input and multi-output production technologies without, however, providing a method for the estimation of the production frontier.

Despite Farrell (1957) developments, efficiency and its measurement only attracted attention again much later. After Farrell (1957), the evolution in the assessment of productive efficiency came via two parallel routes that differ in the way the frontier is specified and estimated. The frontier can be specified as parametric or non-parametric. Both methods can be further divided into stochastic and deterministic. To estimate the frontier, statistical or mathematical programming techniques can be used. Figure 2 shows the various types of production frontiers.

Under the parametric approach, the technology (e.g., represented by a production or cost function) is specified as a function with a precise mathematical form (e.g., the translog or the Cobb-Douglas function). The type of function representing the frontier must be specified *a-priori* and its parameters are estimated from the empirical data. Under the non-parametric approach, the technology is defined by a set of properties that the points in the production possibility set are assumed to satisfy. No function with constant parameters needs to be specified. The specification of production sets is based on the Axiomatic approach (Debreu, 1951; Koopmans, 1951; Shephard, 1970).

A deterministic approach assumes that all the deviations of observed production from the estimated frontier are exclusively explained by inefficiency. It is assumed that there are no random factors affecting the construction of the frontier, such as random noise or errors in the data. Thus, all observations must lie on or below the frontier. The estimation of deterministic frontiers involves the use of mathematical programming techniques. The stochastic approach allows for random noise and measurement error in the data. These factors may affect the DMUs performance and be responsible, together with inefficiency, for observed deviations from the frontier. As a result, the DMUs may lie above or below the frontier, due to either inefficiency or random error. The stochastic approach involves the use of statistical techniques.

3 The Original DEA Model

The Data Envelopment Analysis (DEA) method dates back to the pioneering work by Charnes et al. (1978). DEA is a linear programming technique for measuring the relative efficiency of a homogeneous set of Decision Making Units (DMUs) in their use of multiple inputs to produce multiple outputs. DEA identifies a subset of efficient 'best practice' DMUs and, for the remaining DMUs, their efficiency level is derived by comparison to a frontier constructed from the 'best practice' DMUs.

Instead of trying to fit a hyperplane through the centre of the data, as in statistical regressions, we can imagine a model that floats a segmented linear surface over a set of output possibilities until this flexible surface sits comfortably on top of the observations. The surface thus constructed corresponds to the DEA frontier, which inspired the name of the technique.

> "*A mathematical programming model applied to observational data [that] provides a new way of obtaining empirical estimates of relations such as the production functions and/or efficient production possibility surfaces that are cornerstones of modern economics*" (Charnes et al., 1978)

> "*Hence, the name Data Envelopment Analysis which arises from the procedures (and concepts) applied to observational data which are used to establish the efficiency frontiers via these envelopment procedures*" (Charnes et al., 1981)

DEA derives a summary measure of efficiency for each evaluated DMU aggregating inputs and outputs using optimal weights. This measure can be obtained from two perspectives, corresponding to an input-reduction or output-expansion orientation, as follows:

- Input orientation: Is the DMU using the minimum amount of the inputs given the output levels it is currently producing?
- Output orientation: Is the DMU producing the maximum amount of the outputs from its current input levels?

The choice of orientation will depend on the context of the assessment and the aims of the organisation. According to Farrell's (1957) definition of technical efficiency, the efficiency measure derived with an input orientation corresponds to the minimal factor by which all inputs of the DMU under assessment can be decreased proportionally without decreasing the level of any outputs. Conversely, the efficiency measure with an output orientation is the inverse of the maximum factor by which all outputs can be raised equiproportionally without increasing the level of any inputs.

Beyond the efficiency measure, DEA also provides other sources of managerial information relating to the DMUs' performance. DEA identifies the efficient peers for each inefficient DMU. This is the set of relative efficient DMUs to which an inefficient DMU has been directly compared in the derivation of its efficiency score. Therefore, DEA can be viewed as a benchmarking technique, as it allows decision makers to locate and understand the nature of the inefficiencies of a DMU by comparing it with a selected set of efficient DMUs with a similar profile. In addition, DEA also provides information about the targets that would render an inefficient DMU efficient. These targets correspond to the input reductions and output expansions required for producing on the efficient frontier.

The ratio model

Having defined the efficient frontier and introduced the main features of the DEA approach, the next step is to describe the DEA model. Its mathematical representation will be introduced via the most intuitive formulation, corresponding to a ratio model.

Consider a set of n DMUs ($j = 1, ..., n$), each consuming m inputs x_{ij} ($i = 1 \ldots, m$) to produce s outputs y_{rj} ($r = 1 \ldots, s$). For each DMU_k under assessment, it is possible to obtain a measure of relative efficiency defined by the ratio of all outputs (y_{rk}) to all inputs (x_{ik}). The multiple inputs and outputs are reduced to a single (virtual) input value and a single (virtual) output value by the allocation of weights to each input and output. These weights are not defined *a-priori* and they are chosen so to show the efficiency of DMU_k in the 'best possible light'. This definition leads to the specification of the fractional programming model (1). The next sections describe the linear programming models for computing efficiency within the DEA framework.

DEA ratio model (1)

$$E_k = \max \frac{\sum_{r=1}^{s} u_r y_{rk}}{\sum_{i=1}^{m} v_i x_{ik}}$$

$$\text{s.t.} \quad \frac{\sum_{r=1}^{s} u_r y_{rj}}{\sum_{i=1}^{m} v_i x_{ij}} \leq 1, \forall j = 1, \ldots, n$$

$$u_r \geq 0 \quad \forall r = 1, \ldots, s$$

$$v_i \geq 0 \quad \forall i = 1, \ldots, m$$

u_r and v_i stand for the output and input weights, respectively. To show the efficiency of DMU_k in the 'best possible light', the ratio of weighted outputs to weighted inputs is maximised, subject to the constraints that all efficiency measures for the other DMUs must be less than or equal to unity when evaluated with similar weights. Model (1) is a fractional model but can be converted into linear form through a simple transformation (see Charnes et al., 1978).

The DEA input-oriented model

The DEA model for estimating the efficiency level of DMU k (unit under evaluation) oriented to the reduction of *input* levels and assuming constant returns to scale is presented in (2). This formulation is known as *weights formulation (or multiplier model)*. The dual formulation presented in (3) is known as the *envelopment formulation*.

DEA input-oriented model under CRS Primal (Multiplier) formulation (2)

$$\max \sum_{r=1}^{s} u_r y_{rk}$$

$$\text{s.t.} \sum_{i=1}^{m} v_i x_{ik} = 1$$

$$\sum_{r=1}^{s} u_r y_{rj} - \sum_{i=1}^{m} v_i x_{ij} \leq 0, \forall j = 1, \ldots, n$$

$$u_r \geq 0 \qquad \forall r = 1, \ldots, s$$

$$v_i \geq 0 \qquad \forall i = 1, \ldots, m$$

DEA input-oriented model under CRS Dual (Envelopment) formulation (3)

$$\min \theta$$

$$\text{s.t. } \theta x_{ik} - \sum_{j=1}^{n} \lambda_j x_{ij} \geq 0, \forall i = 1, \ldots, m$$

$$\sum_{j=1}^{n} \lambda_j y_{rj} \geq y_{rk}, \forall r = 1, \ldots, s$$

$$\lambda_j \geq 0 \qquad \forall j = 1, \ldots, n$$

$$\theta \in \mathbb{R}$$

According to the duality theory of Linear Programming (LP), models (2) and (3) provide identical optimal solutions. The value of the objective function varies between 0 (worst) and 1 (best). Thus, if the DMU k under evaluation is radially efficient (meaning that it is located on the frontier), the optimal solution obtained is equal to 1; otherwise, if the value is less than one, the DMU k is considered inefficient.

In the *weights formulation*, the decision variables are v_i e u_r. The LP model (1) seeks to maximize the efficiency of the DMU k under evaluation. Since a LP problem is solved for each DMU, each unit k in evaluation obtains an individual set of weights, which allow it to be shown 'in the best possible light'. This flexibility in determining the weights is a strong argument to conclude that in the case of a DMU being considered inefficient, there is indeed potential for improvement, given that even when it is possible to highlight the strongest aspects and neglect the weakest, there are other DMUs that dominate it in terms of performance.

In the *envelopment formulation*, the decision variables are θ e λ_j. θ is the efficiency level of the DMU k which can also be interpreted as the factor by which the values of the inputs of the DMU under evaluation can be reduced radially (equiproportionally), keeping the outputs at a level equal or higher than the observed one. The variables λ_j are the intensity variables, which can be interpreted as the weights that define a point on the frontier estimated from a convex linear combination of other DMUs in the sample (called *peers* in DEA terminology). This dual formulation is the most widely used for benchmarking purposes, since it allows a direct identification of *peers* (representing best practices to be followed) and *targets* to be reached for each DMU k under assessment to become efficient (through radial projection towards the frontier). The input and output targets for the DMU k under assessment (x_{ik}^T, y_{rk}^T) can be obtained as follows: $x_{ik}^T = \sum_{j=1}^n \lambda_j^* x_{ij}$ and $y_{rk}^T = \sum_{j=1}^n \lambda_j^* y_{rj}$, where the symbol $*$ indicates the value of the decision variable at the optimal solution to model (3).

The ratio model can also be linearised by normalising the value of the numerator of the objective function, leading to an output-oriented formulation of the DEA model, as reported in model (4) and (5). If the evaluated DMU k is radially efficient, the optimal solution obtained is equal to 1. If the value is greater than one, the DMU k is considered inefficient. It should be noted that under CRS the measures of input and output efficiency are equivalent (see Charnes et al., 1978). The output-oriented objective function score is the inverse of the input-oriented score.

DEA output-oriented model under CRS
Primal (Multiplier) formulation (4)

$$\min \sum_{i=1}^{m} v_i x_{ik}$$

$$\text{s.t.} \sum_{r=1}^{s} u_r y_{rk} = 1$$

$$\sum_{i=1}^{m} v_i x_{ij} - \sum_{r=1}^{s} u_r y_{rj} \geq 0, \forall j = 1, \ldots, n$$

$$u_r \geq 0 \qquad \forall r = 1, \ldots, s$$

$$v_i \geq 0 \qquad \forall i = 1, \ldots, m$$

DEA output-oriented model under CRS
Dual (Envelopment) formulation (5)

$$\max \phi$$

$$\text{s.t. } x_{ik} - \sum_{j=1}^{n} \lambda_j x_{ij} \geq 0, \forall i = 1, \ldots, m$$

$$\sum_{j=1}^{n} \lambda_j y_{rj} \geq \phi y_{rk}, \forall r = 1, \ldots, s$$

$$\lambda_j \geq 0 \qquad \forall j = 1, \ldots, n$$

$$\phi \in \mathbb{R}$$

The DEA model with variable returns to scale
The returns to scale is a characteristic of the boundary of the production technology and measures the responsiveness of output to equal proportional changes in all inputs. It assumes that the input mix remains the same whilst the scale size is changed.

The concept of returns to scale can be generalised to the case of multiple inputs and multiple outputs (see Banker et al., 1984). It assumes that the input and output mixes are kept unchanged and can be expressed as follows:

- A DMU exhibits Increasing Returns to Scale (IRS) if a proportional increase (decrease) in the inputs causes a greater than proportionate increase (decrease) to the outputs.
- A DMU exhibits Decreasing Returns to Scale (DRS) if a proportional increase (decrease) in the inputs causes a less than proportionate increase (decrease) in the outputs.

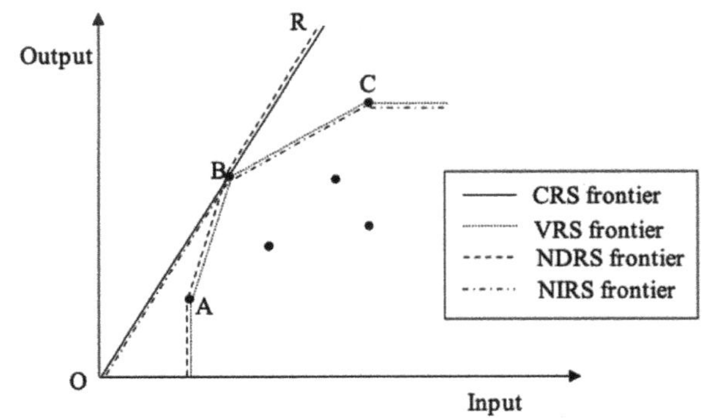

Fig. 3 CRS versus VRS frontiers

- Constant Returns to Scale (CRS) are present when a change in the inputs causes an equally proportionate change in the outputs.

These concepts are illustrated in Fig. 3 for a production frontier defined using DEA.

Under the assumption of CRS, DMU B can be extrapolated to points on the ray OR, such that the change in the input level causes an equally proportional change to the output level. Thus, the CRS frontier is defined by the ray OR.

If the scale extrapolation assumption used in the construction of the CRS frontier is not allowed, the frontier of the PPS must be based on the observed performance of the DMUs given their scale of operation. The efficient frontier in Fig. 3 would be redefined as the segments between A, B and C. This frontier allows for Variable Returns to Scale (VRS) and is made of convex combinations of the extreme points lying on the production surface.

Finally, a frontier of mixed character can be developed where extrapolation is permitted for only a subset of efficient DMUs. Let us consider the frontier defined by the segments between O, B and C. This is defined as a non-increasing returns to scale (NIRS) frontier. Under this assumption, the scale size of the DMUs can be extrapolated for smaller values, although extrapolations for larger scale sizes are not permitted. It is also possible to define a non-decreasing returns to scale (NDRS) frontier, represented in Fig. 3 by the segments linking A, B and R. Note that it is not possible to specify a DRS or an IRS frontier, as there will always be at least one point on the frontier which has constant returns to scale.

Banker et al. (1984) extended the original DEA model to enable the estimation of efficiency under a variable returns to scale context. The VRS models with input orientation are provided in (6) and (7), while with output orientation in (8) and (9).

DEA input-oriented model under VRS
Primal (Multiplier) formulation (6)

$$\max \sum_{r=1}^{s} u_r y_{rk} + \omega$$

$$\text{s.t.} \sum_{i=1}^{m} v_i x_{ik} = 1$$

$$\sum_{r=1}^{s} u_r y_{rj} - \sum_{i=1}^{m} v_i x_{ij} + \omega \leq 0, \forall j = 1, \ldots, n$$

$$u_r \geq 0 \qquad \forall r = 1, \ldots, s$$

$$v_i \geq 0 \qquad \forall i = 1, \ldots, m$$

$$\omega \in \mathbb{R}$$

DEA input-oriented model under VRS
Dual (Envelopment) formulation (7)

$$\min \theta$$

$$\text{s.t.} \ \theta x_{ik} - \sum_{j=1}^{n} \lambda_j x_{ij} \geq 0, \forall i = 1, \ldots, m$$

$$\sum_{j=1}^{n} \lambda_j y_{rj} \geq y_{rk}, \forall r = 1, \ldots, s$$

$$\sum_{j=1}^{n} \lambda_j = 1$$

$$\lambda_j \geq 0 \qquad \forall j = 1, \ldots, n$$

$$\theta \in \mathbb{R}$$

DEA output-oriented model under VRS
Primal (Multiplier) formulation (8)

$$\min \sum_{i=1}^{m} v_i x_{ik} + \overline{\omega}$$

$$\text{s.t.} \sum_{r=1}^{s} u_r y_{rk} = 1$$

$$\sum_{i=1}^{m} v_i x_{ij} - \sum_{r=1}^{s} u_r y_{rj} + \overline{\omega} \geq 0, \forall j = 1, \ldots, n$$

$$u_r \geq 0 \qquad \forall r = 1, \ldots, s$$

$$v_i \geq 0 \qquad \forall i = 1, \ldots, m$$

$$\overline{\omega} \in \mathbb{R}$$

DEA output-oriented model under VRS
Dual (Envelopment) formulation (9)

$$\max \phi$$

$$\text{s.t.} \ x_{ik} - \sum_{j=1}^{n} \lambda_j x_{ij} \geq 0, \forall i = 1, \ldots, m$$

$$\sum_{j=1}^{n} \lambda_j y_{rj} \geq \phi y_{rk}, \forall r = 1, \ldots, s$$

$$\sum_{j=1}^{n} \lambda_j = 1$$

$$\lambda_j \geq 0 \qquad \forall j = 1, \ldots, n$$

$$\phi \in \mathbb{R}$$

In general, under the VRS assumption the orientation of the assessment (input or output) affects the facet of the projection and the resulting DMUs' efficiencies may not be the same, although the subset of efficient DMUs is the same irrespective of the model orientation. The decision variables ω and $\overline{\omega}$ allow the identification of the returns to scale nature of the frontier facet (see for all Banker & Thrall, 1992; Seiford & Zhu, 1999).

4 The Evolution of the DEA Methodology

Since the seminal publication of Charnes et al. (1978) and Banker et al. (1984), there has been an impressive growth of knowledge in this area, both in theoretical aspects and in the applications of these developments to practical situations. Given the vastness of the literature in this area, it is becoming increasingly difficult to follow the developments that have been proposed without having adequate literature reviews. These papers seek to enrich the DEA literature and improve the understanding of this area of knowledge. There are DEA literature reviews focused on theoretical developments (Cook & Seiford, 2009; Cooper et al., 2007, 2011; Dakpo et al., 2016; Seiford, 1996). These papers provide valuable guidance to readers taking their first steps in this area, as well as help to broaden the vision of long-time researchers and practitioners. The resulting insights may encourage reflection on the most promising future directions in this area. In a personal view of the main methodological developments in DEA, from its origin to the present, we would like to highlight four main areas (which naturally do not correspond to an exhaustive overview of the DEA literature, and there are many other relevant areas that are not mentioned here).

1. **Frontier construction and efficiency measurement**.
   This theme includes developments related to:

   - Type of returns to scale of the models (Banker et al., 1984; Banker, 1984; Färe et al., 1985; Banker & Thrall, 1992; Podinovski, 2004);
   - Shape of the frontier, i.e. convex versus non-convex (Deprins et al., 1984; Briec et al., 2004);
   - Stochastic analysis of efficiency estimates, including the use of *boostrapping* (Simar & Wilson, 1998), *partial frontiers* (Aragon et al., 2005; Cazals et al., 2002; Simar, 2003) and *robust conditional models* to incorporate the impact of environmental factors on the performance of the units under evaluation (Daraio & Simar, 2007; De Witte & Marques, 2010; De Witte & Kortelainen, 2013);
   - Non-radial evaluation of the distance to the frontier, including the development of the *additive model* (Charnes et al., 1985), the *Slacks-Based Measure* (SBM) (Tone, 2001), the *Russell measure* (Färe & Lovell, 1978; Pastor et al., 1999), and *Directional Distance Function* (DDF) models (Chambers et al., 1996a);
   - Construction of *Benefit-of-the-Doubt* (BoD) composite indicator models, which aggregate key performance indicators without the productive relation of transformation of inputs into outputs (Cherchye et al., 2007), and *Directional BoD* composite indicators, incorporating undesirable outputs and weight restrictions (Zanella et al., 2015).

   Recent developments in these topics can be found for example in Bădin et al. (2019), Chen and Zhu (2020), Daraio et al. (2020b), Kerstens and Van de Woestyne (2021), Pereira et al. (2021a), Taleb et al. (2022).

2. **Incorporation of decision maker judgements in the models.**
   This theme includes developments related to:

   - Models with weight restrictions, including absolute weight restrictions (Dyson & Thanassoulis, 1988), assurance regions (Thompson et al., 1990) and virtual weight restrictions (Wong & Beasley, 1990);
   - Models for defining objectives to be achieved by organisations (Golany 1988; Sowlati & Paradi, 2004; Thanassoulis & Dyson, 1992; Zhu 1996);
   - Economic efficiency models, such as *cost efficiency* (Färe et al., 1985), *revenue efficiency*, *profit efficiency* (Fare et al., 1994).

   Recent developments in these topics can be found for example in Kuosmanen, Cherchye, and Sipiläinen (2006); Camanho and Dyson (2008); Sotiros, Rodrigues, and Silva (2022).

3. **Comparisons among groups and modelling DMU's internal structure.**
   This theme includes developments related to:

   - Comparative analysis of the performance of groups of DMUs (Charnes et al., 1981; Camanho & Dyson, 2006).
   - *Network* DEA (Charnes et al., 1986; Färe & Grosskopf, 2000; Färe et al., 2007)

   Recent developments in these topics can be found for example in Aparicio et al. (2017), Pastor et al. (2020), Chu and Zhu (2021).

4. **Evolution of productivity over time.**
   This theme includes developments related to:

   - *Malmquist index* for DEA models (Fare et al., 1994)
   - *Luemberger indicator* for DDF models (Chambers et al., 1996b)

   Recent developments in these topics can be found for example in Horta and Camanho (2015), Oliveira et al. (2020), Pereira et al. (2021b).

5 DEA Applications Roadmap

William Cooper, one of the 'fathers' of DEA, has always defended the concept of "*application driven theory*". Effectively, being DEA an operational research/ management science technique, it is natural that the main objective of many methodological developments is to ensure that the models adequately respond to concrete challenges of organisations. Accordingly, the application potential of DEA models is very vast and the information that can be obtained through DEA models to support better decision making is very rich, not only within organisations, but also at sector, regional, national or international level. The practical application of DEA involves several challenges that must examined and resolved, including those relating to the homogeneity of the units under assessment, the input/output set selected and the appropriateness of the measurement scales, the incorporation of exogenous conditions in the performance assessment exercise, or the selection of an adequate formulation of the DEA model, eventually with restricted flexibility of input/output weights. Each of these issues can present difficulties in practice, leading to pitfalls that may compromise the validity and managerial relevance of the results obtained. A detailed analysis of pitfalls and protocols in the application of DEA is available in Dyson et al. (2001).

To illustrate the vastness of the DEA application field, we bring to attention the literature reviews by Gattoufi et al. Gattoufi, Oral, and Reisman (2004), Liu et al. (2013a, 2013b, 2016) together with surveys on specific areas, e.g., on financial institutions (Berger & Humphrey, 1997; Fethi & Pasiouras, 2010), health (Hollingsworth, 2008), energy and environment (Zhou et al., 2008), education (De Witte & López-Torres, 2017). In the last years, environmental efficiency, supply chain, transport, public policy, finance and health are among the main application fields of DEA (Ahn et al., 2018; Emrouznejad & Yang, 2018; Emrouznejad et al., 2019; Daraio et al., 2020a; Rostamzadeh et al., 2021). Recent literature reviews on specific empirical applications cover these topics and many more, as for example DEA models for energy and environmental assessment (Mardani et al., 2018), DEA models to support supplier selection and related management decisions within a sustainability perspective and the adoption of hybrid approaches (Vörösmarty & Dobos, 2020; Dutta et al., 2021), DEA implementation in assessing agricultural productivity with undesirable output (Štreimikis & Saraji, 2021), widely used DEA methods in the efficiency analysis of primary health care (Zakowska & Godycki-Cwirko, 2020), local government efficiency analysis and the role of its determinants with new trends including climate changes and pollution (Milán-García et al., 2021). Table 1 provides a more extensive overview of recent literature reviews, meta-analysis and bibliometric studies.

It is worth emphasising that DEA models are also commonly used in the economic regulation and for policy making, such as in the electricity sector (see Tobiasson et al., 2021 for the Norwegian Energy Regulatory Authority), in the water industry (see Heesche & Bogetoft Pedersen, 2021 for the Danish Water Regulatory Authority), more generally in European regulation regimes (see Agrell et al. 2017; Ennis & Deller, 2019), by the European Commission (see Agasisti et al., 2017) and by the Organisation for Economic Co-operation and Development (see Dutu and Sicari 2020), among others.

The increasing interest in the performance evaluation and the practical use of DEA methods among researchers and policy makers has led to a widespread availability of softwares and codes to implement the empirical analysis. Daraio, Kerstens, Nepomuceno, and Sickles (2019) provide a very recent survey of the options currently available. Just to name a few, the interested users might find dedicated programs (e.g., *DEA-Excel* or *PIM-DEAsoft*), softwares to develop the codes (*Matlab* or *R*), programming languages to call mathematical programming solvers (e.g., *AMPL*, *C++*, *phyton*), or to use packages apt to run DEA models (e.g., "*DEA Toolbox*" in *Matlab* or "*Benchmarking*" in *R*).

Table 1 Recent literature reviews on efficiency, DEA and its applications

Authors	Title
Ahn et al. (2018)	Recent developments on the use of DEA in the **public sector**
Emrouznejad and Yang (2018)	A survey and analysis of the first 40 years of scholarly literature in DEA: 1978–2016
Fall et al. (2018)	DEA and SFA research on the efficiency of **microfinance institutions**: A meta-analysis
Guersola et al. (2018)	**Supply chain** performance measurement: a systematic literature review
Mardani et al. (2018)	Data envelopment analysis in **energy and environmental economics**: an overview of the state-of-the-art and recent development trends
Soheilirad et al. (2018)	Application of data envelopment analysis models in **supply chain management**: A systematic review and meta-analysis
Sassanelli et al. (2019)	**Circular economy** performance assessment methods: A systematic literature review
Tran et al. (2019)	A systematic literature review of efficiency measurement in **nursing homes**
Ahmad et al. (2020)	**Banking** sector performance, profitability, and efficiency: a citation-based systematic literature review
Daraio et al. (2020a)	Empirical surveys of frontier applications: a meta-review
Kaffash et al. (2020)	A survey of data envelopment analysis applications in the **insurance industry** 1993–2018
Mahmoudi et al. (2020)	The origins, development and future directions of data envelopment analysis approach in **transportation systems**
Mohd Chachuli et al. (2020)	**Renewable energy** performance evaluation studies using the data envelopment analysis (DEA): A systematic review
Cvetkoska and Savic (2021)	DEA in **banking**: Analysis and visualization of bibliometric data
Vörösmarty and Dobos (2020)	A literature review of **sustainable supplier** evaluation with Data Envelopment Analysis
Zakowska and Godycki-Cwirko (2020)	Data envelopment analysis applications in **primary health care**: a systematic review
Afsharian et al. (2021)	A review of DEA approaches applying a common set of weights: The perspective of centralized management
Charles et al. (2021)	Data Envelopment Analysis and Big Data: A Systematic Literature Review with Bibliometric Analysis
Dutta et al. (2021)	Applications of data envelopment analysis in **supplier selection** between 2000 and 2020: a literature review
Milán-García et al. (2021)	**Local government** efficiency: reviewing determinants and setting new trends
Rostamzadeh et al. (2021)	Application of DEA in benchmarking: a systematic literature review from 2003–2020
Štreimikis and Saraji (2021)	Green productivity and undesirable outputs in **agriculture**: a systematic review of DEA approach and policy recommendations
Dyckhoff and Souren (2022)	Integrating multiple criteria decision analysis and production theory for performance evaluation: Framework and review
Mergoni and De Witte (2022)	**Policy evaluation** and efficiency: a systematic literature review

6 Opportunities for Future Developments in DEA

More than four decades of research has offered interesting insights on the pervasive and spreading role of efficiency assessment techniques in modern societies. Nevertheless, research in DEA is far from being exhausted and offers many opportunities for future developments. We conclude this chapter discussing three potential avenues for further research as emerged from the outlined literature reviews.

First, addressing sustainability issues might pose new challenges from a methodological development perspective, given the wide arrays of dimensions to be taken into account when exploring this field, and the different, sometimes conflicting, interests involved (Vörösmarty & Dobos, 2020). DEA can make a considerable contribution to the design of public policies and business strategies, leading to better decision making.

Second, there is an emerging literature proposing the combination of DEA models (and more in general frontier methods for efficiency measurement) with econometric policy evaluation techniques to assess policy interventions, both from efficiency and effectiveness perspectives (for a very recent review, see Mergoni & De Witte, 2022). The policy evaluation literature has been focusing on the causal assessment of policy intervention, neglecting the potential mismanagement of the resources involved in the policy implementation process (for a review, see Abadie & Cattaneo, 2018). On the contrary, the efficiency literature has been mostly focusing on correlational evidence, preventing a causal interpretation of the results (D'Inverno et al., 2021). These two streams of literature have been rather separated so far, but the combination of their toolkit represents a promising area for future research. This can allow the development of increasingly refined tools, and provide policy makers with sound evidence-based policy recommendations.

Finally, the emergence of big data paves the way for the development of new tools and algorithms (Charles et al., 2021). As remarked by Zhu (2020), "DEA is evolving into data-enabled analytics". To this extent, challenges might be related to handle the computational burden of large-scale assessments, to uncover spatial patterns or to unravel complex interactions among DMUs and variables.

References

Abadie, A., & Cattaneo, M. D. (2018). Econometric methods for program evaluation. *Annual Review of Economics, 10*, 465–503.

Afsharian, M., Ahn, H., & Harms, S. G. (2021). A review of DEA approaches applying a common set of weights: The perspective of centralized management. *European Journal of Operational Research, 294*(1), 3–15.

Agasisti, T., Hippe, R., Munda, G., et al. (2017). Efficiency of investment in compulsory education: Empirical analyses in Europe. Technical Report. Joint Research Centre (Seville site).

Agrell, P. J., Bogetoft, P., et al. (2017). Regulatory benchmarking: Models, analyses and applications. *Data Envelopment Analysis Journal, 3*(1–2), 49–91.

Ahmad, N., Naveed, A., Ahmad, S., & Butt, I. (2020). Banking sector performance, profitability, and efficiency: A citation-based systematic literature review. *Journal of Economic Surveys, 34*(1), 185–218.

Ahn, H., Afsharian, M., Emrouznejad, A., & Banker, R. (2018). Recent developments on the use of DEA in the public sector. *Socio-Economic Planning Science, 61*, 1–3.

Aigner, D., Lovell, C. K., & Schmidt, P. (1977). Formulation and estimation of stochastic frontier production function models. *Journal of Econometrics, 6*(1), 21–37.

Aigner, D. J., & Chu, S. F. (1968). On estimating the industry production function. *The American Economic Review, 58*(4), 826–839.

Aparicio, J., Crespo-Cebada, E., Pedraja-Chaparro, F., & Santín, D. (2017). Comparing school ownership performance using a pseudo-panel database: A malmquist-type index approach. *European Journal of Operational Research, 256*(2), 533–542.

Aragon, Y., Daouia, A., & Thomas-Agnan, C. (2005). Nonparametric frontier estimation: a conditional quantile-based approach. *Econometric Theory, 21*(2), 358–389.

Bădin, L., Daraio, C., & Simar, L. (2019). A bootstrap approach for bandwidth selection in estimating conditional efficiency measures. *European Journal of Operational Research, 277*(2), 784–797.

Banker, R. D. (1984). Estimating most productive scale size using data envelopment analysis. *European Journal of Operational Research, 17*(1), 35–44.

Banker, R. D., & Thrall, R. M. (1992). Estimation of returns to scale using data envelopment analysis. *European Journal of Operational Research, 62*(1), 74–84.

Banker, R. D., Charnes, A., & Cooper, W. W. (1984). Some models for estimating technical and scale inefficiencies in data envelopment analysis. *Management Science, 30*(9), 1078–1092.

Banker, R. D., Gadh, V. M., & Gorr, W. L. (1993). A monte carlo comparison of two production frontier estimation methods: Corrected ordinary least squares and data envelopment analysis. *European Journal of Operational Research, 67*(3), 332–343.

Berger, A. N., & Humphrey, D. B. (1997). Efficiency of financial institutions: International survey and directions for future research. *European Journal of Operational Research, 98*(2), 175–212.

Briec, W., Kerstens, K., & Eeckaut, P. V. (2004). Non-convex technologies and cost functions: Definitions, duality and nonparametric tests of convexity. *Journal of Economics, 81*(2), 155–192.

Camanho, A., & Dyson, R. (2006). Data envelopment analysis and Malmquist indices for measuring group performance. *Journal of Productivity Analysis, 26*(1), 35–49.

Camanho, A., & Dyson, R. (2008). A generalisation of the farrell cost efficiency measure applicable to non-fully competitive settings. *Omega, 36*(1), 147–162.

Cazals, C., Florens, J. P., & Simar, L. (2002). Nonparametric frontier estimation: A robust approach. *Journal of Econometrics, 106*(1), 1–25.

Chambers, R. G., Chung, Y., & Färe, R. (1996). Benefit and distance functions. *Journal of Economic Theory, 70*(2), 407–419.

Chambers, R. G., Fāure, R., & Grosskopf, S. (1996). Productivity growth in APEC countries. *Pacific Economic Review, 1*(3), 181–190.

Charles, V., Gherman, T., & Zhu, J. (2021). Data envelopment analysis and big data: A systematic literature review with bibliometric analysis. In *Data-enabled analytics* (pp. 1–29).

Charnes, A., Cooper, W. W., & Rhodes, E. (1978). Measuring the efficiency of decision making units. *European Journal of Operational Research, 2*(6), 429–444.

Charnes, A., Cooper, W. W., & Rhodes, E. (1981). Evaluating program and managerial efficiency: An application of data envelopment analysis to program follow through. *Management Science, 27*(6), 668–697.

Charnes, A., Cooper, W. W., Golany, B., Seiford, L., & Stutz, J. (1985). Foundations of data envelopment analysis for Pareto-Koopmans efficient empirical production functions. *Journal of Econometrics, 30*(1–2), 91–107.

Charnes, A., Cooper, W., Golany, B., Halek, R., Klopp, G., Schmitz, E., & Thomas, D. (1986). *Two-phase data envelopment analysis approaches to policy evaluation and management of army recruiting activities: Tradeoffs between joint services and army advertising*. Tex, USA: Center for Cybernetic Studies University of Texas-Austin Austin.

Chen, K., & Zhu, J. (2020). Additive slacks-based measure: Computational strategy and extension to network DEA. *Omega, 91*, 102022.

Cherchye, L., Moesen, W., Rogge, N., & Van Puyenbroeck, T. (2007). An introduction to 'benefit of the doubt' composite indicators. *Social Indicators Research, 82*(1), 111–145.

Chu, J., & Zhu, J. (2021). Production scale-based two-stage network data envelopment analysis. *European Journal of Operational Research, 294*(1), 283–294.

Cobb, C. W., & Douglas, P. H. (1928). A theory of production. *The American Economic Review, 18*(1), 139–165.

Cook, W. D., & Seiford, L. M. (2009). Data envelopment analysis (DEA)-Thirty years on. *European Journal of Operational Research, 192*(1), 1–17.

Cooper, W., Seiford, L., Tone, K., & Zhu, J. (2007). Some models and measures for evaluating performances with DEA: Past accomplishments and future prospects. *Journal of Productivity Analysis, 28*(3), 151–163.

Cooper, W. W., Seiford, L. M., & Zhu, J. (2011). Data envelopment analysis: History, models, and interpretations. In *Handbook on data envelopment analysis* (pp. 1–39). Springer

Cvetkoska, V., & Savic, G. (2021) DEA in banking: Analysis and visualization of bibliometric data. *Data Envelopment Analysis Journal.*

Dakpo, K. H., Jeanneaux, P., & Latruffe, L. (2016). Modelling pollution-generating technologies in performance benchmarking: Recent developments, limits and future prospects in the nonparametric framework. *European Journal of Operational Research, 250*(2), 347–359.

Daraio, C., & Simar, L. (2007). Conditional nonparametric frontier models for convex and nonconvex technologies: A unifying approach. *Journal of Productivity Analysis, 28*(1), 13–32.

Daraio, C., Kerstens, K. H., Nepomuceno, T. C. C., & Sickles, R. (2019). Productivity and efficiency analysis software: An exploratory bibliographical survey of the options. *Journal of Economic Surveys, 33*(1), 85–100.

Daraio, C., Kerstens, K., Nepomuceno, T., & Sickles, R. C. (2020). Empirical surveys of frontier applications: A meta-review. *International Transactions in Operational Research, 27*(2), 709–738.

Daraio, C., Simar, L., & Wilson, P. W. (2020). Fast and efficient computation of directional distance estimators. *Annals of Operations Research, 288*(2), 805–835.

De Witte, K., & Kortelainen, M. (2013). What explains the performance of students in a heterogeneous environment? Conditional efficiency estimation with continuous and discrete environmental variables. *Applied Economics, 45*(17), 2401–2412.

De Witte, K., & López-Torres, L. (2017). Efficiency in education: A review of literature and a way forward. *Journal of the Operational Research Society, 68*(4), 339–363.

De Witte, K., & Marques, R. C. (2010). Incorporating heterogeneity in non-parametric models: A methodological comparison. *International Journal of Operational Research, 9*(2), 188–204.

Debreu, G. (1951). The coefficient of resource utilization. *Econometrica: Journal of the Econometric Society* 273–292

Deprins, D., Simar, L., Tulkens, H. (1984). Measuring labor inefficiency in post offices. In M. Marchand, P. Pestieau, & H. Tulkens (Eds.), *The performance of public enterprises: Concepts and measurements*, (pp. 243–267). Amsterdam, North-Holland.

Dutta, P., Jaikumar, B., Arora, M. S. (2021). Applications of data envelopment analysis in supplier selection between 2000 and 2020: A literature review. *Annals of Operations Research*, 1–56

Dutu, R., & Sicari, P. (2020). Public spending efficiency in the OECD: Benchmarking health care, education, and general administration. *Review of Economic Perspectives, 20*(3), 253–280.

Dyckhoff, H., & Souren, R. (2022). Integrating multiple criteria decision analysis and production theory for performance evaluation: Framework and review. *European Journal of Operational Research, 297*(3), 795–816.

Dyson, R. G., & Thanassoulis, E. (1988). Reducing weight flexibility in data envelopment analysis. *Journal of the Operational Research Society, 39*(6), 563–576.

Dyson, R. G., Allen, R., Camanho, A. S., Podinovski, V. V., Sarrico, C. S., & Shale, E. A. (2001). Pitfalls and protocols in DEA. *European Journal of Operational Research, 132*(2), 245–259.

D'Inverno, G., Smet, M., & De Witte, K. (2021). Impact evaluation in a multi-input multi-output setting: Evidence on the effect of additional resources for schools. *European Journal of Operational Research, 290*(3), 1111–1124.

Emrouznejad, A., & Gl, Yang. (2018). A survey and analysis of the first 40 years of scholarly literature in DEA: 1978–2016. *Socio-Economic Planning Sciences, 61*, 4–8.

Emrouznejad, A., Banker, R. D., & Neralic, L. (2019). Advances in data envelopment analysis: Celebrating the 40th anniversary of DEA and the 100th anniversary of Professor Abraham Charnes' birthday. *European Journal of Operational Research, 278*(2), 365–367.

Ennis, S., & Deller, D. (2019). Water sector ownership and operation: An evolving international debate with relevance to proposals for nationalisation in Italy. CERRE report

Fall, F., Am, Akim, & Wassongma, H. (2018). DEA and SFA research on the efficiency of microfinance institutions: A meta-analysis. *World Development, 107*, 176–188.

Färe, R., & Grosskopf, S. (2000). Network DEA. *Socio-Economic Planning Sciences, 34*(1), 35–49.

Färe, R., & Lovell, C. K. (1978). Measuring the technical efficiency of production. *Journal of Economic theory, 19*(1), 150–162.

Färe, R., Grosskopf, S., & Lovell, C. K. (1985). The measurement of efficiency of production, vol 6. Springer Science & Business Media

Fare, R., Färe, R., Fèare, R., Grosskopf, S., & Lovell, C. K. (1994). *Production frontiers*. Cambridge University Press.

Färe, R., Grosskopf, S., & Whittaker, G. (2007). Network DEA. In: *Modeling data irregularities and structural complexities in data envelopment analysis* (pp. 209–240). Springer

Farrell, M. J. (1957). The measurement of productive efficiency. *Journal of the Royal Statistical Society: Series A (General), 120*(3), 253–281.

Fethi, M. D., & Pasiouras, F. (2010). Assessing bank efficiency and performance with operational research and artificial intelligence techniques: A survey. *European Journal of Operational Research, 204*(2), 189–198.

Gattoufi, S., Oral, M., & Reisman, A. (2004). A taxonomy for data envelopment analysis. *Socio-Economic Planning Sciences, 38*(2–3), 141–158.

Golany, B. (1988). An interactive MOLP procedure for the extension of DEA to effectiveness analysis. *Journal of the Operational Research Society, 39*(8), 725–734.

Greene, W. H. (1980). Maximum likelihood estimation of econometric frontier functions. *Journal of Econometrics, 13*(1), 27–56.

Guersola, M., Lima, E. P. D., & Steiner, M. T. A. (2018). Supply chain performance measurement: A systematic literature review. *International Journal of Logistics Systems and Management, 31*(1), 109–131.

Heesche, E., & Bogetoft Pedersen, P. (2021). Incentives in regulatory DEA models with discretionary outputs: The case of Danish water regulation. Technical Report, IFRO Working Paper.

Hollingsworth, B. (2008). The measurement of efficiency and productivity of health care delivery. *Health Economics, 17*(10), 1107–1128.

Horta, I. M., & Camanho, A. S. (2015). A nonparametric methodology for evaluating convergence in a multi-input multi-output setting. *European Journal of Operational Research, 246*(2), 554–561.

Kaffash, S., Azizi, R., Huang, Y., & Zhu, J. (2020). A survey of data envelopment analysis applications in the insurance industry 1993–2018. *European Journal of Operational Research, 284*(3), 801–813.

Kerstens, K., & Van de Woestyne, I. (2021). Cost functions are nonconvex in the outputs when the technology is nonconvex: Convexification is not harmless. *Annals of Operations Research*, 1–26.

Koopmans, T. C. (1951). An analysis of production as an efficient combination of activities. In T. C. Koopmans (Ed.), *Activity analysis of production and allocation*, Cowles Commission for Research in Economics. Monograph No. 13, Wiley, New York

Kuosmanen, T., & Johnson, A. L. (2010). Data envelopment analysis as nonparametric least-squares regression. *Operations Research, 58*(1), 149–160.

Kuosmanen, T., & Kortelainen, M. (2012). Stochastic non-smooth envelopment of data: Semi-parametric frontier estimation subject to shape constraints. *Journal of Productivity Analysis, 38*(1), 11–28.

Kuosmanen, T., Cherchye, L., & Sipiläinen, T. (2006). The law of one price in data envelopment analysis: Restricting weight flexibility across firms. *European Journal of Operational Research, 170*(3), 735–757.

Liu, J. S., Lu, L. Y., Lu, W. M., & Lin, B. J. (2013). Data envelopment analysis 1978–2010: A citation-based literature survey. *Omega, 41*(1), 3–15.

Liu, J. S., Lu, L. Y., Lu, W. M., & Lin, B. J. (2013). A survey of DEA applications. *Omega, 41*(5), 893–902.

Liu, J. S., Lu, L. Y., & Lu, W. M. (2016). Research fronts in data envelopment analysis. *Omega, 58*, 33–45.

Mahmoudi, R., Emrouznejad, A., Shetab-Boushehri, S. N., & Hejazi, S. R. (2020). The origins, development and future directions of data envelopment analysis approach in transportation systems. *Socio-Economic Planning Sciences, 69*, 100672.

Mardani, A., Streimikiene, D., Balezentis, T., Saman, M. Z. M., Nor, K. M., & Khoshnava, S. M. (2018). Data envelopment analysis in energy and environmental economics: An overview of the state-of-the-art and recent development trends. *Energies, 11*(8), 2002.

Mergoni, A., & De Witte, K. (2022). Policy evaluation and efficiency: A systematic literature review. *International Transactions in Operational Research, 29*(3), 1337–1359.

Milán-García, J., Rueda-López, N., & De Pablo-Valenciano, J. (2021). Local government efficiency: Reviewing determinants and setting new trends. *International Transactions in Operational Research*

Mohd Chachuli, F. S., Ahmad Ludin, N., Mat, S., & Sopian, K. (2020). Renewable energy performance evaluation studies using the data envelopment analysis (DEA): A systematic review. *Journal of Renewable and Sustainable Energy, 12*(6), 062701.

Oliveira, R., Zanella, A., & Camanho, A. S. (2020). A temporal progressive analysis of the social performance of mining firms based on a Malmquist index estimated with a benefit-of-the-doubt directional model. *Journal of Cleaner Production, 267*, 121807.

Pastor, J. T., Ruiz, J. L., & Sirvent, I. (1999). An enhanced DEA Russell graph efficiency measure. *European Journal of Operational Research, 115*(3), 596–607.

Pastor, J. T., Lovell, C. K., & Aparicio, J. (2020). Defining a new graph inefficiency measure for the proportional directional distance function and introducing a new Malmquist productivity index. *European Journal of Operational Research, 281*(1), 222–230.

Pereira, M. A., Camanho, A. S., Figueira, J. R., & Marques, R. C. (2021). Incorporating preference information in a range directional composite indicator: The case of Portuguese public hospitals. *European Journal of Operational Research, 294*(2), 633–650.

Pereira, M. A., Camanho, A. S., Marques, R. C., & Figueira, J. R. (2021). The convergence of the world health organization member states regarding the united nations' sustainable development goal 'good health and well-being'. *Omega, 104*, 102495.

Podinovski, V. V. (2004). Bridging the gap between the constant and variable returns-to-scale models: Selective proportionality in data envelopment analysis. *Journal of the Operational Research Society, 55*(3), 265–276.

Richmond, J. (1974). Estimating the efficiency of production. *International Economic Review*, 515–521.

Rostamzadeh, R., Akbarian, O., Banaitis, A., & Soltani, Z. (2021). Application of DEA in benchmarking: A systematic literature review from 2003–2020. *Technological and Economic Development of Economy, 27*(1), 175–222.

Sassanelli, C., Rosa, P., Rocca, R., & Terzi, S. (2019). Circular economy performance assessment methods: A systematic literature review. *Journal of Cleaner Production, 229*, 440–453.

Seiford, L. M. (1996). Data envelopment analysis: The evolution of the state of the art (1978–1995). *Journal of Productivity Analysis, 7*(2), 99–137.

Seiford, L. M., & Zhu, J. (1999). An investigation of returns to scale in data envelopment analysis. *Omega, 27*(1), 1–11.

Shephard, R. W. (1970). *Theory of cost and production functions*. Princeton University Press.

Simar, L. (2003). Detecting outliers in frontier models: A simple approach. *Journal of Productivity Analysis, 20*(3), 391–424.

Simar, L., & Wilson, P. W. (1998). Sensitivity analysis of efficiency scores: How to bootstrap in nonparametric frontier models. *Management Science, 44*(1), 49–61.

Soheilirad, S., Govindan, K., Mardani, A., Zavadskas, E. K., Nilashi, M., & Zakuan, N. (2018). Application of data envelopment analysis models in supply chain management: A systematic review and meta-analysis. *Annals of Operations Research, 271*(2), 915–969.

Sotiros, D., Rodrigues, V., & Silva, M. C. (2022). Analysing the export potentials of the Portuguese footwear industry by data envelopment analysis. *Omega, 108*, 102560.

Sowlati, T., & Paradi, J. C. (2004). Establishing the "practical frontier" in data envelopment analysis. *Omega, 32*(4), 261–272.

Štreimikis, J., & Saraji, M. K. (2021). Green productivity and undesirable outputs in agriculture: A systematic review of DEA approach and policy recommendations. *Economic Research*, 1–35.

Taleb, M., Khalid, R., Ramli, R., Ghasemi, M. R., & Ignatius, J. (2022). An integrated bi-objective data envelopment analysis model for measuring returns to scale. *European Journal of Operational Research, 296*(3), 967–979.

Thanassoulis, E., & Dunstan, P. (1994). Guiding schools to improved performance using data envelopment analysis: An illustration with data from a local education authority. *Journal of the Operational Research Society, 45*(11), 1247–1262.

Thanassoulis, E., & Dyson, R. (1992). Estimating preferred target input-output levels using data envelopment analysis. *European Journal of Operational Research, 56*(1), 80–97.

Thompson, R. G., Langemeier, L. N., Lee, C. T., Lee, E., & Thrall, R. M. (1990). The role of multiplier bounds in efficiency analysis with application to Kansas farming. *Journal of Econometrics, 46*(1–2), 93–108.

Tobiasson, W., Llorca, M., & Jamasb, T. (2021). Performance effects of network structure and ownership: The Norwegian electricity distribution sector. *Energies, 14*(21), 7160.

Tone, K. (2001). A slacks-based measure of efficiency in data envelopment analysis. *European Journal of Operational Research, 130*(3), 498–509.

Tran, A., Nguyen, K. H., Gray, L., & Comans, T. (2019). A systematic literature review of efficiency measurement in nursing homes. *International Journal of Environmental Research and Public Health, 16*(12), 2186.

Vörösmarty, G., & Dobos, I. (2020). A literature review of sustainable supplier evaluation with data envelopment analysis. *Journal of Cleaner Production, 264*, 121672.

Winsten, C. (1957). Discussion on Mr. Farrell's paper. *Journal of the Royal Statistical Society Series A, 120*, 282–284.

Wong, Y. H., Beasley, J. (1990). Restricting weight flexibility in data envelopment analysis. *Journal of the Operational Research Society, 41*(9), 829–835.

Zakowska, I., & Godycki-Cwirko, M. (2020). Data envelopment analysis applications in primary health care: A systematic review. *Family Practice, 37*(2), 147–153.

Zanella, A., Camanho, A. S., & Dias, T. G. (2015). Undesirable outputs and weighting schemes in composite indicators based on data envelopment analysis. *European Journal of Operational Research, 245*(2), 517–530.

Zhou, P., Ang, B. W., & Poh, K. L. (2008). A survey of data envelopment analysis in energy and environmental studies. *European Journal of Operational Research, 189*(1), 1–18.

Zhu, J. (1996). Data envelopment analysis with preference structure. *Journal of the Operational Research Society, 47*(1), 136–150.

Zhu, J. (2020). DEA under big data: Data enabled analytics and network data envelopment analysis. *Annals of Operations Research*, 1–23.

Stochastic Frontier Analysis: A Review and Synthesis

Mara Madaleno and Victor Moutinho

1 Introduction

Schaltegger and Sturm (1992/94) proposed the term Eco-Efficiency (EE), which was only formally defined and adopted by the World Business Council for Sustainable Development (WBCSD) in 1991 (WBCSD, 1995), as a way of achieving more value from lower inputs of material and energy with reduced emissions. More definitions of EE emerged afterward which may be consulted in Table 1 of Côté et al. (2006). Therefore, to produce with EE means to produce goods and services with minimal environmental damage. As a representative of the EE indicator, usually, a ratio is used, we need the desirable output (for a country, usually the Gross Domestic Product (GDP) is used as a numerator) and the undesirable output (an air emissions representative such as carbon dioxide (CO2) or greenhouse gases (GHG) is used in the denominator) (Moutinho & Madaleno, 2021a, 2021b).

To aggregate environmental pressures to build the EE index, two methods may be used. One which is non-parametric, the Data Envelopment Analysis (DEA), and the parametric one, the Stochastic Frontier Analysis (SFA). Both methods are used to analyze and reach the maximum boundary, turning possible to maximize the metric for EE (the closer to one the higher the EE). The study of EE, thus, joins the study of economic and environmental efficiency. Although we should bear in mind that economic efficiency does not imply environmental efficiency (Robaina-Alves et al.,

M. Madaleno (✉)
GOVCOPP—Research Unit in Governance, Competitiveness and Public Policy, DEGEIT—Department of Economics, Management, Industrial Engineering and Tourism, University of Aveiro, 3810-193 Aveiro, Portugal
e-mail: maramadaleno@ua.pt

V. Moutinho
NECE—Research Unit in Business Sciences, Department of Management and Economics, University of Beira Interior, 6200-209 Covilhã, Portugal
e-mail: ferreira.moutinho@ubi.pt

P. Macedo et al. (eds.), *Advanced Mathematical Methods for Economic Efficiency Analysis*, Lecture Notes in Economics and Mathematical Systems 692,
https://doi.org/10.1007/978-3-031-29583-6_4

Table 1 Reviewed articles and citations

Year	Title	Authors	Journal title	Total
2023	Examining the Relationship Between Eco-efficiency and Energy Poverty: A Stochastic Frontier Models Approach	Moutinho, Leitao, Silva, Serrasqueiro	World-Systems Evolution and Global Futures	0
2022	Understanding water-energy nexus in drinking water provision: An eco-efficiency assessment of water companies	Molinos-Senante, Maziotis, Sala-Garrido, Mocholi-Arce	Water Research	0
2021	Evaluation of the eco-efficiency of territorial districts with seaport economic activities	Quintano, Mazzocchi, Rocca	Utilities Policy	3
2021	(Eco)efficiency in food production: A systematic review of the literature	da Silva, Thomé	Revista em Agronegocio e Meio Ambiente	1
2021	The cost of reducing municipal unsorted solid waste: Evidence from municipalities in Chile	Molinos-Senante, Maziotis	Sustainability (Switzerland)	3
2021	Greenhouse gas emission inefficiency spillover effects in European countries	Kutlu, Wang	International Journal of Environmental Research and Public Health	6
2021a	Assessing eco-efficiency in Asian and African countries using stochastic frontier analysis	Moutinho, Madaleno	Energies	6
2020	Greenhouse gas emission efficiencies of world countries	Kutlu	International Journal of Environmental Research and Public Health	12
2020	The effect of urban air pollutants in Germany: eco-efficiency analysis through fractional regression models applied after DEA and SFA efficiency predictions	Moutinho, Madaleno, Macedo	Sustainable Cities and Society	37

(continued)

Table 1 (continued)

Year	Title	Authors	Journal title	Total
2020	Sustainable land use management for improving land eco-efficiency: a case study of Hebei, China	Deng, Gibson	Annals of Operations Research	25
2019	Improving eco-efficiency for the sustainable agricultural production: A case study in Shandong, China	Deng, Gibson	Technological Forecasting and Social Change	88
2019	Measurement and evolution of eco-efficiency of coal industry ecosystem in China	Wang, Wan, Yang	Journal of Cleaner Production	38
2018	Economic-environmental efficiency of European agriculture–A generalized maximum entropy approach	Moutinho, Robaina, Macedo	Agricultural Economics (Czech Republic)	23
2017	A Parametric Approach to Estimating Eco-Efficiency	Orea, Wall	Journal of Agricultural Economics	17
2016	Measuring eco-efficiency using the stochastic frontier analysis approach	Orea, Wall	International Series in Operations Research and Management Science	7

2015). To study economic efficiency we can go through the technical efficiency study (the ability of a production unit to reach the maximum output from a set of inputs and the production technology) or through allocative efficiency (the ability of a production unit in using an optimal proportion of inputs considering its prices and the production technology). If the production process relies upon fossil fuels or technologies, it is technically efficient and cheap but leads to higher emission levels and environmental inefficiency. However, if technical or economic inefficiency is present, there will be environmental inefficiency provided the waste of resources and increased production levels (waste of raw materials, capital, labor, inefficient use of energy, etc.).

Frontier methods like DEA and SFA are being increasingly recommended by researchers and practitioners to assess the impact of operational strategies and policies on EE performance. Allowing the incorporation of multiple outputs and inputs in performance measurement, frontier methods provide a benchmark, or frontier, against which we may identify areas of "best practices" and "worst practices", close to 1 (high performance) or 0 (worst performance), respectively. Using frontier analysis allows policymakers to identify the gap between their actual performance and optimal performance. Therefore, frontier methods have the advantage to provide a numerical value of performance (the technical efficiency mentioned above) that is objective. Thus, easy to read and interpret, facilitating the understanding of the most efficient resource allocation, and helping decision-makers to measure the outcomes of their various strategies and policies. For a more complete review of frontier methods, we recommend the work of Assaf and Josiassen (2016).

To measure efficiency, we need to start with the estimation of the production or cost frontiers. The main issue is how to reach an estimate of this frontier from a random sample of inputs and outputs. The non-parametric approach to frontier estimation imposes a limited structure on the frontier technology estimation such as the DEA method. Parametric methods like SFA assume a different approach to estimating frontier technology. In the parametric approach, we need to specify a functional form for the relationship between outputs and inputs, being the most common the Cobb–Douglas and the translog. The literature (Pavelescu, 2011) indicates that the translog is better considering it does not assume and implies perfect or smooth substitution between production factors or perfect competition on the production factors market, introducing nonlinear relationships between the outputs and the inputs. To sum up, parametric approaches include a random error, whereas DEA does not, and although the translog is more flexible it can drive incorrect efficiency estimates for observations that are not close to the mean scale (Kumbhakar & Lovell, 2000). Moreover, there is the challenge of the correct selection of the distribution of the inefficiency term. There is no consensus in the literature about which of the half-normal, gamma, exponential, or truncated is the best (Kumbhakar & Lovell, 2000).

EE assessment is usually based on the Data Envelopment Analysis (DEA) approach. However, to construct this index of environmental pressure we may use the Stochastic Frontier Analysis (SFA) (Orea & Wall, 2016, 2017). EE measurement permits advancing research on the environmental impact of economic activity. Its calculus allows us to compare economic outputs obtained through the production of goods and services with the environmental impacts or pressures derived from it. When using solely the DEA technique we need to incorporate bootstrapped truncated regressions in a second stage to analyze the EE determinants (Moutinho & Madaleno, 2021a, 2021b). With SFA, despite dealing with measurement errors in the data, we may directly incorporate the EE determinants in one stage, directly in the stochastic frontier model proposed (Moutinho & Madaleno, 2021a; Orea & Wall, 2016; Reinhard et al., 2000; Song et al., 2018).

The rest of the chapter develops as follows. The next section provides a mathematical specification of the SFA method. In the following section, we summarize the use of frontier methods in the eco-efficiency assessment literature, using data source the Scopus database. With this review of SFA applications in the EE assessment, we can address possible limitations in the current literature and suggest an agenda for future research.

2 Stochastic Frontier Analysis (SFA)

2.1 The SFA Method

As previously stated in the introduction, Stochastic Frontier Analysis (SFA) is a technique for estimating inefficiency. It does that by modeling a frontier concept within a regression framework. Developed by Aigner et al. (1977) and Meeusen and Broeck (1977), the SFA methodology emerged as a theoretical and practical framework for defining and estimating a production function frontier. Since the author's seminal works, the literature has presented numerous reformulations and extensions of their original works. Under SFA (Schmidt, 1986), a data-driven econometric production function is defined as the technological relationship between observed inputs and outputs in use. Thus, it indicates the average levels of outputs that can be produced from a given level of inputs. As with any other production function, we reach the maximum possible produced output provided the set of inputs is considered a boundary or frontier. Any deviations from this frontier are interpreted as inefficiencies. Considering that no economic unit can exceed the maximum frontier, all its deviations are regarded as inefficiencies. In the common regression counterpart method, the output's conditional mean has to be specified, differing from SFA. The SFA approach is analytical using parametric econometric techniques. As also stated in the introduction, the SFA parametric approach differs from the non-parametric approaches as DEA assumes a deterministic frontier. Therefore, in SFA frontier deviations are allowed to represent inefficiency and statistical noise.

Comparing DEA with SFA, the latter is a more realistic approach by including statistical noise in the observed deviation from the estimated frontier. Thus, the frontier specification should be like any other stochastic regression function. Furthermore, SFA observations usually incorporate a random walk by including simultaneously technical inefficiency across production units and a one-sided error term being i.i.d. (independent and identical distribution). Statistically, SFA is implemented by specifying a regression model characterized by a composite error term that incorporates a one-sided disturbance (the inefficiency) and the traditional idiosyncratic disturbance (capturing measurement errors and classical noise). Hence, the SFA goal is to make inferences about frontier parameters and inefficiency usually estimated through likelihood-based methods. This is done independently if we are dealing with a panel or cross-sectional data, with time-varying or constant inefficiency. Traditionally, are used the Cobb–Douglas production functions (Hannesson, 1983), constant elasticity of substitution (CES) production functions (Campbell & Lindner, 1990), and translog production functions (Pascoe & Robinson, 1998; Squires, 1987). These ensure a linkage between the SFA-based efficiency measure and the necessary components to identify energy efficiency determinants.

Following the description of Assaf and Josiassen (2016) of parametric frontier methods as SFA, we can build its empirical framework. It starts with a regression specification involving a logarithmic transformation of the technical efficiency measure, as described in Eq. (1).

$$\ln y_{it} = \ln f\left(X'_{it}, \beta\right) + v_{it} - u_{it} \tag{1}$$

Equation (1) relates the observed output of the *i-th* productive unit at time *t*, y_{it}, to the production frontier, $f\left(x'_{it;}\beta\right)$, x_{it} is a vector of input and explanatory variables, with β_i, a vector of technology parameters. We have $i = 1, \ldots, n; t = 1, \ldots, T$ in Eq. (1), and it is added a random error term, u_{it}, and a measurement error, v_{it}, as presented in (1). We will have $y_{it} = a_{it} f\left(X_{it;}\beta\right)$, with $0 < a_{it} \geq 1$ and $u_{it} = -\ln a_{it} \geq 0$, the random error term, measuring deviations from the frontier technology and representing the technical inefficiency. The output will be bounded by the stochastic frontier $f\left(x'_{it;}\beta\right) \exp(v_{it})$. Considering that inefficiency cannot assume negative values, u_{it} will follow an asymmetric distribution, such as the half-normal, truncated, exponential, or gamma. If panel data is to be considered, Eq. (1) turns a standard unobserved effects model with distributional assumption avoidance.

As noted previously, efficiency scores are computed concerning the maximum feasible possible output given by the stochastic frontier. The model of Eq. (1) accommodates only one output. In the case of multiple outputs, it turns necessary to estimate a distance function. As it happens with DEA, this distance function can be estimated considering either output or input orientation. Furthermore, to estimate Eq. (1) parameters, estimation methods like maximum likelihood (ML), generalized method of moments (GMM), and Bayesian can be used. We just need to specify the functional form of $f\left(x'_{i;}\beta\right)$, meaning, the relationship established between outputs and inputs. As noted previously, the literature proposes several functional forms for *f(.)*, the most common ones the Cobb–Douglas and the translog. So, we need to select the functional production form of the stochastic frontier analysis.

The Cobb–Douglas stochastic frontier specification can be specified as in Eq. (2) whereas the translog stochastic frontier specification can be expressed as in Eq. (3).

$$\ln y_{it} = \beta_0 + \sum_{i=1}^{N} \beta_{it} \ln x_{it} + v_{it} - u_{it}$$

$$\Leftrightarrow y_{it} = \exp(\beta_0 + \sum_{i=1}^{N} \beta_{it} \ln X x_{it}) \times \exp(v_{it}) \times \exp(-u_{it}) + \varepsilon_{it} \tag{2}$$

In Eq. (2), $exp(\beta_0 + \sum_{i=1}^{N} \beta_{it} ln\, x_{it})$ represents the deterministic component, $exp(v_{it})$ represents the noise, and $exp(-u_{it})$ represents inefficiency. The composed error term ε_{it} is the difference between a normally distributed disturbance, where v_{it} symbolizes measurement and specification error, and a one-sided disturbance, u_{it}, denoting inefficiency. Likewise, v_{it} and u_{it} are assumed to be independent of each other and across observations. The disturbance v_{it} accounts for random unobserved factors that can be positive or negative. The translog can be specified as follows:

$$\ln y_{it} = \beta_0 + \sum_{i=1}^{N} \beta_{it} \ln x_{it} + \frac{1}{2} \sum_{i=1}^{N} \sum_{j=1}^{N} \beta_{ijt} \ln x_{it} \ln x_{jt} + v_{it} - u_{it} \tag{3}$$

There are advantages and disadvantages associated with both functional forms, but the translog is considered to be more advantageous as compared to the Cobb–Douglas. Namely, according to Pavelescu (2011, p. 133), it does not assume rigid premises like perfect substitution among production factors, nor even perfect competition in the production factors market. The translog also allows nonlinear relationships between the outputs and inputs whereas the Cobb–Douglas does not.

Most of the existent stochastic frontier models enable the prediction of inefficiency effects. The measure of technical efficiency (TE_{it}) relies on between zero and one and maybe computed as follows Eq. (4):

$$TE_{it} = \exp(-u_{it}) \tag{4}$$

The different model specifications for the production function which can be employed in SFA analysis consider distinct specifications for the technical inefficiency effects or the term u_{it}. In the following, we will present five of these different specifications, having in mind that SFA specifications can be expressed both as time-varying or time-invariant models. The difference is that in the latter the technical inefficiency error term is assumed to be constant over time ($u_{it} = u_{i1} = u_{i2} = \cdots = u_{iT}$), whereas in the latter, time-varying models, it is allowed the technical inefficiency error term varies over time.

The model in Eq. (2) has been extended by Pitt and Lee (1981) to longitudinal data considering time-invariant models. They used MLE (maximum likelihood estimation) of the half-normal stochastic frontier model considering that $u_i \sim N(0, \sigma_u^2)$ and $v_i \sim N^+(0, \sigma_v^2)$. Thus, the probability density function of each u_i is a truncated version of a normal random variable having zero mean and variance σ_u^2, and the error terms v_i are independently and identically distributed normal random variables with zero means and variances σ_v^2.

Battese and Coelli (1988) generalized this model to the normal-truncated case, being this time-invariant model estimated through the MLE method, and the long run (or persistent inefficiency) is given by Eq. (4) with $E(u_{it}|v_{it} - u_{it})$. Besides the use of panel data in SFA for the first time by Pitt and Lee (1981), the link between panel data and frontier literature was only established by Schmidt and Sickles (1984). Schmidt and Sickles (1984) have demonstrated the advantages of panel data estimation provided that fewer assumptions are required, leading to consistent technical inefficiency, and permitting the researcher to appraise if inefficiency embodied by the one-sided part of the error term is time varying or constant. However, the disadvantage is the assumption that inefficiency is constant over time.

A stochastic frontier model with time-invariant inefficiency can be estimated using traditional fixed-effects estimation techniques where the individual technical inefficiency is treated as an unknown fixed parameter to be estimated. For Schmidt and Sickles (1984), the general fixed-effects production frontier model, when we assume that technical inefficiency is time invariant, can be written as in Eq. (5):

$$\ln y_{it} = \beta_{0i} + \ln f\left(X'_{it;}\beta\right) + v_{it}$$

$$\text{where} \quad \beta_{0i} = \overline{\beta_0} - u_i \tag{5}$$

In Eq. (5), the error term v_{it} is uncorrelated with the regressors. No distributional assumptions are required for the time-invariant u_i being only assumed that it is correlated with v_{it} or with the regressors. The Feasible Generalized Least Squares (FGLS) model provides consistent estimators of the parameters, being valid even if either N or T grows to infinity. We recover u_i as $\widehat{u_i} = \widehat{\beta_0} - \widehat{\beta_{0i}}$, where the individual time-invariant technical inefficiency is obtained through $\widehat{\beta_0} = max_i\left\{\widehat{\beta_{0i}}\right\}$ guaranteeing that $\widehat{u_i} \geq 0$. To sum up, assuming fixed effects in the time-invariant model, there is no distribution assumption for the error termu_i, being allowed u_i to be correlated with the regressors and withv_{it}. The parameters are estimated through OLS, and β estimates are consistent as either $N \to \infty$ or $T \to \infty$ (being this assumption a requirement for consistency).

Besides the fixed-effects model, we can employ the random-effects model. In this situation, the error term u_i is assumed as randomly distributed, being uncorrelated with the regressors and with v_{it}. To estimate the model, a two-step generalized least squares (GLS) method may be used. However, assuming that technical inefficiency is time invariant can be very strong or even unrealistic. In the '90s, Battese and Coelli (1992, 1995) proposed models that consider a stochastic frontier production function under panel data with a time-varying inefficient error term. In Battese and Coelli (1992), the inefficient error term is given by $u_{it} = G(t)u_i$, being G a function for a time, under the assumptions $u_i \sim N(0, \sigma_u^2)$ and $v_i \sim N^+(0, \sigma_v^2)$. Respectively, we have that the probability density function of each u_i is a truncated version of a normal random variable having zero mean and variance σ_u^2, and that the error terms v_i are independently and identically distributed normal random variables with zero means and variances σ_v^2. Afterward, Battese and Coelli (1995) specified the technical inefficiency effect as $u_{it} = z_{it}\delta + W_{it}$, being the random variable W_{it} assumed to be independently distributed and obtained by truncation (at zero) of the normal distribution with mean $z_{it}\delta$ and variance σ^2. Therefore, the point of truncation is $W_{it} \geq z_{it}\delta$.

Cornwell et al. (1990) assume the model presented in Eq. (2) and that the technical inefficiency error term is time varying. The authors tackled the problem from the perspective of a panel regression model with individual-specific slope coefficients, where $u_{it} = \omega_i + \omega_{i1}t + \omega_{i1}t^2$. The quadratic specification allows for a unit-specific temporal pattern of inefficiency. Using this, the model parameters are estimated by extending the conventional fixed or random-effects panel data estimators. If distribution and independence assumptions are valid, we may use MLE to estimate the parameters (Battese & Coelli, 1995).

Afterward, Kumbhakar and Lovell (2000) proposed that if v_{it} is heteroscedastic, the consistency of parameter estimates and technical inefficiency is maintained beneath both time-invariant fixed-effects and random-effects approaches. However, the maximum likelihood method is impractical unless T is large relative to N. Though heteroscedastic problems can occur in time-varying models. When x is heteroscedastic, the Cornwell et al. (1990) methodology can be used to correct the inaccurate estimation. When the error term is heteroscedastic only the random-effects model is considered once the error term can't be heteroscedastic if the error term is fixed effects. In the presence of heteroscedasticity for both error term parameters, to have consistent estimates, we may only use the generalized method of moments (GMM). But, heteroscedastic problems can occur in time-varying models. When v_{it} is heteroscedastic, the Cornwell et al. (1990) method can be used to correct the imprecise estimation. Thus, when u_{it} is heteroscedastic, only the random-effects model is taken into account once it is not possible for the error term u_{it} to be heteroscedastic if the error term is fixed effects.

2.2 SFA Approach in EE

Kuosmanen and Kortelainen (2005) defined EE, in a frontier setting, as a ratio between economic value added (α) and environmental damage or pollution (p). The pressure-generating technology set $T = \left\{(\alpha, p) \in R^{(1+k)} | \alpha can be generated by p\right\}$ proposed by the authors can describe all the feasible combinations of economic value, α, and environmental pressure, p. Total environmental damage, defined by D(p), is measured by aggregating the K environmental pressures $(p_1, \ldots, p_k)$ which are due to production activity.

If individually assessed, for producer i, individual EE scores can be obtained through Eq. (6).

$$EE = \frac{Economic\ Value\ Added}{Environmental\ Pressure} = \frac{\alpha_i}{D_i(p)} \tag{6}$$

Total environmental damage, defined by D(p), is a function aggregating all environmental pressures in a unique indicator, by a linear weighted average of the individual environmental pressures (Eq. (7)), being w_k assigned to p_k.

$$D_i(p) = \sum_{k=1}^{K} w_{k,i}\, p_{k,i} \tag{7}$$

Both Kuosmanen and Korteinen (2005), Picazo-Tadeo et al. (2012) use DEA as a non-subjective weighting method. For each firm i, using DEA, the EE score can be computed as in Eq. (8), by solving this programming problem:

$$max_{w_{ki}} EE_i = \frac{\alpha_i}{\sum_{k=1}^{K} w_{k,i} p_{k,i}} \tag{8}$$

Subject to

$$\frac{\alpha_j}{\sum_{k=1}^{K} w_{k,i} p_{k,j}} \leq 1; j = 1, \ldots, N$$

And

$$w_{k,i} \geq 0; k = 1, \ldots, K$$

These optimization problem constraints force the weights to be non-negative and make EE scores take values between 0 and 1, namely,

$$EE_i = \frac{\alpha_i}{\sum_{k=1}^{K} w_{k,i} p_{k,i}} \leq 1, \forall i = 1, \ldots, N \tag{9}$$

Equation (9) is used to derive a stochastic frontier EE model. Under the parametric approach, the coefficients of environmental pressures are estimated and represent marginal contributions to value added. To ensure non-negativity we may reparametrize the weights as $w_k = e^{\beta_k}$ and taking logs in (9) results in Eq. (10).

$$lnEE_i = ln\left(\frac{\alpha_i}{\sum_{k=1}^{K} e^{\beta_k}.p_{k,i}}\right) \leq 0 \tag{10}$$

Equation (10) may be rewritten as

$$ln(\alpha_i) = ln\left(\sum_{k=1}^{K} e^{\beta_k}.p_{k,i}\right) - u_i \tag{11}$$

Being $u_i = -lnEE_i \geq 0$, which can be interpreted as a non-negative random term capturing firm i's eco-inefficiency. By extending the model in (11) by adding a symmetric random noise term, v_i, and a non-zero intercept θ, we are incorporating the effects of random shocks to economic added value (Eq. (12)).

$$ln(\alpha_i) = \theta + ln\left(\sum_{k=1}^{K} e^{\beta_k}.p_{k,i}\right) + v_i - u_i \tag{12}$$

In Eq. (12), deviations from the frontier due to random noise are integrated jointly with eco-inefficiency. Having a non-zero intercept guarantees unbiased parameter estimates if random noise components have a level effect on the firm's economic value added. As previously discussed, the error term $\varepsilon_i = v_i - u_i$ in Eq. (12) is composed of two independent parts, namely, v_i, a two-sided random noise term assumed to be normally distributed with mean zero and constant standard deviation ($\sigma_v = e^{\delta}$) and u_i is a one-sided error term that captures the underlying eco-inefficiency, which in accordance to Aigner et al. (1977) follows a half-normal distribution, being the truncation at zero of a normally distributed random variable of mean zero and constant standard deviation given by $\sigma_u = e^{\gamma}$.

Considering these distributional assumptions, the density function of the composed error term $\varepsilon_i = v_i - u_i$ in (12) is equal to the density function of a standard normal-half-normal frontier model. Following Kumbhakar and Lovell (2000), Orea and Wall (2017), the log-likelihood function for a sample of N firms/producers can be written as in Eq. (13).

$$\ln L(\theta, \beta, \delta, \gamma) = -\frac{N}{2} ln\left[\sigma_v^2 + \sigma_u^2\right] + \sum_{i=1}^{N} \ln \Phi\left[-\frac{\varepsilon_i(\theta, \beta).{}^{\sigma_u}/_{\sigma_v}}{\left(\sigma_v^2 + \sigma_u^2\right)^{1/2}}\right] - \frac{1}{2\left(\sigma_v^2 + \sigma_u^2\right)} \sum_{i=1}^{N} \varepsilon_i(\theta, \beta)^2 \tag{13}$$

where $\beta = (\beta_1, \ldots, \beta_K)$ and $\varepsilon_i(\theta, \beta) = ln(\alpha_i) - \theta - ln\left(\sum_{k=1}^{K} e^{\beta_k}.p_{k,i}\right)$. The likelihood function in (13) can be maximized concerning parameters $(\theta, \beta, \delta, \gamma)$ to obtain their consistent estimates of our EE model. The error term $\varepsilon_i(\theta, \beta)$ is simply $\varepsilon_i = v_i - u_i$. Following Jondrow et al. (1982) we may use the conditional expectation $E(u_i|\varepsilon_i)$ for the asymmetric random term u_i estimation and compute the firm's EE score as $exp(-u_i)$.

For traditional SFA production models, $\varepsilon_i(\theta, \beta)$ will be a simple linear function of the parameters to be estimated, but $\varepsilon_i(\theta, \beta) = ln(\alpha_i) - \theta - ln\left(\sum_{k=1}^{K} e^{\beta_k}.p_{k,i}\right)$ in the EE frontier model is a nonlinear function of the β parameters. Moreover, the SFA model based on $\varepsilon_i(\theta, \beta) = ln(\alpha_i) - \theta - ln\left(\sum_{k=1}^{K} e^{\beta_k}.p_{k,i}\right)$ can accommodate heteroscedastic inefficiency and noise terms if we model the variances σ_v^2 and σ_u^2 as functions of some exogenous variables.

As noticed by its developers and applicants, compared to the DEA EE model, the SFA approach mitigates the effects of outliers and measurement errors in the data on the EE scores. It should be emphasized that the "technology" in the EE model is a simple index aggregating all environmental pressures into a unique value. Therefore, the parametric specification of a functional form is not as tricky as it might be in a traditional production function with multiple inputs and outputs simultaneously.

Considering that we have few parameters to be estimated, like in DEA, the SFA model can be used even when we have a small number of observations to work with.

The β parameters have an interesting interpretation under the parametric model. Provided the EE score is constant and equal to 1 along the EE frontier if we differentiate (6) and use the reparameterization of the pressure weights in (10) we will get Eq. (14).

$$\frac{\partial \alpha_i / \partial p_j}{\partial \alpha_i / \partial p_k} = \frac{e^{\beta_j}}{e^{\beta_k}} \quad (14)$$

Equation (14) represents the marginal rate of technical substitution of environmental pressures providing important information on the possibilities of substitutions among pressures and the effects for firms of legislation requiring a reduction in individual pressures. Thus, once we compute the β parameters, e^{β_k} represents the marginal contribution of pressure p_k to firm i's economic value added, or else, it is the monetary loss in the economic value if the pressure p_k was reduced by one unit.

3 Applications of SFA in EE Assessment

A Scopus documents database search within the Article Title, Abstract, and Keywords using "Stochastic Frontier Analysis" and "Eco-Efficiency", on 20 December 2022, yields 18 documents. The first was published in 2016 and the most recent was already accepted to be published in 2023. We had to remove three of these works from the analysis considering that one uses DEA for EE assessment, which is not our goal, the other considers a completely different setting, and a third was impossible for the authors to download the article. The 15 articles to be presented next have an h-index of 8, meaning that of the 16 documents considered for the h-index, 8 have been cited at least 8 times. Table 1 summarizes the total citations of the 15 documents. In total there have been 289 citations for the 18 collected articles at the date the collection was made. The highest number of citations has been made from 2017 onward, respectively, 2019 with 17, 2020 with 41, 2021 with 88, and finally 2022 with 136 citations made to these works.

Not all documents are journal articles, we have considered all in the search and the most recent is, for example, a book chapter. The most cited article is the one from Deng and Gibson (2019) about "Improving eco-efficiency for sustainable agricultural production: A case study in Shandong,China" which was published by Technological Forecasting and Social Change (Table 1). The most recent articles are those with fewer citations, but still, we observe that several applications of the SFA methodology have been made in the context of EE. In the following, we will concentrate on the details of each of these articles and comment on the directions of future research besides the use of their applications in practical terms. The structure of the literature analysis will follow the decreasing date of publication. As such, we will start from the most recent article to the latter.

Moutinho et al. (2023) made use of a stochastic frontier analysis log-linear function and comparable data for 26 European economies over the 2008–2020 period. They select to analyze the relationship between selected socioeconomic indicators of energy poverty and eco-efficiency measures. Results were derived from the estimations of a time-varying fixed-effects model, a true random-effects model, and an inefficiency effects model under truncated-normal. Regarding selected energy poverty determinants, the main empirical results point to a significant interaction between those same socioeconomic determinants and relevant measures of eco-efficiency. The authors conclude that in the period of the first Kyoto commitment (2008–2012), Sweden, Hungary, France, Latvia, and Lithuania stand out in the TOP5 in terms of technical eco-efficiency. But in the second Kyoto period (2013–2018) in this TOP5, we have Sweden, Hungary, France, Latvia, and Slovenia. Thus only one country changed position between the two periods.

For the water sector, Molinos-Senante et al. (2022) estimated hyperbolic distance functions to compute eco-efficiency. The empirical application focused on the English and Welsh water companies from 2011 to 2019. With this, the potential reduction in greenhouse gas emissions (GHG) by water companies was estimated, being concluded that water companies could expand water delivered by 8.75% and reduce emissions by 8.0%. Moreover, the authors state that average environmental efficiency and eco-efficiency scores were 0.920 and 0.962, respectively, and that the English and Welsh water industry operates under decreasing economies of scale. Notice that due to the economies of scale, the cost of reducing GHG emissions was higher for water and sewerage companies than for water-only companies. Thus, environmental efficiency and eco-efficiency are useful to understand the water provision/energy costs/GHG emissions nexus.

Following Quintano et al. (2021), research papers connected to environmental efficiency using radial and non-radial (SBM—slacks-based measure) SFA have been only two. Serra et al. (2014) studied the environmental and technical efficiency (of Catalan arable crop farms) and Trujillo and Tovar (2007) studied the economic efficiency in the European port industry. Quintano et al. (2021) assessed the performance of 24 European ports from the perspective of EE. SFA was used for port performance evaluation. Results suggest that recent management efforts to boost the volume of the ports' activity do not worsen their eco-efficiency.

In the agriculture sector, EE has been explored by da Silva and Thomé (2021), Moutinho et al. (2018), and Orea and Wall (2017). A systematic literature review on (eco)efficiency in agricultural production was conducted by da Silva and Thomé (2021). From the 100 articles collected, 69 were extensively analyzed to conclude that 14% of the articles were published by the Journal of Cleaner Production. The authors denoted a considerable growth in studies on (eco)efficiency and that for 75% of the articles the research method adopted was the DEA or SFA. Moutinho et al. (2018) aimed to estimate the agricultural economic-environmental efficiency (eco-efficiency) for European countries (from 2005 until 2010) using DEA and SFA with a generalized maximum entropy (GME) approach. The EE measure used is the

GVA/GHG ratio, with agriculture gross value added (GVA) deemed as the desired output and greenhouse gas (GHG) emissions as the unwanted output. The inputs used were capital, labor, land, energy, and nutrients. Positions in the EE ranking changed between 2005 and 2010, with Portugal maintaining its position under those revealing higher EE levels. Finally, Orea and Wall (2017) noticed that DEA was solely used in the literature up to that date to measure agriculture producers' EE and used the SFA approach using a sample of 50 Spanish dairy farms. The authors conclude that the SFA model yields nearly identical EE scores to those calculated by DEA. They suggest extending the stochastic frontier model to include the determinants of EE, where the explanatory variables are integrated right into the inefficiency component. Previously, and using the same dataset, Orea and Wall (2016) emphasized the usefulness of the SFA approach to measuring eco-efficiency. The authors defend that in addition to dealing with measurement errors in the data, the SFA allows determinants of eco-efficiency to be incorporated in one stage. Additionally, it permits an analysis of the potential substitutability between environmental pressures.

For China, environmental problems and environmental issues are becoming increasingly urgent. Thus, Wang et al. (2019) analyzed the EE of the coal industry ecosystem to contribute toward the healthy development of coal-mining cities (28 were considered), using SFA based on the assumptions of a translog production function. The authors declare significant regional differences with ownership structure and environmental regulations exerting significant inhibitory effects on the improvement of EE, whereas marketization level, technological innovation, and energy price are revealed to have significant positive impacts. Also for China, Deng and Gibson (2019) estimated land productivity in Shandong, China during 1990–2010 and analyzed the EE based on SFA. Results pointed out that regional EE in Shandong was mostly over 0.9, except for cities located far from the political or economic centers. Overall, after the pointed existent trade-offs between agricultural production and urbanization, the authors argue to be necessary to adjust its agricultural technological measures according to local specific conditions to improve eco-efficiency for sustainable agricultural development. One year later, the same authors Deng and Gibson (2020) studied the relationship between land use management and land EE in Hebei, China using SFA to analyze land use conversions and land EE. Results revealed that land use output is the key factor connecting land use management and land EE and demonstrate a corresponding decrease in land EE with a reduction in the distance to the city center.

For a specific country, Chile, Molinos-Senante and Maziotis (2021) study quantify the marginal cost of reducing any unsorted waste using SFA, allowing them to estimate the EE of the waste sector. Results imply substantial eco-inefficiency in the sector once the average eco-efficiency score is roughly 0.5. Thus, the municipalities could approximately halve their operational costs and unsorted waste to produce the same level of output. The average marginal cost of reducing unsorted waste computed by the authors was 32.28 Chilean pesos per ton, even if prominent differences are revealed among the waste utilities evaluated. For a specific country as well, Moutinho et al. (2020) analyze the effect of urban air pollution for the specific years 2007, 2010, and 2013 in 24 German cities. Both DEA and SFA approaches are used to predict eco-efficiency scores. Afterward, fractional regression is applied to infer the influencing factors of the eco-efficiency scores, at the city level. Results suggest a significant impact on the eco-efficiency of excess PM10, the average temperature, the average NO2 concentration, and rainfall.

Considering several countries, Kutlu (2020) uses SFA to infer the efficiencies of countries in terms of achieving the lowest GHG emission levels per GDP output in the years between 1990 and 2015. The authors conclude that compared to the 1990–1997 period, 92.50%, 79.51%, and 59.84% of the countries improved their GHG emission efficiencies in the 1998–2007, 2008–2012, and 2013–2015 periods, respectively. Hence, even if the Kyoto protocol helped in increasing GHG emissions efficiency, this efficiency-boosting effect faded away over time. For 22 Asian and 22 African countries, Moutinho and Madaleno (2021a) evaluate the economic and environmental efficiency using as a measure of EE the ratio GDP/CO2 emissions (the output in the model). Capital, labor, fossil fuels, and renewable energy consumption are regarded as inputs, using SFA and a log-linear Translog production function, during 2005 and 2018. Results demonstrate cross-countries heterogeneity among production inputs, namely, labor, capital, and type of energy use and its efficiency. Whereas labor and renewable energy share were found to increase technical eco-efficiency, fixed capital decreases it under time-variant models. However, technical improvements in EE are validated through time considering the time variable in the model estimations, by replacing fossil fuels with renewable sources. Finally, the authors found an inverted U-shaped eco-efficiency function concerning the share of fossil fuel consumption. In the same year, but for European countries, Kutlu and Wang (2021) examine whether spatial spillover effects exist for GHG emission efficiency for 38 European countries between 2005 and 2014. Results point out that inefficiencies of other countries would lead to lower efficiency levels for a country. However, this negative inefficiency spillover effect decreased till 2008, increasing then till 2011, turning relatively stable after 2011. Thus, any strategy to reduce the inefficiencies of other countries could potentially improve a country's efficiency levels. Additionally, the authors conclude for a positive and significant impact of the human development index on GHG emission efficiency levels (it is pointed out that a one standard deviation increase in the human development index would lead to an 11.12 percentage points increase in the GHG emission efficiencies on average). Finally, it is stated that despite the different countries evidencing distinct efficiency levels and efficiency growth patterns over time, the pattern of spatial spillover is quite analogous among all countries over time.

Although very limited this brief literature review is, from these studies we may infer economic and environmental econometric studies of EE assessment, which are very recent. In the following section, we will discuss some policy implications that were and may be derived from the SFA approach to EE assessment and evidence of some possible avenues of future research using SFA, or more advanced methodologies.

4 Policy Recommendations and Possible Future Research Applications of SFA

First, we would highlight the fact that empirical assessment of EE is usually done at the country level but it can be used at the industry or firm level as well. However, the methodology proposed in these studies could be applied to different companies operating in other countries and could also be extended to other different economic activity sectors. Second, it is noted that the methodology of these studies could be used to quantify the environmental impact of any other undesirable product in different industries. In need of analysis is, for example, water leakage or any other industry such as supply interruptions in the energy sector (as suggested by Molinos-Senante et al., 2022).

Considering that regulators recognize the need for and importance of reducing environmental damage, like GHG emissions, in the processes to set specific sector tariffs, EE assessment through SFA needs to be done appropriately. The SFA approach can also be used to infer the percentage increase to be achieved in the desired output and the percentage decrease of the undesired output, or costs, or else, we can deduce from the approach the possible percentages to be reached to be over the efficient production frontier. Therefore, through SFA we may detect higher levels of performance whenever EE is assessed. Moreover, a general conclusion from these explored articles is that the more complex the treatment necessary for production and the involved costs are (energy and others), the more undesirable outputs are to be produced (increases environmental damage) and the lower the EE scores attained. Therefore, it is suggested for policymakers, managers, producers, and stakeholders force lowering costs, improve processes efficiency, bet in technological development and sophistication able to answer challenges more promptly, more efficient and rational use of resources and inputs to keep sustainable growth of outputs.

By using the SFA approach for EE assessment, the firm's managers will be able to evaluate how efficient companies are when expanding production and curtailing GHG emissions at the same time, lowering the environmental damage and reducing associated costs. It also allows for determining how efficiency changes when firms want to reduce costs. It has also been demonstrated that energy costs and other costs are strong determinants of firms' efficiency (Molinos-Senante et al., 2022). But other factors may be identified and corrected, such as technical requirements and their consequent influence on efficiency improvements. Compensations for good practices in the form of financial rewards, lower taxes, green financing access, and others would even help firms to become more eco-efficient.

Studies demonstrate that policy actions can refer to the translog SFA function (Moutinho et al., 2020; Quintano et al., 2021) to verify the potential impact of specific measurements when the undesirable output indicator (GHG, CO2 emissions, etc.) becomes a critical part of the production process and competitive strategy. Moreover, authority managers and policymakers should consider initiatives targeting environmental impact, such as the improvement of transition toward renewable energy sources, by focusing on the firms/sectors/countries that present the strongest SFA efficiency scores (Quintano et al., 2021). Thus, management and legal authorities can design policy actions through the results attained through the SFA model proposed.

Usually, homogeneity among the entities studied to determine EE is considered, but the (observed or unobserved) heterogeneity in the data could also be investigated to identify the clusters having similar behavior patterns. Management can consider specific tools to manage the functional activities, and must not exclude the impact on the efficiency of the dimensions connected to different contextual factors and the national legislative environment (Kutlu, 2020). Economic dimensions such as costs, investment in assets, and quality of services must also be considered even though they are more challenging to measure (Quintano et al., 2021).

The research results explored above also highlight the need to allow generalization that can be applied to different economic sectors. Considering that one of the main limits of the above-presented research is that the models consider only a small number of indicators, an extended investigation of both the involved dimensions and additional dimensions not considered are required, for instance, the incidence of the most prevalent firm dimension. Therefore, additional proxies—and different NACE selections—may be required in further investigation. Future research could also be relevant to add new data covering several years while monitoring the annual variations in efficiency. Furthermore, a spatial econometric analysis could bring valuable interaction efforts in terms of network analysis and its correspondent impact on EE.

The recurrent addition of a second stage when using the DEA demonstrates the relevance of analyzing the influence of non-controllable factors on the (eco)efficiency and turns SFA into a more suitable technique. Topics of future research include a detailed econometric analysis to better study the specific determinants of the eco-efficiency indicators, including the variables of taxes, subsidies, or information about collaboration with other institutions. Another useful approach could be the use of the decomposition analysis to identify the most relevant factors in the eco-efficiency assessment. Provided there remain some uncertainties related to the parameters' estimated values, and considering it reduces the accuracy of the estimated results, it is suggested a joint consideration of interaction factors. For example, land productivity is influenced by natural factors and human activities, which should be considered together.

Results obtained through the application of the SFA approach are of extreme importance to policymakers. They allow stakeholders to evaluate how eco-efficient services, firms, companies, regions, and/or countries are. They allow us to understand how much it costs to deal with any unsorted product. It is fundamental to use tools such as eco-efficiency to support decision-making which integrates not only costs but also environmental impacts. Moreover, policymakers can now understand the factors that affect the desired outputs through EE improvements. This information is essential for adopting strategies such as mergers or eco-taxes to enhance eco-efficiency. Municipalities need to act to improve economic and environmental performance and EE assessment at this Municipal level and by the economic activity sector will help policymakers to understand the needed intervention areas. As well, more periods, and historical data, or even projections allow for the measurement of productivity analysis and its drivers, efficiency change, and technical change. Moreover, the inclusion of undesirable outputs in the analysis could enable us to measure their impact on the elements of productivity change. This type of analysis could recognize, on one side, how the less eco-efficient entities have improved/worsened their performance compared to the best ones (efficiency change). On the other way, how the more eco-efficient entities improved/worsened their performance (technical change). This knowledge could be used by managers to identify the best procedures that could boost productivity and sustainability and move toward a greener economy with massive advantages to the environment and society.

Eco-efficiency can help us measure an entity's sustainability as the above-exposed articles made us realize. Several companies incorporate eco-efficiency into their business strategy, incorporating their operational, product innovation, and marketing strategies. However, it stills remain unanswered the question of how to improve EE. A possibility is through less environmental damage and the reduction of emissions. Another is by exploring the determinants of EE. Moreover, EE is usually measured using a representative of emissions that does not reflect environmental pollution. Using GHG or CO_2 emissions is common but we could use other pollutants like PM2.5, PM10, SO_2, or Nox, to represent the undesirable outputs in EE assessment. As suggested by Moutinho et al. (2020), the development of low-emission zones in urban centers should be established by including targets as limit values and implementing stricter regulations for the inclusion of all relevant air pollutants and emitters. The promotion of electric vehicle circulation in urban areas is also able to reduce emissions despite the improvement in traffic flows in cities, as well as the promotion of renewable energy sources in electricity production, which can be seen as a valuable way to increase EE and ensure sustainable economic growth. For all these suggestions, we should bear in mind that we need to ensure appropriate development in technological terms and policy options need to be available for the correct implementation. Mandatory regulatory instruments and target setting that should be readjusted over time as energy transition processes develop as well as industrial or domestic equipment able to provide correct responses to increased populations and corresponding needs are crucial.

For future research and readers' ease, a debate about the interpretation of what the environmental variables are assumed to affect should be completed. Only one of the above-mentioned articles (Kutlu & Wang, 2021) considers GHG efficiency rather than total GHG emission amounts. Therefore, for given input levels, as long as the GHG emission grows slower than real GDP, the efficiency would enhance. Thus, despite that in general, the GHG emission levels could increase over time, it is possible to observe some increase in efficiency levels. Policymakers should be aware that a potential way to improve GHG emission efficiency levels is by investing in environmentally friendly production technologies or similar measures allowing to lower the bad output keeping or increasing the amount of good output. Imposing environmental taxes may be a way to achieve sustainability goals but these policies must be weighted properly considering the present living costs of the population due to higher inflation levels, decreasing standard living costs and increasing poverty levels. On the other way, these environmental taxes could be transferred to firms to research and invest in environmentally friendly technologies. Therefore, improving the environment cannot be achieved at the expense of individual efforts. Governmental financing available has to be managed in the most appropriate way to reach the general sustainability goals and funds must be applied suitably. Partnerships with universities and research centers could be a way to reach goals and variables like investment in environmentally friendly technologies investment, investment in research and development, financial support received, and types of financing sources should be included as variables able to explain EE levels. As for individuals or families, collected environmental taxes may be redeployed as subsidies in the form of grants, loans with low interest, procurement mandates, better tax treatments, etc., or similar applications allowing families to contribute to EE levels and scores improvement. Individual well-being and public awareness may stimulate greater demand for energy obtained from renewable sources (Moutinho & Madaleno, 2021a; Moutinho et al., 2020; Quintano et al., 2021). This would enhance GHG emission efficiency turning urgent that policymakers should not only use incentive/disincentive tools like subsidies and taxes but also promote public awareness about the need for renewable energy economies.

Some policy advice has been provided by Moutinho and Madaleno (2021a) concerning Asian and African economies, which can be suitable for other realities. When EE is assessed through scores we can observe different behaviors, development stages, heterogeneous growth, different levels of renewables use and production, as well as different levels of environmental efficiency, not only for countries but for firms as well. The authors advised a deep reform of government subsidies which are still granted for fossil fuel production and consumption that the use of renewable energy should be encouraged, and that eco-innovation and energy efficiency should be promoted to ensure environmentally friendly technology investments. Moreover, present energy efficiency projects should be enhanced and promoted, which might help reduce emissions from highly polluting sectors, whereas implementing strategies to maximize benefits from renewable energy technology transfers happening whenever capital goods are to be imported. More concise reforms of stacked regulatory frameworks, learning by doing in what respects decreasing fossil fuel energy consumption, encouraging the creation of renewable energy projects like energy communities to reduce energy poverty levels, fight general poverty levels, and increase fair access to electricity for all. All these recommendations need to be in line with the sustainable development goals proposed by the United Nations. Changing the way of thinking and individuals' behavior is harder, but would increase the good output level (like GDP growth) while simultaneously decreasing the bad output levels (environmental damage). By doing so, policymakers would be able to ensure the necessary living standards and respectable life quality levels, reduce energy poverty and poverty, increase employment, increase income per capita, and ensure better levels of education and awareness. These can only be achieved with effective, direct, and meaningful policy measures and regulations, fighting simultaneously corruption and rent-seeking behaviors or disadvantageous effects that may also emerge in the context.

From these studies, we might also take that it is necessary to account for heterogeneity in the data, which should be included in the EE assessment model. Additionally, provided there are multiple ways to define and measure similar concepts of EE, studies should highlight the theories behind their studies and provide appropriate definitions of what is to be analyzed. Thus, it is necessary to reinforce the difference between eco-efficiency and concepts such as sustainability, environmental effectiveness, or environmental performance indicators, even if the differences between them are shaky. Another suggestion is to build EE indicators able to reflect the real situation of regions, countries, sectors, and firms, such that researchers could use similar data to derive reasonable and comparable results. It is not easy to provide reasonable explanations for the EE levels obtained through the SFA application solely, and more important than doing the EE assessment is to be able to quantify and explain the different levels reached for EE.

DEA has a lot of strengths but does not consider statistical noise and all the production frontier deviations are computed as technical inefficiency, turning DEA more sensitive to outliers. As pointed out in the literature (Kontodimopoulos et al., 2011; Song & Chen, 2019), DEA suffers from measurement errors in the variables included and the omission of unobserved and potentially relevant variables in the EE assessment that might lead to erroneous DEA efficiency measurements. This, in turn, would lead to uninformed policymaking decisions which are to be avoided. Therefore, SFA emerged to overcome these shortcomings assuming that the distance of a production unit from the best practice frontier is the sum of two components: (1) the true inefficiency; (2) random fluctuations. Other SFA advantages include the possibility of specification in the case we are using panel data and given that most of the freely available data by its nature are in the form of panel data (firms, countries, sectors, with year observations) it becomes useful, allowing formal statistical testing of hypothesis and that confidence intervals are constructed. Despite the SFA's usefulness in the EE assessment, very few studies analyze and evaluate EE using SFA as highlighted by Song and Chen (2019) and which we were trying to evidence in the present chapter. A possible explanation for this underutilization of SFA highlighted by Robaina-Alves et al. (2015) is that the functional form used in SFA needs to be correctly specified and the fact that SFA can only be used to estimate the production efficiency with only one output. Thus, to choose between DEA and SFA we need to be aware if we pretend flexibility in the mean structure or if we wish precision in the noise separation. As Coelli (1995) suggests, the SFA is recommended for use in agriculture applications once measurement error, weather, and missing variables are likely to play a significant role in the sector. The studies explored above evidence the SFA approach's usefulness as well in other contexts.

It should also be important to highlight that using SFA for EE assessment, considering as inputs capital, labor, and production-related resources, the estimated EE indicator provides the overall outcome of the economic, resource, and environmental efficiency of the joint use of all production factors. Using the SFA approach, the production frontier is deemed by technical inefficiency, measurement error, statistical noise, and other non-systematic influences, dodging the possibility that a large amount of random noise is possibly mistaken for inefficiency as in DEA. The SFA approach can also explain the variations in the EE effects in terms of other variables in a single-stage approach. Notice that a determinant analysis of EE can also be performed with the DEA technique but in a two-stage process. Finally, the output elasticity of each factor (capital, labor, and other production resources) can be assessed using the estimated parameters of the production frontier, as to reveal the factor liable for the good or bad performance of the region/country/sector/firm in terms of eco-efficiency. In sum, whether a lot has already been done, a lot more remains to be and the present section's suggestions intend to be a starting point for those who pretend to advance research using SFA for the EE assessment.

5 Concluding Remarks

This chapter is an attempt to highlight the usefulness of applying the SFA approach for the EE assessment. For that, initially, it is presented the application of frontier models in EE assessment, and for that, an appropriate definition of EE was provided, while setting the framework for its correct application. Afterward, the SFA methodology was mathematically described, in general settings and considering its application in the EE assessment. Doing a brief search in the literature for SFA approach applications in the EE assessment, the present chapter provides a literature review. As observed, there are currently more applications of SFA in EE assessment, but a lot more remains to be done, especially considering the use of its application. Finally, we end the chapter by evidencing future research trends and by providing policy directions that might be useful to be pursued given the information conveyed by the EE scores obtained through the application of the SFA approach. Policy-makers, managers, investors, and stakeholders can learn from the results presented in this chapter and research can be enhanced by following some of the suggestions presented above, whereas clear future research avenues and applications' usefulness was highlighted and described. Results allow us to favor the SFA methodology as compared to the DEA approach for its easiness of application and superior advantages when assessing EE and establishing rankings or even inferring about its determinants as demonstrated above.

Acknowledgements This work was supported by the Research Unit on Governance, Competitiveness and Public Policies (GOVCOPP), and the Research Unit in Business Science and Economics (NECE-UBI) through the Portuguese Foundation for Science and Technology (FCT—Fundação para a Ciência e a Tecnologia), references UIDB/04058/2020 and UID/GES/04630/2021, respectively.

References

Aigner, D., Lovell, C. A. K., & Schmidt, P. (1977). Formulation and estimation of stochastic frontier production function models. *Journal of Econometrics, 6*(1), 21–37. https://doi.org/10.1016/0304-4076(77)90052-5

Assaf, A. G., & Josiassen, A. (2016). Frontier analysis: A state-of-the-art review and meta-analysis. *Journal of Travel Research, 55*(5), 612–627. https://doi.org/10.1177/0047287515569776

Battese, G. E., & Coelli, T. J. (1988). Prediction of firm-level technical efficiencies with a generalized frontier production function and panel data. *Journal of Econometrics, 38*(3), 387–399. https://doi.org/10.1016/0304-4076(88)90053-X

Battese, G. E., & Coelli, T. J. (1992). Frontier production functions, technical efficiency and panel data: With application to paddy farmers in India. *Journal of Productivity Analysis, 3*(1), 153–169.

Battese, G. E., & Coelli, T. J. (1995). A model for technical inefficiency effects in a stochastic frontier production function for panel data. *Empirical Economics, 20*, 325–332. https://doi.org/10.1007/BF01205442

Campbell, H. F., & Lindner, R. K. (1990). The production of fishing effort and the economic performance of licence limitation programs. *Land Economics, 66*(1), 56–66. https://doi.org/10.2307/3146683

Coelli, T. J. (1995). Recent developments in frontier modelling and efficiency measurement. *Australian Journal of Agricultural Economics, 39*, 219–245. https://doi.org/10.1111/j.1467-8489.1995.tb00552.x

Cornwell, C., Schmidt, P., & Sickles, R. C. (1990). Production frontiers with cross-sectional and time-series variation in efficiency levels. *Journal of Econometrics, 46*(1–2), 185–200. https://doi.org/10.1016/0304-4076(90)90054-W

Côté, R., Booth, A., & Louis, B. (2006). Eco-efficiency and SMEs in Nova Scotia. *Canada. Journal of Cleaner Production, 14*(6–7), 542–550. https://doi.org/10.1016/j.jclepro.2005.07.004

da Silva, J. V. B., & Thomé, K. M. (2021). (Eco)efficiency in food production: A systematic review of the literature. [(Eco)eficiência da produção alimentar: Uma revisão sistemática da literatura]. Revista Em Agronegocio e Meio Ambiente 14(3). https://doi.org/10.17765/2176-9168.2021V14N3E8055

Deng, X., & Gibson, J. (2019). Improving eco-efficiency for the sustainable agricultural production: A case study in Shandong, China. *Technological Forecasting and Social Change, 144*, 394–400. https://doi.org/10.1016/j.techfore.2018.01.027

Deng, X., & Gibson, J. (2020). Sustainable land use management for improving land eco-efficiency: A case study of Hebei, China. *Annals of Operations Research, 290*(1–2), 265–277. https://doi.org/10.1007/s10479-018-2874-3

Hannesson, R. (1983). Bioeconomic production function in fisheries: Theoretical and empirical analysis. *Canadian Journal of Fisheries and Aquatic Sciences, 40*(7), 968–982. https://doi.org/10.1139/f83-123

Jondrow, J., Lovell, C. K., Materov, I. S., & Schmidt, P. (1982). On the estimation of technical inefficiency in the stochastic frontier production function model. *Journal of Econometrics, 19*(2–3), 233–238. https://doi.org/10.1016/0304-4076(82)90004-5

Kontodimopoulos, N., Papathanasiou, N. D., Flokou, A., Tountas, Y., & Niakas, D. (2011). The impact of non-discretionary factors on DEA and SFA technical efficiency differences. *Journal of Medical Systems, 35*(5), 981–989. https://doi.org/10.1007/s10916-010-9521-0

Kumbhakar, S. C., & Lovell, C. A. K. (2000). *Stochastic frontier analysis*. Cambridge University Press. https://doi.org/10.1017/cbo9781139174411

Kuosmanen, T., & Kortelainen, M. (2005). Measuring eco-efficiency of production with data envelopment analysis. *Journal of Industrial Ecology, 9*, 59–72. https://doi.org/10.1162/108819805775247846

Kutlu, L. (2020). Greenhouse gas emission efficiencies of world countries. *International Journal of Environmental Research and Public Health, 17*(23), 1–11. https://doi.org/10.3390/ijerph17238771

Kutlu, L., & Wang, R. (2021). Greenhouse gas emission inefficiency spillover effects in European countries. *International Journal of Environmental Research and Public Health, 18*(9). https://doi.org/10.3390/ijerph18094479

Meeusen, W., & van Den Broeck, J. (1977). Efficiency estimation from Cobb-Douglas production functions with composed error. *International Economic Review, 18*, 435–444. https://doi.org/10.2307/2525757

Molinos-Senante, M., & Maziotis, A. (2021). The cost of reducing municipal unsorted solid waste: Evidence from municipalities in Chile. *Sustainability (Switzerland), 13*(12). https://doi.org/10.3390/su13126607

Molinos-Senante, M., Maziotis, A., Sala-Garrido, R., & Mocholi-Arce, M. (2022). Understanding water-energy nexus in drinking water provision: An eco-efficiency assessment of water companies. *Water Research*, 225. https://doi.org/10.1016/j.watres.2022.119133

Moutinho, V., Leitão, J., Silva, P. M., & Serrasqueiro, J. (2023). Examining the relationship between eco-efficiency and energy poverty: A stochastic frontier models approach. . Retrieved from www.scopus.com. https://doi.org/10.1007/978-3-031-16477-4_7

Moutinho, V., & Madaleno, M. (2021a). A two-stage DEA model to evaluate the technical eco-efficiency indicator in the EU countries. *International Journal of Environmental Research and Public Health, 18*, 3038. https://doi.org/10.3390/ijerph18063038

Moutinho, V., & Madaleno, M. (2021b). Assessing eco-efficiency in Asian and African countries using stochastic frontier analysis. *Energies, 14*(4), 1168. https://doi.org/10.3390/en14041168

Moutinho, V., Madaleno, M., & Macedo, P. (2020). The effect of urban air pollutants in Germany: Eco-efficiency analysis through fractional regression models applied after DEA and SFA efficiency predictions. *Sustainable Cities and Society, 59*. https://doi.org/10.1016/j.scs.2020.102204.

Moutinho, V., Robaina, M., & Macedo, P. (2018). Economic-environmental efficiency of European agriculture–A generalized maximum entropy approach. *Agricultural Economics (czech Republic), 64*(10), 423–435. https://doi.org/10.17221/45/2017-AGRICECON

Orea, L., & Wall, A. (2016). Measuring Eco-efficiency using the stochastic frontier analysis approach. In J. Aparicio, C. Lovell, & J. Pastor (Eds.), *Advances in efficiency and productivity* (Vol. 249). International Series in Operations Research & Management Science. Springer. https://doi.org/10.1007/978-3-319-48461-7_12

Orea, L., & Wall, A. (2017). A parametric approach to estimating eco-efficiency. *Journal of Agricultural Economics, 68*(3), 901–907. https://onlinelibrary.wiley.com/doi/full/10.1111/1477-9552.12209

Pascoe, S., & Robinson, C. (1998). Input controls, input substitution and profit maximization in the English channel beam trawl fishery. *Journal of Agricultural Economics, 49*, 16–33. https://doi.org/10.1111/j.1477-9552.1998.tb01249.x

Pavelescu, F. M. (2011). Some aspects of the translog production function estimation. *Romanian Journal of Economics, 32*, 131–150.

Picazo-Tadeo, A. J., Beltrán-Esteve, M., & Gómez-Limón, J. A. (2012). Assessing eco-efficiency with directional distance functions. *European Journal of Operational Research, 220*(3), 798–809. https://doi.org/10.1016/j.ejor.2012.02.025

Pitt, M., & Lee, L. (1981). The measurement and sources of technical efficiency in the indonesian weaving industry. *Journal of Development Economics, 9*, 43–64.

Quintano, C., Mazzocchi, P., & Rocca, A. (2021). Evaluation of the eco-efficiency of territorial districts with seaport economic activities. *Utilities Policy, 71*. https://doi.org/10.1016/j.jup.2021.101248

Reinhard, S., Knox Lovell, C. A., & Thijssen, G. J. (2000). Environmental efficiency with multiple environmentally detrimental variables; estimated with SFA and DEA. *European Journal of Operational Research, 121*(2), 287–303. https://doi.org/10.1016/S0377-2217(99)00218-0

Robaina-Alves, M., Moutinho, V., & Macedo, P. (2015). A new frontier approach to model the eco-efficiency in European countries. *Journal of Cleaner Production, 103*, 562–573. https://doi.org/10.1016/j.jclepro.2015.01.038

Schaltegger, S., & Sturm, A. (1992/94) Environmentally oriented decisions in companies (in German: Ökologieorientierte Entscheidungen in Unternehmen) (2nd ed.) Haupt, Bern/Stuttgart.

Schmidt, P. (1986). Frontier production functions. *Econometric Reviews, 4*, 289–328. https://doi.org/10.1080/07474938608800089

Schmidt, P., & Sickles, R. C. (1984). Production frontiers and panel data. *Journal of Business & Economic Statistics, 2*(4), 367–374. https://doi.org/10.2307/1391278

Serra, T., Chambers, R. G., & Lansink, A. O. (2014). Measuring technical and environmental efficiency in a state-contingent technology. *European Journal of Operational Research, 236*(2), 706–717. https://doi.org/10.1016/j.ejor.2013.12.037

Song, J., & Chen, X. (2019). Eco-efficiency of grain production in China based on water footprints: A stochastic frontier approach. *Journal of Cleaner Production, 236*, 117685. https://doi.org/10.1016/j.jclepro.2019.117685

Song, M., Peng, J., Wang, J., & Zhao, J. (2018). Environmental efficiency and economic growth of China: A Ray slack-based model analysis. *European Journal of Operational Research, 269*(1), 51–63. https://doi.org/10.1016/j.ejor.2017.03.073

Squires, D. (1987). Fishing effort: Its testing, specification, and internal structure in fisheries economics and management. *Journal of Environmental Economics and Management, 14*(3), 268–282. https://doi.org/10.1016/0095-0696(87)90020-9

Trujillo, L., & Tovar, B. (2007). The European port industry: An analysis of its economic efficiency. *Maritime Economics & Logistics, 9*, 148–171. https://doi.org/10.1057/palgrave.mel.9100177

Wang, D., Wan, K., & Yang, J. (2019). Measurement and evolution of eco-efficiency of coal industry ecosystem in China. *Journal of Cleaner Production, 209*, 803–818. https://doi.org/10.1016/j.jclepro.2018.10.266

WBCSD. (1995). Achieving Eco-efficiency in Business, Report of the World Business Council for Sustainable Development, 14–15 March, Second Antwerp Eco-efficiency Workshop, WBCSD, Conches.

Part 2

Combining Directional Distances and ELECTRE Multicriteria Decision Analysis for Preferable Assessments of Efficiency

Thyago C. C. Nepomuceno and Cinzia Daraio

1 Introduction

Since the introduction of the efficiency measure on the multiple input–output spaces by Charnes et al. (1978), Data Envelopment Analysis (DEA) has consolidated itself as the leading scientific tool for measuring efficiency and productivity in their many facets with countless empirical applications, innovative models, reviews and computational developments (Adler et al., 2002; Aldamak & Zolfaghari, 2017; Daraio et al., 2019, 2020). Nevertheless, one issue that emerges from this highly adopted methodology is how to cope with tradeoffs of compensation among the many possible production possibilities. Surely there must be more than one way that Decision Making Units (DMUs) can resort to production resources to generate goods and services, some of which are preferable to others. In addition, when there is an explicit preference or value judgment from a manager or policymaker, the quantitative measure for technical efficiency might not represent the production scenario as well as a subjective and more flexible measure of efficiency.

Another question that comes up in this discussion is how much efficient do we really need to be? A typical tradeoff between efficiency and quality is reported in many frontier applications (Daraio et al., 2021; Dismuke & Sena, 2001; Nepomuceno et al., 2021; Shimshak & Lenard, 2007), especially in healthcare (Clement et al., 2008; Khushalani & Ozcan 2017; Nayar & Ozcan, 2008; Nepomuceno et al. 2020a, 2020b, 2022a; Piubello Orsini et al., 2021). The evidence is that most inefficient units, e.g., healthcare units with excess beds for hospitalization, staff, or longer service time,

T. C. C. Nepomuceno · C. Daraio
DIAG, Sapienza University of Rome, Rome, Italy
e-mail: daraio@diag.uniroma1.it

T. C. C. Nepomuceno (✉)
Campus Agreste, Federal University of Pernambuco, Caruaru, Brazil
e-mail: thyago.nepomuceno@ufpe.br

P. Macedo et al. (eds.), *Advanced Mathematical Methods for Economic Efficiency Analysis*, Lecture Notes in Economics and Mathematical Systems 692,
https://doi.org/10.1007/978-3-031-29583-6_5

are also high-quality institutions, and technically efficient hospitals serving as many patients as possible have poor quality, sometimes unacceptable healthcare services. In addition, sometimes efficiency will not be the primary goal for a business or a public administration if it means dismissing employees or cutting off important legal benefits. Thus, in an input-oriented production scenario under such circumstances, if there is a possibility of attaining efficiency by contracting other less preferable resources, it would represent a more viable managerial framework keeping in mind that such efficiency prospect may not represent the most *economical* way, but the most *preferable* way for evaluation.

An interesting approach is presented by Nepomuceno et al. (2020c). The authors employed a weighting scheme defined by the manager of a public bank institution in Brazil based on a sustainable banking preference structure for directional efficiency instead of efficiency based on technical aspects or profitability. They considered as inputs the electricity consumption, printing services and employees used to produce business transactions. The input preference elicitation was made on four decision criteria (Cost, Environmental Impact, Availability and Accessibility to the inputs). The authors report differences among the methodologies with some units becoming more inefficient due to different targets on the frontier that different directions aim to achieve. A similar analogy can be made considering an output-oriented scenario. The vast majority of businesses and industries produce more than one output, and they may have different opinions on which should be produced more based on profits, costs, simplicity, externalities, economies of scale and scope, market expansion expectations, supply chain, social impact, or personal affinities.

Many theoretical developments have addressed this issue in the efficiency analysis literature. Examples are weights restrictions to degrees of desirability for potential production configurations (Allen et al., 1997; Dyson & Thanassoulis, 1988; Halme & Korhonen, 2000); value efficiency analysis with or without preference weighting structures (Halme & Korhonen, 2000; Halme et al., 1999); including virtual (artificial or pseudo) DMUs (Nepomuceno & Costa, 2019; Thanassoulis & Allen, 1998) and input or output target settings that may reflect the decision maker preference over alternative ways of reaching efficiency (Golany, 1988; Nepomuceno et al., 2022b; Thanassoulis & Dyson, 1992; Zhu, 1996; Zhu et al., 2022). Multiple Criteria Decision Aid (MCDA) models and methodologies can be a useful tool for developments on preference modeling (Belton & Stewart, 2002; Da Silva et al., 2022; de Almeida, et al., 2017; de Almeida Filho et al., 2018; Vieira Junior, 2008; Vincke, 1992).

One way of integrating the subjective multicriteria decision outlook into a consistent preference modeling for efficiency analysis is through the advances in non-radial Direction Distance Functions (Chambers et al., 1996a, 1996b). When it comes to frontier models, an additional parameter called *directional vector* (for single orientations) or *directional matrix* (for both orientations) $g = (g_{(x)}, g_{(y)})$ is imposed in the linear formulation for contracting the $i = (1, 2, 3, \ldots, m)$ inputs in the direction $g_{(x)}$ or expanding the $r = (1, 2, 3, \ldots, s)$ in the direction $g_{(y)}$, so that the optimal solution $\sup\{\beta \geq 0 | (x_i - \beta g_{(x_i)}, y_r + \beta g_{(y_r)})$ is feasible and may contain a possible *subjective* evaluation of a decision maker (for instance, considering a policymaker's preferences information on decision criteria).

There are many multicriteria methods and ways of combining and exploring pairwise comparisons to produce the directions based on organizational needs. For its simplicity and straightforward processing, we propose to use in this chapter the outranking method called ELECTRE (*Elimination and Choice Translating Algorithm*), formerly designed by Benayoun, et al. (1966) and Roy (1968). ELECTRE can be a useful ground for any practitioner to reproduce in combination with other more sophisticated evaluations. The next sections explain the approach in detail and provide a numerical application for analyzing the efficiency of Higher Education Institutions (HEIs) considering the preference structure coming from an international ranking such as the Times Higher Education (THE) World University Ranking.

2 Using ELECTRE Outranking to Set a "Preferable Direction"

Multicriteria models and approaches are meant to provide classifications, ordering choices or preference descriptions that require a subjective knowledge on basic concepts, variables and domains, such as criteria, alternatives, decision units and weights, so that alternatives can be evaluated in a compensatory or non-compensatory way according to multiple decision criteria weighted by the decision maker (de Carvalho et al., 2015; Roy, 1996). The outranking methods regard the non-compensatory perspective. They construct pairwise relations on a set of alternatives based on a decision marker's preferences instead of a single value for each alternative as it is made by aggregative functions in compensatory approaches (Belton & Stewart, 2002; Vincke, 1992).

In this panorama, a decision making unit (DMU) 'a', for instance, is said to outrank another DMU' b' if, and only if, 'a' is at least as good as 'b' and there is no other stronger argument to contradict this assertion (Belton & Stewart, 2002). The ELECTRE (Elimination and Choice Translating Algorithm) method, formerly designed by Benayoun, Roy and Sussman (1966) and Roy (1968), outstands among non-compensatory outranking approaches, side-by-side with the PROMETHEE (*Preference Ranking Organization Method for Enrichment of Evaluations*) procedures developed by Brans et al. (1986) due to the relative easiness and lower operating assumptions to construct the pairwise aggregations (Belton & Stewart, 2002).

The methods of ELECTRE family (ELECTRE I, II, III, IV, IS and TRI) differ from each other based on the volume of information required, the problematicity of the decision (whether it is a choice, ranking and sorting problem), and according to the decision maker's preference structure, i.e., complete order, preorder, semiorder, interval order or partial order, preorder or semiorder (de Carvalho et al. 2018; Vinke 1992, Triantaphyllou, 2000). The ELECTRE II (Roy & Bertier, 1971, 1973) is intended to rank problems appropriate to the context of subjective efficiency evaluations, and it demands an assignment of concordance and discordance indexes to pairwise compare alternatives (Roy, 1996).

In the case of defining directions for an efficiency evaluation, the alternatives are the inputs to be used or the outputs to be produced under some subjective preference structure. Considering an output-oriented scenario, the concordance index measures the weight proportion required for an output r to outrank another output r', i.e., to evidence that alternative r is at least as good as r'. . The discordance index stands against this assumption (Olson, 1996; Vincke, 1992). Thus:

$$C_{(r,r')} = \frac{\sum_{q}^{q'} w_q}{\sum_{Q=1}^{Q'} w_Q} \tag{1}$$

is the concordance index, where: w_q : $P_q\ (r) \geq P_q\ (r')$, $P_{q+1}\ (r) \geq P_{q+1}\ (r')$, ..., $P_{Q-1}\ (r) \geq P_{Q-1}\ (r')$, $P_Q\ (r) \geq P_Q\ (r')$ is the set of weights from output r where the preference value $P\ (r)$ for criteria q is equal to or greater than the decision's marker preference for the same criteria considering alternative r', , inside a decision universe containing a total of ($Q = 1, 2, 3, \ldots, Q'$) criteria, and the *discordance index* is

$$D_{(r,r')} = \begin{cases} 0 \\ \frac{\max(\mathrm{w})_q [P_q(\mathrm{r}') - P_q(r)]}{\max(\mathrm{w})_{\mathrm{Q}=1}^{Q'} [P_Q(\mathrm{b}) - P_Q(\mathrm{a})]} \end{cases} \tag{2}$$

for which output (alternative) r is better than r'. It evidences arguments against the assumption that r performs as well as r' given the relative disadvantage between the alternatives being evaluated, i.e., the ratio of the maximum difference between outputs r and r' by the maximum difference between any two other outputs, 'a' and 'b', in the decision universe A(Q, ..., Q′). The procedure follows two outranking relations by the usage of four preference constraints, namely, a strong concordance threshold (C*), weak concordance threshold (c*), strong discordance threshold (D*) and a weak discordance threshold (d*) in order to construct the final ranking relation, determined by two other ascending and descending rankings of alternatives (Belton & Stewart, 2002; Roy & Bertier, 1973).

The descending order ranking begins with the best outputs B(Q, ..., Q′) ϵ A(Q, ..., Q′) which are not strongly outranked by any other output, and goes downwards by selecting a subgroup B′ (Q, ..., Q′) ϵ B(Q, ..., Q′) of outputs which are not weakly outranked by any other output in B(Q, ..., Q′). Then, B′ (Q, ..., Q′) is removed from A(Q, ..., Q′) to become the first class in the descending ranking, and the procedure continues until each output is adequately evaluated. The ascending order ranking has similar reasoning to the descending ranking; however, it represents an upward relation with the worst outputs W(Q, ..., Q′) ϵ A(Q, ..., Q′) which do not strongly outrank any other output and goes up by selecting a subgroup, say W′ (Q, ..., Q′) ϵ B(Q, ..., Q′) of alternatives, which do not weakly outrank any other alternative of W(Q, ..., Q′). W′ (Q, ..., Q′) is removed from A(Q, ..., Q′) to become the first class in the ascending ranking, and the procedure continues until each alternative is evaluated.

For a multicriteria application, it is sufficient to consider the intersection of ascending and descending ranking of alternatives to determine a partial order of units. In the approach for defining an appropriate direction for expanding outputs based on a prior preference elicitation, one can follow Nepomuceno et al. (2020c) non-compensatory approach inspired by the PROMETHEE method for aggregating the preference functions of concordance (see Nepomuceno et al., 2020c for the exact notation):

$$g_{(y_r)} = \sum_{r=1}^{s} \left(\frac{\sum_{a=1}^{A} w_a C_a(r', r)}{\sum_{a=1}^{A} w_a} \right), \forall (r \neq r') \in R' | C_a(r', r) = 1\}$$

where: *Concordance index*: $C_a(i', i) = \begin{cases} 0\, if\ f_a(i) + q_a \geq f_a(i') \\ 1\, if\ f_a(i) + p_a < f_a(i') \end{cases}$ for any i .

Or use the sum of the gains in the pairwise comparisons, which provides richer and more relevant value judgments on the set of alternatives and avoids many unrealistic mathematical hypotheses. In the numerical example ahead we have opted for the second approach instead of Nepomuceno et al. (2020c), combining the ELECTRE perspective in defining the directions.

3 Numerical Example

The *Times Higher Education (THE)* is a weekly magazine that reports statistics and discussions about higher education. THE produces also a world university ranking for a comprehensive and balanced comparison that academics, stakeholders, and governments can use in which 13 performance indicators are used. These performance indicators evaluate research-intensive universities across their core missions and are sorted into five groups: teaching (the learning environment), research (volume, income and reputation), citations (research influence), knowledge transfer and international outlook (staff, students and research). Table 1 reports the weighting criteria and individual indicators related to the evaluation of THE engineering and technology rankings in 2016.

Let us conduct an efficiency analysis of 392 universities in Europe using expenditure data as input and $r = 5$ outputs, products of Higher Education such as graduates and revenues. The data used in this example can be accessed from the European Tertiary Education Register (https://eter-project.com/). This European-level database reports data at the institutional level on HEIs' activities and outputs. Let us assume that universities have a subjective preference on what products they should improve to attain efficiency based on the THE evaluation (criteria listed in Table 1). In this approach, we can either set a different direction for each output of each decision unit (university) or set the same direction for all the outputs. For example, a university manager may prefer expanding one product (e.g., graduates) more than another (e.g., citations) in a different perspective than another university manager of another

Table 1 THE weighted subcriteria on worldwide university ranking (subject area: Engineering & Technology)

Criteria	Subcriteria	Weight
Teaching (0.3)	Staff-to-student ratio	0.1
	Doctoral awards	0.05
	Doctorate-to-bachelor's ratio	0.15
	Reputation survey	0.65
	Institutional income	0.05
Research (0.3)	Scholarly papers	0.15
	Research income	0.15
	Reputation survey	0.7
Citations per paper (0.275)	Research influence	1
Industry income (0.05)	Knowledge transfer and innovation	1
International outlook (0.075)	Ratio of international to domestic staff	0.33
	International co-authorship	0.33
	Ratio of international to domestic students	0.34

institution, or they both may agree that one output should consistently be expanded more than the other in the same perspective, which means that the industry has an overall preference, or maybe a public figure deciding for all institutions.

Table 2 considers a production process for one university using financial resources (e.g., public expenditures) to generate five different outputs, and each output has a different impact on five decision criteria related to the HEI. Higher Education. This information was adapted from 2016 THE rankings for Higher Education, considering the evaluation information of top universities as variables. Each output has a different impact (contribution) toward the teaching, international outlook, research, citations and industry income decision criteria, and each of those criteria has a different weight preference according to Table 1. After this elicitation process that is usually carried out together with the decision maker, we can apply ELECTRE pairwise comparisons, restricting or abolishing compensations (if necessary) to construct the concordance matrix reported in Table 3.

The non-compensatory multicriteria methodology demands the inclusion of four unbiased constraints to pairwise compare universities: a strong (C^+) and weak (C^-) concordance thresholds, defined as 0,7 and 0,5, respectively, such that $C^+ > C^-$, and a strong (D^+) and weak (D^-) discordance thresholds, defined as 30 and 50 (points in the former original ranking score) respectively, such that D- > D + . The concordance indexes for each comparison are computed on the proportional score the criterion has each weight instead of an overall weighting value. For instance, if output 1 beats output 3 by 95.2 against 94.9 in the Teaching criteria, weighted by 0.3,

Table 2 Example of output index impact based on 2016 THE university ranking (subject area: Engineering & Technology)

Output	Teaching	International outlook	Research	Citations	Industry income
Output 1	95.2	75.7	98.2	100	91.2
Output 2	98.3	54.8	97.4	99.9	99.8
Output 3	94.9	78.5	91.8	99.4	98.0
Output 4	93.9	86.8	96.0	95.2	81.8
Output 5	91.6	69.1	97.8	99.9	77.2

Table 3 Example of concordance matrix after pairwise evaluation

	Output 1	Output 2	Output 3	Output 4	Output 5	Output directions
Output 1	–	0.654	0.893	0.940	1	3.487
Output 2	0.362	–	0.957	0.957	0.650	2.925
Output 3	0.114	0.062	–	0.645	0.417	1.239
Output 4	0.069	0.069	0.376	–	0.413	0.929
Output 5	0.000	0.349	0.608	0.608	–	1.567

instead of adding 0.3 to their concordance index, we consider adding the proportion of 0.952*0.3 = 0.2856 to aggregate the real contribution of that output performance into the analysis.

Table 3 reports the concordance matrix. As an example, the concordance index of "0.654" from comparing output 1 with output 2 comes from the proportion of criteria (weighted as from Table 1) in which output 1 beats output 2: (75.7 * 0.075) + (98.2 * 0.3) + (100 * 0.275) = 0.654126. The resulting directions in this simple exercise are the sum of the pairwise comparisons (gains) for each output.

In this chapter, we report a comparison of the efficiency scores calculated with a traditional directional distance employing a unit direction with the efficiency scores obtained by applying a "preferable direction" obtained by applying the ELECTRE directional approach described in the previous section.

Table 4 reports the main results and differences from our ELECTRE directional efficiency (Model B and Eff_Range (B)) to a traditional directional distance model employing the common unit vector (model A and Eff_Range (A)).

About 47% of units (185 universities) have an average efficiency between 0.5 and 0.99 in the first model compared to 61.2% (240 universities) in the second model. The summary of results indicates that there is no change in the location of the efficient frontier, only in the distribution of inefficiencies from one perspective to the other. About 5.35% (21 universities) have an efficiency score below 0.2 in the second model (B) compared to about 13% (51 units) in the first model (A). From an output-aggregate perspective, all the inefficient units become less inefficient. The sum of the efficiency scores decreases from 6931.744 to 4022.9. From a specific output perspective, only the first output (revenues) is bigger in the second model,

Table 4 Efficiency results

Eff_Range (A)	*Units*	*Percentage*	*Eff_Range (B)*	*Units*	*Percentage*
E = = 1	18	4.592%	E = = 1	18	4.592%
1 < E ≤ 0,80	71	18.112%	1 < E ≤ 0.80	112	28.571%
0,80 < E ≤ 0,50	114	29.082%	0.80 < E ≤ 0.50	128	32.653%
0,50 < E ≤ 0,20	138	35.204%	0.50 < E ≤ 0.20	113	28.827%
0,20 < E	51	13.010%	0.20 < E	21	5.357%
Sum of Eff. Scores (Model A)			*Sum of Eff. Scores (Model B)*		
Aggr. Score (Out1)	1097.179		Aggr. Score (Out1)	1134.889	
Aggr. Score (Out2)	560.6619		Aggr. Score (Out2)	536.2802	
Aggr. Score (Out3)	741.5463		Aggr. Score (Out3)	517.4122	
Aggr. Score (Out4)	3990.744		Aggr. Score (Out4)	1373.677	
Aggr. Score (Out5)	541.6126		Aggr. Score (Out5)	460.6417	
Aggregate Score	**6931.744**		**Aggregate Score**	**4022.9**	

highlighting the different dynamics in which evaluations can be considered changing the preferences about the direction to reach the efficient frontier.

Figure 1, panels (a), (b), (c) and (d) illustrate the direction differences in a two-dimensional isotrans visualization considering outputs 1 and 4 (revenues in higher education and graduates at ISCED 8), and outputs 1 and 5 (revenues in higher education and the total graduates at any level) as examples in the axis of the graphs. Panels (a) and (c) were constructed using the unit vector. Panels (b) and (d) were constructed using the directions from the ELECTRE pairwise evaluations (reported in Table 3). The linear combination generating the production frontier does not change from one representation to its equivalent, but directions in which inefficient units must expand the outputs change considerably. Depending on the institution's strategic objectives, the evaluation may significantly change the ranking positions of universities and intrinsic incentives for managerial purposes.

4 Conclusions

In this chapter, we discuss on the critical issue of considering *subjective values* and *preferences* in the quantitative efficiency analysis computed using frontier models. Among the many developments in the field, combining multicriteria decision analysis with directional distances to construct coherent directions for the expansion of outputs or contraction of inputs is a practical and prominent approach that has not been explored in its full potential. Here we have offered just one perspective on how to advance in this avenue by developing multicriteria non-compensatory pairwise evaluation for setting *preferable* directions.

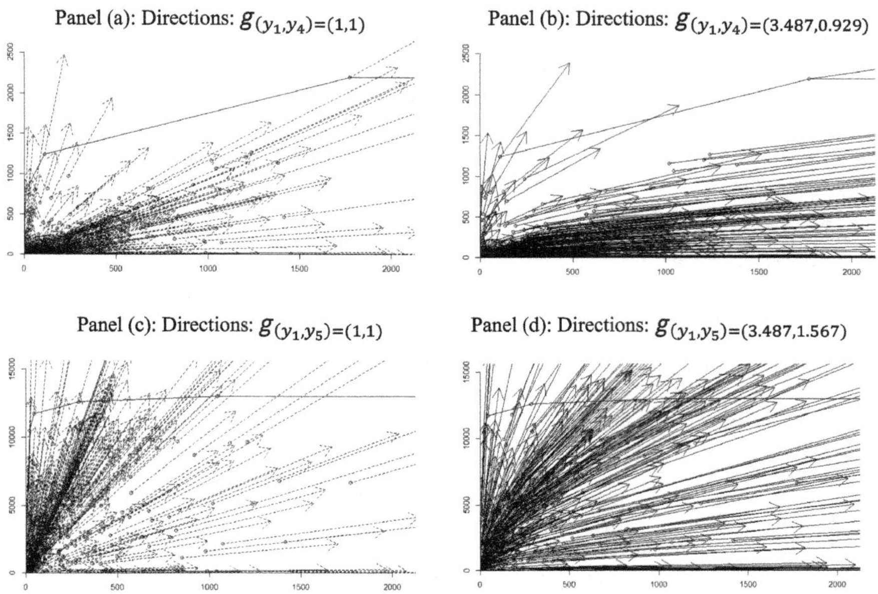

Fig. 1 Directions illustrations

We have discussed and applied as an example the ELECTRE outranking procedure which utilizes four preference constraints (a strong concordance threshold C*, weak concordance threshold c*, strong discordance threshold D* and a weak discordance threshold d*) in order to construct the final ranking relation for the output directions. Nevertheless, many other frameworks can be appropriate for different problems depending on the required information volume and the decision problem can be extended from this point of view.

In the numerical example, the elicitation was made considering the THE World Ranking of universities decision criteria (teaching, research, citations, knowledge transfer and international outlook), but any other decision criteria or policy instrument can be considered in this flexible approach. The results highlight significant differences from applying a traditional unit vector direction compared with the application of the *preferable* direction determined by applying the ELECTRE outranking approach. One can be more appropriate than the other, depending on the context and institutional objectives. We hope to advance and contribute to these prospects in future extensions of combinations of frontier methods with multicriteria procedures.

References

Adler, N., Friedman, L., & Sinuany-Stern, Z. (2002). Review of ranking methods in the data envelopment analysis context. *European Journal of Operational Research, 140*(2), 249–265.

Aldamak, A., & Zolfaghari, S. (2017). Review of efficiency ranking methods in data envelopment analysis. *Measurement, 106*, 161–172.

Allen, R., Athanassopoulos, A., Dyson, R. G., & Thanassoulis, E. (1997). Weights restrictions and value judgements in data envelopment analysis: Evolution, development and future directions. *Annals of Operations Research, 73*, 13–34.

Belton V., & Stewart T. J. (2002). Multiple criteria decision analysis: An integrated approach. Kluwer Academic Publisher.

Benayoun, R., Roy, B., & Sussman, N. (1966). Manual de Reference du Programme Electre, Note De Synthese et Formaton, No.25, Direction Scientifque SEMA, Paris, France.

Brans, J. P., Vincke, P., & Mareschal, B. (1986). How to select and how to rank projects: The PROMETHEE method. *European Journal of Operational Research, 24*(2), 228–238.

Chambers, R. G., Chung, Y., & Färe, R. (1996a). Benefit and distance functions. *Journal of Economic Theory., 70*, 407–419.

Chambers, R., Färe, R., & Grosskopf, S. (1996b). Productivity growth in APEC countries. *Pacific Economic Review., 1*(3), 181–190.

Charnes, A., Cooper, W. W., & Rhodes, E. (1978). Measuring the efficiency of decision making units. *European Journal of Operational Research, 2*(6), 429–444.

Clement, J. P., Valdmanis, V. G., Bazzoli, G. J., Zhao, M., & Chukmaitov, A. (2008). Is more better? An analysis of hospital outcomes and efficiency with a DEA model of output congestion. *Health Care Management Science, 11*(1), 67–77.

da Silva, A. L. C. de Lima, Cabral Seixas Costa, A. P., & de Almeida, A. T. (2022). Analysis of the cognitive aspects of the preference elicitation process in the compensatory context: a neuroscience experiment with FITradeoff. *International Transactions in Operational Research.*

Daraio, C., Kerstens, K. H., Nepomuceno, T. C. C., & Sickles, R. (2019). Productivity and efficiency analysis software: An exploratory bibliographical survey of the options. *Journal of Economic Surveys, 33*(1), 85–100.

Daraio, C., Kerstens, K., Nepomuceno, T., & Sickles, R. C. (2020). Empirical surveys of frontier applications: A meta-review. *International Transactions in Operational Research, 27*(2), 709–738.

Daraio, C., Simar, L., & Wilson, P. W. (2021). Quality as a latent heterogeneity factor in the efficiency of universities. *Economic Modelling, 99*, 105485.

de Almeida, A. T., Alencar, M. H., Garcez, T. V., & Ferreira, R. J. P. (2017). A systematic literature review of multicriteria and multi-objective models applied in risk management. *IMA Journal of Management Mathematics, 28*(2), 153–184.

de Almeida Filho, A. T., Clemente, T. R., Morais, D. C., & de Almeida, A. T. (2018). Preference modeling experiments with surrogate weighting procedures for the PROMETHEE method. *European Journal of Operational Research, 264*(2), 453–461.

de Carvalho, V. D. H., Poleto, T., Camara, L., & Costa, A. P. C. S. (2015). Abordagem multicritério de apoio a decisões estrategicamente sustentáveis nas organizações. *Revista Produção Online, 15*(3), 925–947.

de Carvalho, V. D. H., Poleto, T., & Seixas, A. P. C. (2018). Information technology outsourcing relationship integration: A critical success factors study based on ranking problems (P. γ) and correlation analysis. *Expert Systems, 35*(1), e12198.

Dismuke, C. E., & Sena, V. (2001). Is there a trade-off between quality and productivity? The case of diagnostic technologies in Portugal. *Annals of Operations Research, 107*(1), 101–116.

Dyson, R. G., & Thanassoulis, E. (1988). Reducing weight flexibility in data envelopment analysis. *Journal of the Operational Research Society, 39*(6), 563–576.

Golany, B. (1988). An interactive MOLP procedure for the extension of DEA to effectiveness analysis. *Journal of the Operational Research Society, 39*(8), 725–734.

Halme, M., Joro, T., Korhonen, P., Salo, S., & Wallenius, J. (1999). A value efficiency approach to incorporating preference information in data envelopment analysis. *Management Science, 45*(1), 103–115.

Halme, M., & Korhonen, P. (2000). Restricting weights in value efficiency analysis. *European Journal of Operational Research, 126*(1), 175–188.

Khushalani, J., & Ozcan, Y. A. (2017). Are hospitals producing quality care efficiently? An analysis using Dynamic Network Data Envelopment Analysis (DEA). *Socio-Economic Planning Sciences, 60*, 15–23.

Nayar, P., & Ozcan, Y. A. (2008). Data envelopment analysis comparison of hospital efficiency and quality. *Journal of Medical Systems, 32*(3), 193–199.

Nepomuceno, T. C., & Costa, A. P. C. (2019). Resource allocation with time series DEA applied to Brazilian federal saving banks. *Economics Bulletin, 39*(2), 1384–1392.

Nepomuceno, T. C., Silva, W. M., Nepomuceno, K. T., & Barros, I. K. (2020a). A DEA-based complexity of needs approach for hospital beds evacuation during the COVID-19 outbreak. *Journal of Healthcare Engineering*.

Nepomuceno, T., Silva, W. M. D. N., & Silva, L. G. D. O. (2020b). PMU7 efficiency-based protocols for BEDS evacuation during the COVID-19 pandemic. *Value in Health, 23*, S604.

Nepomuceno, T. C. C., Daraio, C., & Costa, A. P. C. S. (2020c). Combining multicriteria and directional distances to decompose non-compensatory measures of sustainable banking efficiency. *Applied Economics Letters, 27*(4), 329–334.

Nepomuceno, T. C., Daraio, C., & Costa, A. P. C. (2021). Multicriteria ranking for the efficient and effective assessment of police departments. *Sustainability, 13*(8), 4251.

Nepomuceno, T. C. C., Piubello Orsini, L., de Carvalho, V. D. H., Poleto, T., & Leardini, C. (2022a). The core of healthcare efficiency: A comprehensive bibliometric review on frontier analysis of hospitals. In *Healthcare* (Vol. 10, No. 7, p. 1316).

Nepomuceno, T. C. C., de Carvalho, V. D. H., Nepomuceno, K. T. C., & Costa, A. P. C. (2022b). Exploring knowledge benchmarking using time-series directional distance functions and bibliometrics. *Expert Systems*, e12967.

Olson, D. L. (1996). *Decision aids for selection problems*. Springer.

Piubello Orsini, L., Leardini, C., Vernizzi, S., & Campedelli, B. (2021). Inefficiency of public hospitals: A multistage data envelopment analysis in an Italian region. *BMC Health Services Research, 21*(1), 1–15.

Roy, B. (1968). Classement et choix en présence de points de vue multiples (la méthode ELECTRE). *La Revue D'informatique Et De Recherche Opérationelle (RIRO), 8*, 57–75.

Roy, B. (1996). Multicriteria methodology for decision aiding (Vol. 12). Springer Science & Business Media.

Roy, B., & Bertier, P. (1971). La methode ELECTRE II: Une methode de classement en presence de critteres multiples. SEMA (Metra International), Direction Scientifique, Note de Travail No. 142, Paris (p. 25).

Roy, B., & Bertier, P. (1973). La methode ELECTRE II: Une methode au media-planning. In M. Ross (Ed.), *Operational research 1972* (pp. 291–302). North-Holland Publishing Company.

Shimshak, D. G., & Lenard, M. L. (2007). A two-model approach to measuring operating and quality efficiency with DEA. INFOR: *Information Systems and Operational Research, 45*(3), 143–151.

Thanassoulis, E., & Dyson, R. G. (1992). Estimating preferred target input-output levels using data envelopment analysis. *European Journal of Operational Research, 56*(1), 80–97.

Thanassoulis, E., & Allen, R. (1998). Simulating weights restrictions in data envelopment analysis by means of unobserved DMUs. *Management Science, 44*(4), 586–594.

Triantaphyllou, E. (2000). *Multi-criteria decision making methods: A comparative study*. Kluwer Academic Publishers, Boston, MA, U.S.A.

Vieira Junior, H. (2008). Multicriteria approach to data envelopment analysis. *Pesquisa Operacional, 28*, 231–242.

Vincke, P. (1992). *Multicriteria decision-aid*. Wiley.

Zhu, J. (1996). Data envelopment analysis with preference structure. *Journal of the Operational Research Society, 47*(1), 136–150.

Zhu, Q., Aparicio, J., Li, F., Wu, J., & Kou, G. (2022). Determining closest targets on the extended facet production possibility set in data envelopment analysis: Modeling and computational aspects. *European Journal of Operational Research, 296*(3), 927–939.

Benefit-of-the-Doubt Composite Indicators and Use of Weight Restrictions

Ana S. Camanho, Andreia Zanella, and Victor Moutinho

1 Introduction

A composite indicator (CI) is given by the aggregation of several key performance indicators. CIs are intended to reflect multidimensional concepts in a single measure. Their main benefits are the capacity to summarize information, the facility to interpret results compared with a battery of separate indicators, and the capacity to reduce the visible size of a set of indicators without dropping the underlying base information (Nardo et al., 2008). Examples of well-established composite indicators are the Environmental Performance Index (Emerson et al., 2012), Climate Change Performance Index (Burck et al., 2012), and the Human Development Index (United Nations, 2013).

The Organisation for Economic Co-operation and Development (OECD) and the European commission provide a handbook for the construction of composite indicators that discusses the range of methodological approaches available to construct CIs (Nardo et al., 2008). The handbook highlights the growing interest in composite indicators by the academic circles, media and policymakers.

Although a considerable amount of research related to the construction of composite indicators has been developed in recent years (as reviewed in Nardo et al. (2008)), they are not effective in providing managerial information to guide improvements.

A. S. Camanho (✉)
Faculdade de Engenharia, Universidade do Porto, Porto, Portugal
e-mail: acamanho@fe.up.pt

A. Zanella
Universidade Federal de Santa Catarina, Florianópolis, Brazil
e-mail: andreia.zanella@ufsc.br

V. Moutinho
Universidade da Beira Interior, Covilhã, Portugal
e-mail: ferreira.moutinho@ubi.pt

P. Macedo et al. (eds.), *Advanced Mathematical Methods for Economic Efficiency Analysis*, Lecture Notes in Economics and Mathematical Systems 692,
https://doi.org/10.1007/978-3-031-29583-6_6

Furthermore, they are prone to criticism regarding the subjectivity inherent in the specification of the relative importance given to the individual indicators in the construction of the CI. The handbook points Data Envelopment Analysis (DEA) as an interesting weighting and aggregation procedure to reduce the inherent subjectivity associated with the specification of weights. As the indicator weights result from an optimizing process based on linear programming, they are less prone to subjectivity and controversy. The use of DEA to construct CIs was popularized by Cherchye et al. (2007). This approach is known as 'Benefit-of-the-Doubt' (BoD) construction of composite indicators.

The next two sections present the BoD CI formulations derived from traditional DEA models and from Directional Distance Functions (DDF). Directional Distance Functions allow the construction of CIs that can simultaneously accommodate desirable and undesirable output indicators in the performance assessment.

2 BoD Composite Indicators Based on DEA Models

DEA is a linear programming technique that measures the relative efficiency of a homogeneous set of Decision Making Units (DMUs) in their use of multiple inputs to produce multiple outputs. It derives a single summary measure of efficiency for each DMU, based on direct comparisons with other DMUs in the sample. This feature makes DEA particularly interesting for benchmarking purposes, as it evaluates performance by comparison with what was actually observed. Another important feature of the DEA models is related with the specification of the weights assigned for each input and output. DEA allows to specify the weights recurring to optimization, avoiding the criticism of subjectivity and disagreement often present in the specification of weights. In addition, DEA is able to handle data measured in different scales, without requiring prior normalization of data.

The DEA technique can also be applied to construct composite indicators that look only at the achievements (results obtained) by a set of DMUs, providing a measure of effectiveness. The use of DEA for performance assessments that disregard the conversion of inputs to outputs, focusing only on the outcomes, was first proposed by Cook and Kress (1990), with the purpose to construct a preference voting model (for aggregating votes in a preferential ballot).

The CI can be derived both from the input or output oriented DEA models. The output oriented model introduced by Charnes et al. (1978), assuming constant returns to scale, can be formulated as follows:

$$
\begin{aligned}
&\frac{1}{E(y,x)} = \min \sum_{i=1}^{m} v_i\, x_{ij_0} \\
&s.t. \sum_{r=1}^{s} u_r\, y_{rj_0} = 1 \\
&\sum_{r=1}^{s} u_r\, y_{rj} - \sum_{i=1}^{m} v_i\, x_{ij} \leq 0 \quad j = 1, \ldots, n \\
&u_r \geq 0 \qquad\qquad r = 1, \ldots, s \\
&v_i \geq 0 \qquad\qquad i = 1, \ldots, m
\end{aligned} \tag{1}
$$

In the formulation above, x_{ij} are the inputs ($i = 1, \ldots, m$) consumed by the DMU j ($j = 1, \ldots, n$) to produce the outputs y_{rj} ($r = 1, \ldots, s$). u_r and v_r are the weights attached to the outputs and inputs in the performance assessment, and these are the decision variables of the model. The relative efficiency score $E(y, x)$ estimated for DMU j_0 is given by the inverse of the objective function value of model (1), which is equivalent to $\sum_{r=1}^{s} u_r\, y_{rj_0} / \sum_{i=1}^{m} v_i\, x_{ij_0}$. This efficiency score ranges between 0 (worst) and 1 (best).

In composite indicators constructed using DEA models, all individual key performance indicators have to be specified as outputs. As originally suggested by Lovell et al. (1995) and Lovell (1995), an identical input value should underlie the evaluation of every DMU, which for simplicity can be assumed to be equal to one. Following Koopmans (1951), this unitary input can be interpreted as a 'helmsman' associated with every DMU, who attempts to steer the units towards the maximization of outputs.

By considering a unitary input level for all DMUs in the output oriented model (1) we obtain the CI model presented in (2).

$$
\begin{aligned}
&\frac{1}{E(y,\mathbb{1})} = \min v \\
&s.t. \sum_{r=1}^{s} u_r\, y_{rj_0} = 1 \\
&\sum_{r=1}^{s} u_r\, y_{rj} - v \leq 0 \quad j = 1, \ldots, n \\
&u_r \geq 0 \qquad\qquad r = 1, \ldots, s \\
&v \geq 0
\end{aligned} \tag{2}
$$

In addition to the performance score $E(y, \mathbb{1})$, the CI constructed based on DEA is also able to provide other managerial information. This feature is particularly interesting for benchmarking purposes. For each DMU j_0, with a composite indicator score smaller than one, it is possible to identify the peers and the targets that it should aim to achieve. The identification of the peers and targets for the inefficient DMUs can be done through the envelopment formulation of model (2), shown in (3). The objective function value at the optimal solution of the model (3) corresponds to the factor θ by which all outputs of the DMU under assessment can be proportionally improved to reach the target output values. The performance score, or composite indicator value for DMU j_0 under assessment, is the reciprocal of this value, i.e., $CI = E(y, \mathbb{1}) = \frac{1}{\theta}$. Therefore, the DMUs with the best performance are those for which there is no evidence that it is possible to expand their outputs, such that the value of θ^* is equal to 1.

$$
\begin{aligned}
& \frac{1}{E(y, \mathbb{1})} = \max \theta \\
& s.t. \;\; \theta \; y_{rj_0} - \sum_{j=1}^{n} \lambda_j \; y_{rj} \leq 0 \;\; r = 1, \ldots, s \\
& \sum_{j=1}^{n} \lambda_j \leq 1 \qquad\qquad\qquad j = 1, \ldots, n \\
& \lambda_j \geq 0 \qquad\qquad\qquad\qquad j = 1, \ldots, n
\end{aligned}
\tag{3}
$$

The peers for the DMU j_0 under assessment are the DMUs with values of λ_j^* greater than zero at the optimal solution of model (3) (the symbol * signals the optimal value of a decision variable). For the DMUs with a composite indicator score smaller than one, the targets to improve performance are given by $\sum_{j=1}^{n} \lambda_j^* \; y_{rj}$, for $r = 1, \ldots, s$.

The formulation of BoD CI proposed by Cherchye et al. (2007) differs from the formulation of the BoD CI model presented in (3) in the orientation of the assessment, as the latter has an output orientation.

The original input oriented model introduced by Charnes et al. (1978), assuming constant returns to scale, can be formulated as follows:

$$
\begin{aligned}
& E(y, x) = \max \sum_{r=1}^{s} u_r \; y_{rj_0} \\
& s.t. \sum_{i=1}^{m} v_i \; x_{ij_0} = 1 \\
& \sum_{r=1}^{s} u_r \; y_{rj} - \sum_{i=1}^{m} v_i \; x_{ij} \leq 0 \;\; j = 1, \ldots, n \\
& u_r \geq 0 \qquad\qquad\qquad r = 1, \ldots, s \\
& v_i \geq 0 \qquad\qquad\qquad i = 1, \ldots, m
\end{aligned}
\tag{4}
$$

The BoD CI proposed by Cherchye et al. (2007) is shown in (5). It is equivalent to the input oriented DEA model (4) with a unitary input level for all DMUs.

$$
\begin{aligned}
E(y, \mathbb{1}) &= \max \sum_{r=1}^{s} u_r \, y_{rj_0} \\
s.t. \sum_{r=1}^{s} u_r \, y_{rj} &\leq 1 \qquad j = 1, \ldots, n \\
u_r &\geq 0 \qquad r = 1, \ldots, s
\end{aligned}
\tag{5}
$$

The dual of model (5) is shown in (6).

$$
\begin{aligned}
E(y, \mathbb{1}) &= \min \sum_{j=1}^{n} \lambda_j \\
s.t. \; \sum_{j=1}^{n} \lambda_j \, y_{rj} &\geq y_{rj_0} \quad r = 1, \ldots, s \\
\lambda_j &\geq 0 \qquad j = 1, \ldots, n
\end{aligned}
\tag{6}
$$

As both BoD CI formulations (models (2) and (5)) assume constant return to scale, the performance scores obtained from these two models are the same. The advantage of using the output oriented formulation is that it facilitates the incorporation of weight restrictions. Furthermore, as noted by Färe and Karagiannis (2014), the output oriented formulation may have a more appealing interpretation in terms of efficiency measurement when there is a single unitary input.

One important assumption of the CI derived from DEA models is that higher values of the output indicators correspond to better performance. However, real world applications may involve both desirable and undesirable outputs indicators. For example, in environmental performance assessments, we may have an output indicator related to quality of the water, for which more output corresponds to better performance and an other output indicator related to the level of CO_2 emissions, for which less output corresponds to better performance. Thus, the literature in this field has evolved to propose enhanced models that can accommodate both desirable and undesirable outputs in BoD assessments.

The advantages and limitations of different approaches to handle undesirable outputs in DEA models were discussed by Scheel (2001), Dyson et al. (2001), Seiford and Zhu (2002) and Zanella et al. (2015). The most common approaches suggest the incorporation of undesirable outputs in the form of their additive inverses ($-y_{und}$) or multiplicative inverses ($1/y_{und}$), or to add to the additive inverses of the undesirable outputs a sufficient large positive number ($-y_{und} + c$). This third option is the most frequently used in the literature (Cook & Green, 2005; Oggioni et al., 2011). It has the advantage of enabling a simple interpretation of results, but it is sensitive to the choice of the constant c, impacting both the performance scores and the ranking of the DMUs, as discussed by Zanella et al. (2015).

In addition to the above mentioned approaches, in Cherchye et al. (2011) the transformation in the measurement scale of the undesirable outputs was performed based on a normalization procedure, which was applied both to desirable and undesirable outputs. This procedure results in indicators varying between 0 and 1. As data normalization leads to a loss of information, this approach is rarely used in DEA studies. It does not take advantage of the ability of DEA to deal with data measured in different scales.

In the next section we discuss the BoD CI model specified using Directional Distance Functions (DDFs). The DDF allows including both desirable and undesirable output indicators in the CI model considering their original measurement scales, avoiding the need to normalise the indicators or to adjust their scales.

3 BoD Composite Indicators Based on Directional Distance Functions

The efficiency evaluation using the directional distance function (DDF), developed by Chambers et al. (1996), allows to simultaneously expand outputs and contract inputs according to a directional vector. The constant returns to scale model of Chambers et al. (1996) is specified as shown in (7).

$$\begin{aligned} &\vec{D}(y, x; g) = \max \delta \\ &s.t. \sum_{j=1}^{n} y_{rj}\, \lambda_j \geq y_{rj_0} + \delta\, g_y \quad r = 1, \ldots, s \\ &\sum_{j=1}^{n} x_{ij}\, \lambda_j \leq x_{rj_0} - \delta\, g_x \qquad i = 1, \ldots, m \\ &\lambda_j \geq 0 \qquad\qquad\qquad\qquad j = 1, \ldots, n \end{aligned} \tag{7}$$

In formulation (7), x_{ij} $(i = 1, \ldots, m)$ are the inputs used by the DMU j $(j = 1, \ldots, n)$ to produce y_{rj} $(r = 1, \ldots, s)$ outputs. The λ_j are the intensity variables. The components of vector $g = (g_y, -g_x)$ indicate the direction of change for the outputs and inputs. Positive values for the components are associated to expansion of desirable outputs and negative values are associated to contraction of inputs. The factor δ indicates the extend of DMU's inefficiency. It corresponds to the maximal feasible expansion of outputs and contraction of inputs that can be achieved simultaneously. For the particular case in which the directional vector is specified as $g = (g_y, -g_x) = (y_{rj_0}, -x_{ij_0})$, the inefficiency measure reflects the magnitude of the proportional change to the original levels of outputs and inputs that are required to achieve the efficient frontier. Boussemart et al. (2003) provided an exact equivalence between the DDF and the DEA radial efficiency score as:

$$E(y, x) = \frac{1 - \vec{D}(y, x; g)}{1 + \vec{D}(y, x; g)} = \frac{1 - \delta}{1 + \delta} \tag{8}$$

Chung et al. (1997) extended this approach to allow including undesirable outputs in the efficiency evaluation. The constant returns to scale model of Chung et al. (1997) is specified as shown in (9).

$$
\begin{aligned}
&\vec{D}(y, b, x; g) = \max \delta \\
&s.t. \sum_{j=1}^{n} y_{rj}\, \lambda_j \geq y_{rj_0} + \delta\, g_y \quad r = 1, \ldots, s \\
&\sum_{j=1}^{n} b_{kj}\, \lambda_j = b_{kj_0} - \delta\, g_b \qquad k = 1, \ldots, l \\
&\sum_{j=1}^{n} x_{ij}\, \lambda_j \leq x_{rj_0} - \delta\, g_x \qquad i = 1, \ldots, m \\
&\lambda_j \geq 0 \qquad\qquad j = 1, \ldots, n
\end{aligned}
\tag{9}
$$

In model (9), the components of the vector $g = (g_y, -g_b, -g_x)$ indicate the direction of change for the inputs and desirable and undesirable outputs. While inputs and desirable outputs are assumed to be strongly disposable, the undesirable outputs b_{kj} $(k = 1, \ldots, l)$ are assumed to be weakly disposable, as it is shown by the equality in the constraint associated to the undesirable outputs. When imposing weak disposability of undesirable outputs we assume that they are by-products of the desirable outputs and cannot be reduced without cost, which implies that abatement of an undesirable output is possible if accompanied by a reduction in a desirable output or an increase in an input. The decision on whether to assume strong disposability or weak disposability for the variables of a DEA model depends on the nature of the application under analysis (Liu et al., 2009).

The CI proposed by Zanella et al. (2015), referred as Directional BoD CI model, is shown in (10). It is an adaptation of the model (9).

$$
\begin{aligned}
&\vec{D}(y, b, \mathbb{1}; g) = \max \delta \\
&s.t. \sum_{j=1}^{n} y_{rj}\, \lambda_j \geq y_{rj_0} + \delta\, g_y \quad r = 1, \ldots, s \\
&\sum_{j=1}^{n} b_{kj}\, \lambda_j \leq b_{kj_0} - \delta\, g_b \qquad k = 1, \ldots, l \\
&\sum_{j=1}^{n} \lambda_j = 1 \\
&\lambda_j \geq 0 \qquad\qquad j = 1, \ldots, n
\end{aligned}
\tag{10}
$$

In formulation (10), b_{kj} $(k = 1, \ldots, l)$ are the indicators that should be reduced for DMU j $(j = 1, \ldots, n)$, and y_{rj} $(r = 1, \ldots, s)$ are the indicators that should be increased. The components of vector $g = (g_y, -g_b) = (y_{rj_0}, -b_{kj_0})$ indicate the direction of change for the indicators. When the directional vector is specified as the current value of the indicators for the DMU under assessment, the desirable and undesirable outputs are expanded and contracted, respectively, according to a direction that corresponds to proportional changes to the original levels.

The factor δ indicates the extent of DMU's inefficiency. It corresponds to the maximal feasible expansion of desirable indicators and contraction of undesirable indicators that can be achieved simultaneously. Using expression (8), the value of the Directional BoD CI that corresponds to an efficiency measure can be obtained as $(1-\delta)/(1+\delta)$ when the directional vector is specified as the current value of the outputs for the DMU under assessment.

The most important difference between the Directional CI model (10) and model (9) is the redesign of the efficient frontier. In order to avoid downward-sloping segments in the frontier, the authors replaced the equality constraint related to the undesirable outputs in model (9) by an inequality constraint "less than or equal to", as shown in model (10). This constraint becomes similar to the one imposed to inputs in traditional directional distance function models, where proportional reductions to the levels of these variables are sought, and ensures that a DMU will only be classified as efficient when no further improvements to both desirable and undesirable outputs are possible.

In addition, in the CI model (10) the authors modified the input restriction. Using a unitary level of input for all DMUs and setting the directional vector as $g = (g_y, -g_b, 0)$ (a vector that allows to, simultaneously, expand the desirable outputs and contract undesirable ones by keeping inputs fixed) in model (9), the input restriction would become $\sum_{j=1}^{n} \lambda_j \leq 1$. This restriction was modified to $\sum_{j=1}^{n} \lambda_j = 1$ in model (10). The equality constraint intends to obtain a frontier that is appropriate for assessments involving aggregation of performance indicators expressed as ratios, as is the case of composite indicators. This constraint allows obtaining a frontier identical to the one used in the BoD CI presented in Sect. 2 (models (2) and its dual model (3) or model (5) and its dual model (6)).

Hollingsworth and Smith (2003) explains that when the variables are expressed as ratios in DEA models, the formulation proposed by Banker et al. (1984) (known as the BCC model) is needed. This model (with the sum of lambdas equal to one) ensures that the frontier is constructed based only on interpolation among observed DMUs. Consequently, extrapolation of the production possibility set to areas corresponding to infeasible activity is not allowed. Hollingsworth and Smith (2003) also explains that, in this context, the BCC model should not be associated with a variable returns to scale assumption, as the use of ratios leads to a loss of information about the size of the DMU, and implicitly assumes constant returns to scale in the operation of the DMUs under analysis.

As by-products of the assessment using the Directional CI model, it is possible to identify the peers and targets for the inefficient DMUs. The peers are the DMUs with λ_j^* greater than zero at the optimal solution of model (10), and the targets are given by the expressions shown in (11).

$$\begin{cases} y_{rj_0}{}^{target} = \sum_{j=1}^{n} \lambda_j^* \, y_{rj} & r = 1, \ldots, s \\ b_{kj_0}{}^{target} = \sum_{j=1}^{n} \lambda_j^* \, b_{kj} & k = 1, \ldots, l \end{cases} \tag{11}$$

The dual of model (10), corresponding to the multiplier formulation, is shown in (12). The multiplier formulation facilitates the incorporation of weight restrictions in the model.

$$\begin{aligned} \min \quad & -\sum_{r=1}^{s} y_{rj_0}\, u_r + \sum_{k=1}^{l} b_{kj_0}\, p_k + v \\ s.t. \quad & \\ & \sum_{r=1}^{s} g_y\, u_r + \sum_{k=1}^{l} g_b\, p_k = 1 \\ & -\sum_{r=1}^{s} y_{rj}\, u_r + \sum_{k=1}^{l} b_{kj}\, p_k + v \geq 0 \qquad j = 1, \ldots, n \\ & u_r \geq 0 \qquad r = 1, \ldots, s \\ & p_k \geq 0 \qquad k = 1, \ldots, l \\ & v \in \Re \end{aligned} \tag{12}$$

4 Incorporating Value Judgments in BoD Composite indicators

The CI models presented in the previous sections allow the specification of the indicator weights recurring to optimization, such that each DMU tries to maximise its performance score. This total flexibility in the selection of weights is important for the identification of under-performing DMUs. Under these conditions, if a DMU does not achieve the maximum score, even when evaluated with a set of weights that intends to maximise its performance score, it provides irrefutable evidence that other DMUs are performing better. However, one can argue that the weights assigned to the indicators are not realistic, and thus the robustness of the performance measure and its applicability to a real-world context is questionable. Therefore, in some cases it may be important to incorporate in the model expert opinion (or value judgments) on the importance that each individual indicator should have in the assessment. This can be done by imposing restrictions to weights in the BoD CI model. As noted by Cherchye et al. (2011), the ability to add extra information related to the importance of the individual indicators enables enhancing credibility and acceptance of CIs in practical applications. This section discusses the implementation of weight restrictions in the context of assessments involving composite indicators.

The first attempt to restrict the flexibility of the weights in DEA models was made in the mid 80's. The use of DEA by Thompson et al. (1986) to support the selection of potential sites for a nuclear research laboratory in Texas raised a problem of lack of discrimination between efficient DMUs. The authors improved the discrimination between the DMUs' efficiency scores by defining ranges of acceptable weights. Since the work of Thompson et al. (1986), several types of weight restrictions have been proposed in the DEA literature. Allen et al. (1997) and Thanassoulis et al. (2004) reviewed the literature on weight restrictions and discussed the advantages and limitations of the different approaches.

The most prevalent types of weight restrictions are restrictions to virtual weights and direct restrictions to weights. While the direct restrictions to weights are more often used in DEA assessments that involve an input-output framework, the restrictions to virtual weights are more prevalent in assessments with composite indicators. Cherchye et al. (2007) presented and discussed different ways of implementing the virtual weight restrictions in this context.

4.1 Restrictions to Virtual Weights in CIs

The virtual weights are the product of the output value of the DMU_j with the corresponding weight. The restrictions to virtual weights in DEA models were originally proposed by Wong and Beasley (1990). Such restrictions assume the form presented in (13). They restrict the importance attached to an output indicator y_r, expressed in percentual terms, ranging between a lower and an upper bound (ϕ_r and ψ_r, respectively).

$$\phi_r \leq \frac{u_r\, y_{rj}}{\sum_{r=1}^{s} u_r\, y_{rj}} \leq \psi_r \quad j = 1, \ldots, n \tag{13}$$

Cherchye et al. (2007) pointed out that CIs are often composed by individual indicators that can be classified in mutually exclusive categories $C_z,\ z = 1, \ldots, q$. In this case, one may want to impose restrictions on a category of indicators rather than on individual output indicators, specially when it is difficult to define weights for individual indicators. The restrictions to categories of indicators are natural extensions of the restrictions presented in (13). Instead of restricting the relative importance allocated to the output indicator r, they restrict the relative importance allocated to the set of output indicators from category C_z, as shown in (14).

$$\phi_z \leq \frac{\sum_{r \in C_z} u_r\, y_{rj}}{\sum_{r=1}^{s} u_r\, y_{rj}} \leq \psi_z \quad j = 1, \ldots, n \tag{14}$$

An important advantage of the restrictions to virtual weights is that they are independent of the units of measurement of the inputs and outputs. However, as they are DMU-specific, they may be computationally expensive and lead to infeasible solutions when the bounds are loosely specified.

In order to overcome these problems, Wong and Beasley (1990) suggested to apply the above restrictions only to the virtual outputs of the DMU under assessment (j_0), as shown in (15). This procedure has been often used in the literature (e.g. Morais and Camanho (2011); Lins et al. (2012); Rogge (2012)).

$$\phi_r \leq \frac{u_r \; y_{rj_0}}{\sum_{r=1}^{s} u_r \; y_{rj_0}} \leq \psi_r \tag{15}$$

When the DEA model is output oriented, if the restrictions to the virtual weights are only imposed to the DMU under assessment, the denominator of expression (15) corresponds to the normalization constraint of the DEA model, which is always equal to 1 (see models (1) and (2) in Sect. 2). Thus, expression (15) can be simplified, as shown in (16).

$$\phi_r \leq u_r \; y_{rj_0} \leq \psi_r \tag{16}$$

The restrictions imposed only to the DMU under assessment also have drawbacks. According to Dyson et al. (2001), if the restrictions are imposed only to the virtual outputs of the DMU under assessment, they compromise the symmetry of the model with respect to all DMUs, as each DMU is assessed based on a different feasible region (DMU-specific frontiers). Consequently, some DMUs considered inefficient when evaluated against their specific frontier may appear as peers for others. Note that the weights imposed in their own evaluation may be more restrictive that the weights allowed using a different specification of the virtual weight restrictions, corresponding to the assessment of other DMUs. Sarrico and Dyson (2004) added that these restrictions imposed only to the DMU under assessment might impose unreasonable restrictions on the virtual weights of the other DMUs.

4.2 Direct Weight Restrictions in CIs

The most prevalent type of direct weight restrictions used in DEA applications are Assurance Regions type I (ARI), proposed by Thompson et al. (1990). They usually incorporate information concerning marginal rates of substitution between the outputs (or between the inputs). The formulation of ARI restrictions is shown in (17), where r_a and r_b are outputs in the set $r = 1, \ldots, s$ considered in the estimation of the composite indicator.

$$\phi \leq \frac{u_{r_a}}{u_{r_b}} \leq \psi \tag{17}$$

As pointed out by Allen et al. (1997) and Sarrico and Dyson (2004), a disadvantage of the direct restrictions on weights is that they are sensitive to the units of measurement of the variables, as a result it is often difficult to specify meaningful marginal rates of substitution.

In the context of the composite indicators, an enhanced formulation of ARI weight restrictions was proposed by Zanella et al. (2015). This formulation enables expressing the relative importance of the indicators in percentual terms, instead of specifying marginal rates of substitution. This requires the use of an "artificial" DMU representing the average values of the outputs in the sample analyzed. This type of formulation for the weight restrictions recurring to the use of an "artificial" DMU equal to the sample average was originally proposed by Wong and Beasley (1990), as a complement of DMU-specific virtual weight restrictions.

If instead of restricting the virtual outputs of a DMU_j, as shown in (13), the restrictions are imposed in relation to the average DMU ($\bar{y}_r$), as shown in (18), all DMUs are assessed with identical restrictions. Thus, these weight restrictions in fact work as ARIs.

$$\phi_r \leq \frac{u_r \, \bar{y}_r}{\sum_{r=1}^{s} u_r \, \bar{y}_r} \leq \psi_r \tag{18}$$

The bounds of the restrictions become independent of the units of measurement of the variables, as the numerator and denominator are given by the product of the raw weights with the output quantities. Thus, the bounds ϕ_r or ψ_r of expression (18) may be interpreted as the percentual importance of the indicator y_r in the assessment. A value of ϕ_r and ψ_r equal to one means that output y_r is the only one to be considered in the assessment, whereas values of these bounds equal to zero mean that the corresponding output should be ignored.

The restrictions shown in (18) have the advantage of avoiding comparisons against DMU-specific frontiers, as they reflect a system of weights that is imposed only for the "artificial" DMU. As a result, the evaluation is performed against a frontier that envelops all DMUs under assessment. This feature prevent inefficient DMUs of being selected as peers for others DMUs in the assessment.

It is worth noting that using this type of weight restrictions does not imply that the virtual weights of the DMUs under evaluation are between the percentage bounds specified for the 'artificial' DMU. As these bounds are only imposed for the 'artificial' DMU (corresponding to the average value observed in the sample for the indicators considered), for some of the DMUs the bounds of the virtual weights used in the assessment are outside of the range defined by ϕ_r and ψ_r, as these bounds are not formulated as DMU-specific restrictions.

These weight restrictions can be used with different formulations of composite indicator models, as well as models involving both inputs and outputs. However, it is important to ensure that specification of output-oriented models, as the denominator of the weight restrictions should be aligned with the normalisation constraint of the model.

Furthermore, it is also possible to impose this type of restrictions on a category of indicators, rather than on individual output indicators, as shown in (19).

$$\phi_z \leq \frac{\sum_{r \in C_z} u_r \, \bar{y}_r}{\sum_{r=1}^{s} u_r \, \bar{y}_r} \leq \psi_z \tag{19}$$

The impact of the specification of this type of restrictions in a BoD composite indicator model is illustrated in the next section, using a small contrived example that allows a visual representation of the shape of the production possibility set.

5 Graphical Interpretation of a Directional BoD Model

This section illustrates the estimation of a Directional BoD CI model considering a small example with one desirable output and one undesirable output. This graphical illustration intends to enhance the understanding of how the CI models estimate relative performance, as well as the impact of the model specification and weight restrictions on the shape of the frontier.

The illustrative example is adapted from Zanella et al. (2015), and consists of a set of 6 DMUs, whose data are shown in Table 1. These DMUs are assessed considering two output indicators: y (a desirable output) and b (an undesirable output).

Figure 1 shows the production frontier that would be obtained for our illustrative example reported in Table 1 with an assessment based on model (10) or its dual model (12). The efficient frontier is defined by the segments linking A, C and E. Note that the production possibility set frontier also includes the vertical and horizontal extensions of the efficiency frontier defined from DMUs A and E, respectively.

By setting the directional vector as $g = (-g_b, g_y) = (-b_{kj_0}, y_{rj_0})$, i.e. the current value of the outputs for the DMU under assessment, it is possible to simultaneously expand the desirable outputs and contract the undesirable outputs through a path that allows proportional interpretation of improvements. In order to facilitate the interpretation of the DMUs' projection to the frontier, Fig. 1 illustrates the projection of the inefficient DMUs B, D and F on the frontier.

Table 2 shows the results obtained using the Directional BoD CI model. The value of δ^* can be interpreted as the inefficiency of a given DMU.

For example, for DMU D the value of $\delta^* = 0.381$ corresponds to $\frac{DD^*}{O'D} = \frac{7.842}{20.591}$ in Fig. 1.

Table 1 Indicators of the illustrative example

DMU	y (Desirable)	b (Undesirable)
A	5	7
B	25	28
C	22	15
D	10	18
E	30	30
F	25	40

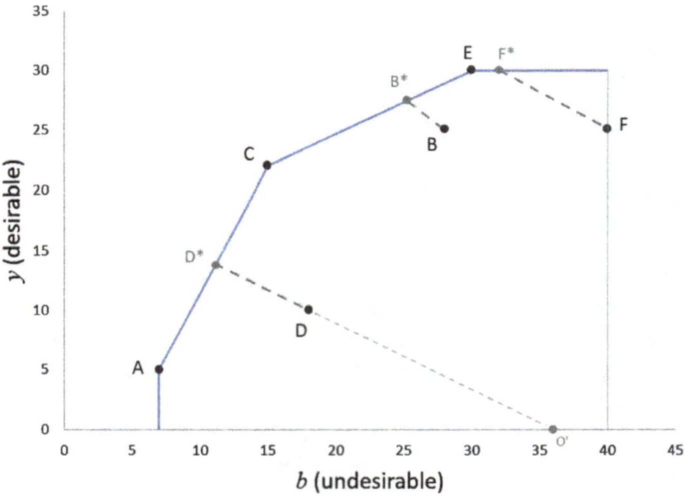

Fig. 1 Production possibility set

Table 2 Composite indicator, peers and targets obtained from the directional BoB CI model

DMU	δ	CI	Peers (λ)	Target y	Target b
A	0	1	A (1)		
B	0.098	0.821	C (0.317); E (0.683)	27.462	25.242
C	0	1	C (1)		
D	0.381	0.448	A (0.482); C (0.518)	13.808	11.145
E	0	1	E (1)		
F	0.200	0.667	E (1)	30	32

The CI score for the DMU D is interpreted as an efficiency score, and corresponds to the ratio $\frac{1-\delta}{1+\delta} = \frac{1-0.381}{1+0.381} = 0.448$. The graphical interpretation of the CI score for DMU D is as follows:

$$CI = \frac{1 - \frac{DD^*}{O'D}}{1 + \frac{DD^*}{O'D}} = \frac{O'D - DD^*}{O'D^*} = \frac{20.591 - 7.842}{28.433} = 0.448 \quad (20)$$

The point D* is the target that the DMU D should achieve to become efficient, i.e., to operate at the frontier (obtained from expression (11)), which corresponds to the value 13.808 for the output indicator y and 11.145 for the output indicator b. The peers for DMU D are DMUs A and C, with values of λ_A and λ_C equal to 0.482 and 0.518, respectively.

5.1 Directional BoD Model with Virtual Weight Restrictions

Next we illustrate the estimation of the directional BoD CI model (10), or its dual (12), using DMU-specific virtual weight restrictions that allow the incorporation of the relative importance of the indicators in percentage terms.

Consider that the virtual weight of each output presented in Table 1 should be at least 40% of the total virtual weights. Following the expression (16), the corresponding weight restrictions to be imposed in model (12) become $u_1\ y_{1j_0} \geq 0.4$ for the desirable output y, and $p_1\ b_{1j_0} \geq 0.4$ for the undesirable output b. Note that as the directional vector of the model (12) was specified as the current value of the outputs of the DMU under assessment, the total virtual weights correspond to the normalization constraint of the model, which is equal to 1. Thus, the denominator of the virtual weight restrictions could be omitted as in (16).

Different bounds could have been used (both upper or lower limits), provided they allow reaching a feasible solution with the sum of virtual weights for the DMU under assessment equal to 100%.

The values of the CI and the targets that the inefficient DMUs must achieve to become efficient are presented in Table 3. Note that by restricting the virtual weights, DMUs A and E are no longer efficient.

The efficient frontiers for the evaluation of DMU D and DMU B with virtual weight restrictions are shown in Figs. 2 and 3, respectively.

Considering the assessment of DMU D, the weight restriction imposed to the desirable output ($u_1\ 10 \geq 0.4$) determines the segment labelled I, whereas segment II represents the restriction associated with the undesirable output ($p_1\ 18 \geq 0.4$). This implies that the slope of the frontier must be between the slopes of segments I and II to ensure that both outputs are given a weight representing at least 40% of the total virtual weight of DMU D. The resulting frontier is defined by the segments in bold in Fig. 2. Note that this is a value frontier, than extends the production possibility set to areas beyond the original technology of production spanned by the DMUs observed in the sample under analysis. The point D* is the target that the DMU D should achieve to operate on the value frontier, which corresponds to the value 15.8 for the output indicator y and 7.56 for the output indicator b.

Table 3 Composite indicator and targets obtained from the directional BoD CI model with DMU-specific virtual weight restrictions

DMU	δ	CI	Target y	Target b
A	0.674	0.195	8.371	2.28
B	0.114	0.796	27.843	24.816
C	0	1		
D	0.58	0.266	15.8	7.56
E	0.04	0.923	31.2	28.8
F	0.22	0.639	30.5	31.2

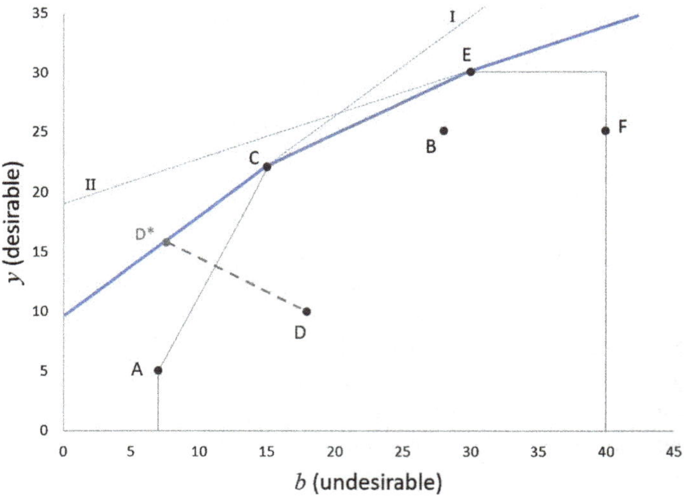

Fig. 2 Assessment of DMU D with DMU-specific virtual weight restrictions

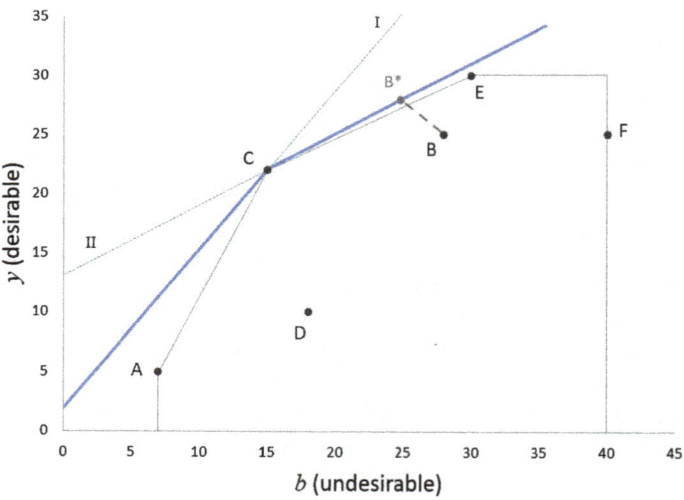

Fig. 3 Assessment of DMU B with DMU-specific virtual weight restrictions

Regarding the assessment of DMU B, it can be seen in Fig. 3 that the frontier against which this DMU is assessed is different from the frontier underlying the evaluation of DMU D. As explained in Sect. 4.1, this happens because the virtual weight restrictions are DMU-specific and thus different restrictions lead to the specification of different frontiers, one for each DMU assessed.

5.2 Directional BoD Model with ARI Restrictions

Next we illustrate the use of ARI weight restrictions (18) for the small example involving 6 DMUs presented in Table 1. With the aim to impose that each output is assigned an importance of at least 40% in the assessment, the weight restrictions imposed to the directional BoD CI model (12) are given by $\frac{u_1\ \bar{y}_1}{u_1\ \bar{y}_1+p_1\ \bar{b}_1} \geq 0.4$ for the desirable output y and $\frac{p_1\ \bar{b}_1}{u_1\ \bar{y}_1+p_1\ \bar{b}_1} \geq 0.4$ for the undesirable output b, with $\bar{y}_1 = 19.5$ and $\bar{b}_1 = 23$.

Table 4 reports the values of CI and the targets for each inefficient DMU, obtained from the directional BoD CI model with ARI weight restrictions.

In Fig. 4, the restrictions on the desirable and undesirable outputs are represented by the segments I and II, respectively. The slope of the efficient frontier must be between the slopes of these segments to ensure that both outputs are weighted at least 40% in the assessment. The efficient frontier used for the assessment of all DMUs is defined by the segments in bold.

Figure 4 also illustrates the projection of the inefficient DMUs on the frontier. The points A*, B*, D*, E* and F* are the targets that the DMUs should achieve to become efficient, whose values are reported in Table 4. Note that the efficiency scores and targets result from an assessment against a value frontier that extends the original production possibility set to areas beyond those defined by the technology of production corresponding to the actual observations.

Unlike what happened with the virtual weight restrictions, which led to assessments against DMU-specific frontiers, using the ARI type of restrictions all DMUs are assessed against a unique frontier, as the weight restrictions are identical for all DMUs.

Table 4 Composite indicator and targets obtained from the directional BoB CI model with ARI weight restrictions

DMU	δ	CI	Target y	Target b
A	0.491	0.341	7.455	3.563
B	0.106	0.806	27.662	25.018
C	0	1		
D	0.481	0.351	14.808	9.345
E	0.01	0.98	30.306	29.694
F	0.234	0.621	30.845	30.648

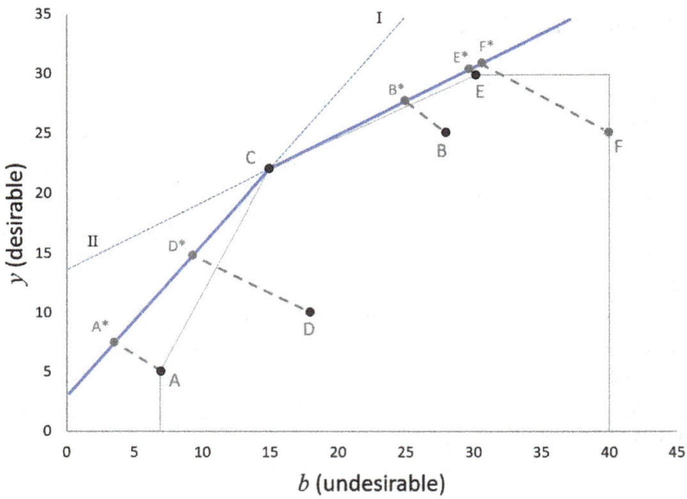

Fig. 4 Assessment of all DMUs with ARI weight restrictions

6 Conclusions

This chapter reviewed the literature on the construction of composite indicator models based on the DEA technique. The motivation for the specification of BoD models and Directional BoD models was presented, as well as the different alternatives concerning the imposition of weight restrictions in this context.

It is shown that BoD CI model allows estimating a frontier identical to the one obtained in assessments using the traditional DEA model, assuming CRS and a dummy input equal to one. Thus, BoD assessments estimate a theoretically sound frontier in the presence of indicators measured as ratios. The Directional BoD model preserves the advantages of estimating a frontier that is identical to the DEA frontier and BoD frontier, with the advantage of accommodating undesirable indicators without any transformations in their measurement scale. As the Directional BoD model is specified using a DDF, it also can seek proportional improvements in all indicators (an increase of desirable outputs and reduction of undesirable outputs).

This chapter also explored different ways to incorporate information on decision-maker preferences about the relative importance of individual indicators aggregated in the CI. The specification of two different types of weight restrictions that can be used in this context (virtual weight restrictions and ARI weight restrictions) was reviewed.

Finally, the use of a Directional BoD model was illustrated using a small contrived example. The impact of the specification of virtual DMU-specific weight restrictions and ARI restrictions on the shape of the frontier was also shown. These restrictions allow incorporating in the assessment the relative importance of indicators, expressed in percentage terms, and enhance the discrimination among DMUs in terms of performance scores.

References

Allen, R., Athanassopoulos, A. D., Dyson, R. G., & Thanassoulis, E. (1997). Weights restrictions and value judgments in data envelopment analysis: Evolution, development and future directions. *Annals of Operations Research, 0*(73), 13–64.

Banker, R. D., Charnes, A., & Cooper, W. W. (1984). Some models for estimating technical and scale inefficiencies in data envelopment analysis. *Management Science, 30*(9), 1078–1092.

Boussemart, J. P., Briec, W., Kerstens, K., & Poutineau, J.-C. (2003). Luenberger and Malmquist productivity indices: Theoretical comparisons and empirical illustration. *Bulletin of Economic Research, 55*, 391–405.

Burck, J., Hermwille, L., & Krings, L. (2012). *Climate Change Performance Index*. Technical report, Germanwatch and Climate Action Network Europe, Berlin, Germany.

Chambers, R. G., Chung, Y., & Fare, R. (1996). Benefit and distance functions. *Journal of Economic Theory, 70*(2), 407–419.

Charnes, A., Cooper, W. W., & Rhodes, E. (1978). Measuring the efficiency of decision making units. *European Journal of Operational Research, 2*(6), 429–444.

Cherchye, L., Moesen, W., Rogge, N., & Van Puyenbroeck, T. (2007). An introduction to 'benefit of the doubt' composite indicators. *Social Indicators Research, 82*(1), 111–145.

Cherchye, L., Moesen, W., Rogge, N., & Van Puyenbroeck, T. (2011). Constructing composite indicators with imprecise data: A proposal. *Expert Systems with Applications, 38*(9), 10940–10949.

Chung, Y., Fare, R., & Grosskopf, S. (1997). Productivity and undesirable outputs: A directional distance function approach. *Journal of Environmental Management, 51*(3), 229–240.

Cook, W. D., & Green, R. H. (2005). Evaluating power plant efficiency: A hierarchical model. *Computers & Operations Research, 32*(4), 813–823.

Cook, W. D., & Kress, M. (1990). A data envelopment model for aggregating preference rankings. *Management Science, 36*(11), 1302–1310.

Dyson, R. G., Allen, R., Camanho, A. S., Podinovski, V. V., Sarrico, C. S., & Shale, E. A. (2001). Pitfalls and protocols in DEA. *European Journal of Operational Research, 132*(2), 245–259.

Emerson, J. W., Hsu, A., Levy, M. A., de Sherbinin, A., Mara, V., Esty, D. C., & Jaiteh, M. (2012). *Environmental Performance Index and Pilot Trend Environmental Performance Index*. Technical report, Yale Center for Environmental Law and Policy, New Haven.

Färe, R., & Karagiannis, G. (2014). Benefit-of-the-doubt aggregation and the diet problem. *Omega, 47*, 33–35.

Hollingsworth, B., & Smith, P. (2003). Use of ratios in data envelopment analysis. *Applied Economics Letters, 10*(11), 733–735.

Koopmans, T. (1951). Analysis of production as an efficient combination of activities. *Activity Analysis of Production and Allocation* (pp. 33–97). New York: Wiley.

Lins, M. E., Oliveira, L. B., da Silva, A. C. M., Rosa, L. P., & Pereira, A. O., Jr. (2012). Performance assessment of alternative energy resources in Brazilian power sector using data envelopment analysis. *Renewable and Sustainable Energy Reviews, 16*(1), 898–903.

Liu, W. B., Meng, W., Li, X. X., & Zhang, D. Q. (2009). DEA models with undesirable inputs and outputs. *Annals of Operations Research, 173*(1), 177–194.

Lovell, C. A. K. (1995). Measuring the macroeconomic performance of the Taiwanese economy. *International Journal of Production Economics, 39*(1–2), 165–178.

Lovell, C. A. K., Pastor, J. T., & Turner, J. A. (1995). Measuring macroeconomic performance in the OECD: A comparison of European and non-European countries. *European Journal of Operational Research, 87*(3), 507–518.

Morais, P., & Camanho, A. S. (2011). Evaluation of performance of European cities with the aim to promote quality of life improvements. *Omega, 39*(4), 398–409.

Nardo, M., Saisana, M., Saltelli, A., Tarantola, S., Hoffmann, A., & Giovannini, E. (2008). *Handbook on constructing composite indicators: Methodology and user guide*. Organisation for Economic Co-operation and Development.

Oggioni, G., Riccardi, R., & Toninelli, R. (2011). Eco-efficiency of the world cement industry: A data envelopment analysis. *Energy Policy, 39*(5), 2842–2854.

Rogge, N. (2012). Undesirable specialization in the construction of composite policy indicators: The environmental performance index. *Ecological Indicators, 23*, 143–154.

Sarrico, C. S., & Dyson, R. G. (2004). Restricting virtual weights in data envelopment analysis. *European Journal of Operational Research, 159*(1), 17–34.

Scheel, H. (2001). Undesirable outputs in efficiency valuations. *European Journal of Operational Research, 132*(2), 400–410.

Seiford, L. M., & Zhu, J. (2002). Modeling undesirable factors in efficiency evaluation. *European Journal of Operational Research, 142*(1), 16–20.

Thanassoulis, E., Portela, M. C., & Allen, R. (2004). Incorporating value judgments in DEA. In *Handbook on data envelopment analysis*. Boston: Kluwer Academic Publishers.

Thompson, R. G., Langemeier, L. N., Lee, C. T., Lee, E., & Thrall, R. M. (1990). The role of multiplier bounds in efficiency analysis with application to Kansas farming. *Journal of Econometrics, 46*(1–2), 93–108.

Thompson, R. G., Singleton, F. D., Thrall, R. M., & Smith, B. A. (1986). Comparative site evaluations for locating a high-energy physics lab in Texas. *Interfaces, 16*(6), 35–49.

United Nations. (2013). *2013 Human development report*. Technical report, United Nations Development Programme, New York.

Wong, Y. H. B., & Beasley, J. E. (1990). Restricting weight flexibility in data envelopment analysis. *The Journal of the Operational Research Society, 41*(9), 829–835.

Zanella, A., Camanho, A. S., & Dias, T. G. (2015). Undesirable outputs and weighting schemes in composite indicators based on data envelopment analysis. *European Journal of Operational Research, 245*(2), 517–530.

Multidirectional Dynamic Inefficiency Analysis: An Extension to Include Corporate Social Responsibility

Magdalena Kapelko, Alfons Oude Lansink, and Spiro E. Stefanou

1 Introduction

The remedies to address the presence of inefficiency will depend on the nature of the factors. With variable inputs being those that are readily measured, invoices delivered on a weekly or monthly basis, the remedies are to assess and monitor these factors; for example, better record keeping, monitoring and supervision of variable input allocations. On the other hand, remedies for dynamic factors involve the management of structures, equipment and land that relate to asset management practices. Having insights into input-specific inefficiencies can provide the decision maker with direction remedies that can take on.

Several streams of research focus on the input-specific inefficiency measure to identify paths to overall efficiency improvement.[1] Bogetoft and Hougaard (1999) and Asmild et al. (2003) propose the multidirectional inefficiency analysis (MEA) to address input-specific inefficiency measurement. The advantage of MEA over other input-specific approaches is that it is able to select benchmarks for inputs reduction

M. Kapelko (✉)
Department of Logistics, Wroclaw University of Economics and Business, ul. Komandorska 118/120, 53-345 Wroclaw, Poland
e-mail: magdalena.kapelko@ue.wroc.pl

A. Oude Lansink
Business Economics, Wageningen University, Wageningen, The Netherlands

S. E. Stefanou
Economic Research Service, United States Department of Agriculture, Washington D.C, USA

[1] Early attempts to input-specific inefficiency measurement were based on the concept of input subvector technical efficiency (Färe et al., 1994). Another stream of literature developed input-specific inefficiency measures based on the Russell measure (Färe & Lovell, 1978) or slacks-based measure (Tone, 2001) and directional slacks-based measure (Färe & Grosskopf, 2010; Fukuyama & Weber, 2009).

P. Macedo et al. (eds.), *Advanced Mathematical Methods for Economic Efficiency Analysis*, Lecture Notes in Economics and Mathematical Systems 692,
https://doi.org/10.1007/978-3-031-29583-6_7

which are not proportional to the actual production, but to the potential improvements related to each input variable separately. In addition, in differentiation to other measures, MEA ensures the characteristic of technological monotonicity. Further methodological extensions of MEA include Bogetoft and Hougaard (2004) (super-efficiency in MEA), Asmild and Pastor (2010) (slack-free MEA), Asmild et al. (2016) (program efficiency in MEA), Baležentis and De Witte (2015) (conditional MEA), and Soltani et al. (2021) (ranking in MEA). Applications of MEA have followed, for example, Holvad et al. (2004), Asmild and Matthews (2012) or Manevska-Tasevska et al. (2021).

This chapter uses MEA in the context of the dynamic production framework. Stefanou (2022) presents a review of the evolution of the dynamic production framework. The approach taken here is based on the theory of adjustment costs that is formalized initially by Epstein (1981) and developed more fully in Silva and Stefanou (2003, 2007), Silva et al. (2015, 2020) and Kapelko et al. (2014). Kapelko and Oude Lansink (2017) introduce the dynamic extension of MEA, which was further developed to dynamic program MEA in Kapelko and Oude Lansink (2018).

The empirical context for the dynamic MEA is to measure inefficiency as firms address the increasingly high awareness of simultaneously being profitable and socially responsible. The past two decades have seen an increasing corporate awareness to address the undesirable consequences (often referred to as undesirable or bad outputs) of firm activity. Being responsive to environmental, health, governance, and societal impacts entails firm strategies to mitigate undesirable outputs as firms address profit maximization. Chambers and Serra (2018) and Puggioni and Stefanou (2019) are the early contributions that extend the production framework to include corporate and social responsibility (CSR), based on by-production model (Murty et al., 2012) that poses that the firm is simultaneously engaging two production technologies; namely, one for the desirable output and the other for the undesirable output. Engida et al. (2020) build on the by-production framework presented in Dakpo and Oude Lansink (2019) by introducing dynamics of the firms' production process that allows for CSR. Ait Sidhoum et al. (2020) address the measurement of inefficiency in the assessment of sustainability. Generalizing the technology to allow for indivisibility, Kapelko et al. (2021) address the firm's challenge that a firm can face the input indivisibility as it seeks to integrate socially responsible activity in its core business activities with the specialization of factors and the introduction of new processes and installations.

However, no efforts have explored the prospect of input-specific inefficiency in a CSR context using the power of the MEA framework. This chapter presents the methodological framework for the dynamic MEA approach that specifically incorporates the CSR activity as part of the firm's decision making. The dynamic DEA approach is used that assumes variable returns to scale and the input-specific inefficiency measures for variable and dynamic factors are presented. The dynamic MEA approach is applied to a sample of 201 European firms for the period 2010–2017, using firm-level CSR and financial data from three industries (capital, consumption, and other category). Across all three industries, the results identify the dynamic factor of investment as the one being the greatest contributor to firm inefficiency.

This chapter proceeds as follows. Section 2 describes the methodology of dynamic MEA with CSR incorporated. Section 3 presents the dataset, while Sect. 4 describes the empirical results. Section 5 offers concluding comments.

2 Dynamic Multidirectional Inefficiency Analysis with CSR Incorporated

This section develops the MEA approach for measuring dynamic inefficiency in the context of firms that are conducting CSR activities. The approach used in this paper builds on earlier work of Kapelko and Oude Lansink (2017, 2018) who extended the MEA approach to the dynamic context, though without consideration of CSR activities. Essentially, the dynamic MEA framework is based on two sequential steps. In the first step, the input coordinates of an ideal point are determined by solving linear programmes for each input, output, and investment separately. The ideal reference point expresses the largest possible reduction in each input or expansion of output and investment keeping everything else constant. The second step solves a DEA model of which the outcome is used to compute a vector of input-, investment- and output-specific dynamic inefficiencies. Below we extend the dynamic MEA model with consideration of CSR activities.

Consider a set of $j = 1,\ldots,J$ firms using a vector of N variable inputs $x = (x_1,\ldots,x_N)$, a vector of F gross investments in quasi-fixed inputs $I = (I_1,\ldots,I_F)$, a vector of F quasi-fixed inputs $k = (k_1,\ldots,k_F)$ and producing a vector of M marketable outputs $y = (y_1,\ldots,y_M)$ and a vector of R CSR outputs $z = (z_1,..,z_R)$. The dynamic production technology transforms variable inputs and gross investments into marketable outputs and CSR outputs at a given level of quasi-fixed inputs and can be defined as:

$$P = \{(x, I, y, z, k) : \ x, I \ can\ produce\ y, z\ given\ k\} \tag{1}$$

The first step in the dynamic MEA is to find an ideal reference point with regard to variable inputs x_n^* for the DMU^0 under analysis (x_n^0, I_f^0, y_m^0, z_r^0). Hence, for each variable input n (=1,…,N), the following linear programming model needs to be solved using DEA, assuming Variable Returns to Scale (VRS):

$$\begin{aligned} x_n^* = \min_{x_n, \lambda^j} & \ x_n \\ \text{s.t.} & \\ & \sum_{j=1}^{J} \lambda^j y_m^j \geq y_m^0, \qquad m = 1, \ldots, M \\ & \sum_{j=1}^{J} \lambda^j z_r^j \geq z_r^0, \qquad r = 1, \ldots, R \end{aligned} \tag{2}$$

$$\sum_{j=1}^{J} \lambda^j x_n^j \leq x_n$$

$$\sum_{j=1}^{J} \lambda^j x_{-n}^j \leq x_{-n}^0 \qquad - n = 1, \ldots, n-1, n+1, \ldots, N$$

$$\sum_{j=1}^{J} \lambda^j (I_f^j - \delta^j k_f^j) \geq I_f^0 - \delta^j k_f^0 \qquad f = 1, \ldots, F$$

$$\sum_{j=1}^{J} \lambda^j = 1$$

$$\lambda^j \geq 0 \qquad j = 1, \ldots, J$$

where (x_{-n}^0) is a vector of all variable inputs except input n, λ^j are intensity weights to form the linear combinations of J observed DMUs, and δ^j denotes depreciation rates that are firm specific. x_n^* denotes the optimal solutions of model (2), and x_n is a target value for the nth input reduction. The condition $\sum_{j=1}^{J} \lambda^j = 1$ imposes VRS.

Next, an ideal reference point for investments I_f^* for the DMU^0 (x_n^0, I_f^0, y_m^0, z_r^0) is identified using DEA and solving the following linear programming models for each investment f(=1,…,F):

$$\begin{aligned}
I_f^* = &\max_{I_f, \lambda^j} I_f \\
&\text{s.t.} \\
&\sum_{j=1}^{J} \lambda^j y_m^j \geq y_m^0, && m = 1, \ldots, M \\
&\sum_{j=1}^{J} \lambda^j z_r^j \geq z_r^0, && r = 1, \ldots, R \\
&\sum_{j=1}^{J} \lambda^j x_n^j \leq x_n^0 && n = 1, \ldots, N \\
&\sum_{j=1}^{J} \lambda^j (I_f^j - \delta^j k_f^j) \geq I_f - \delta^j k_f^0 \\
&\sum_{j=1}^{J} \lambda^j (I_f^j - \delta^j k_f^j) \geq I_{-f}^0 - \delta^j k_{-f}^0 && - f = 1, \ldots, f-1, f+1, \ldots, F \\
&\sum_{j=1}^{J} \lambda^j = 1 \\
&\lambda^j \geq 0 && j = 1, \ldots, J
\end{aligned} \tag{3}$$

where I^0_{-f} is a vector of all investments except investment f. I^*_f denotes the optimal solutions of model (3), and I_f is a target value for the fth investment expansion.

Next, an ideal reference point for marketable output y^*_m for the DMU^0 (x^0_n, I^0_f, y^0_m, z^0_r) is identified by using DEA and solving the following linear programming models for each marketable output m (=1,…,M):

$$
\begin{aligned}
y^*_m = \max_{y_m,\lambda^j} y_m \\
\text{s.t.} \\
&\sum_{j=1}^{J}\lambda^j y^j_m \ge y_m, \\
&\sum_{j=1}^{J}\lambda^j y^j_m \ge y^0_{-m}, && m = 1,\ldots,m-1,m+1,\ldots,M \\
&\sum_{j=1}^{J}\lambda^j z^j_r \ge z^0_r, && r = 1,\ldots,R \\
&\sum_{j=1}^{J}\lambda^j x^j_n \le x^0_n && n = 1,\ldots,N \\
&\sum_{j=1}^{J}\lambda^j (I^j_f - \delta^j k^j_f) \ge I^0_f - \delta^j k^0_f && f = 1,\ldots,F \\
&\sum_{j=1}^{J}\lambda^j = 1 \\
&\lambda^j \ge 0 && j = 1,\ldots,J
\end{aligned}
\tag{4}
$$

where y^0_{-m} is a vector of all marketable outputs except output m. y^*_m denotes the optimal solution of model (4), i.e. the maximum output that can be obtained, keeping all inputs, investments, and CSR output constant, and y_m is a target value for the mth output expansion.

Finally, an ideal reference point for CSR output z^*_r for DMU^0 (x^0_n, I^0_f, y^0_m, z^0_r) is identified using DEA by solving the following linear programming model for each CSR output r (=1,…,R):

$$
\begin{aligned}
z^*_r = \max_{z_r,\lambda^j} z_r \\
\text{s.t.} \\
&\sum_{j=1}^{J}\lambda^j y^j_m \ge y^0_m, && m = 1,\ldots,m
\end{aligned}
\tag{5}
$$

$$
\begin{aligned}
&\sum_{j=1}^{J} \lambda^j z_r^j \geq z_r, \\
&\sum_{j=1}^{J} \lambda^j z_r^j \geq z_{-r}^0, && r = 1, \ldots, r-1, r+1, \ldots, R \\
&\sum_{j=1}^{J} \lambda^j x_n^j \leq x_n^0 && n = 1, \ldots, N \\
&\sum_{j=1}^{J} \lambda^j (I_f^j - \delta^j k_f^j) \geq I_f^0 - \delta^j k_f^0 && f = 1, \ldots, F \\
&\sum_{j=1}^{J} \lambda^j = 1 \\
&\lambda^j \geq 0 && j = 1, \ldots, J
\end{aligned}
$$

As before, z_{-r}^0 is a vector of all CSR outputs except output r. z_r^* denotes the optimal solution of model (5), i.e. the maximum CSR output that can be obtained, keeping all inputs, investments, and marketable output constant, and z_r is a target value for the rth CSR output expansion.

With all ideal reference points $(x_n^*, I_f^*, y_m^*, z_r^*)$ determined for DMU^0, the next step is to use a DEA model to determine marketable output-, CSR output-, input- and investment-specific inefficiency scores:

$$
\begin{aligned}
\beta^* = \max_{\beta, \lambda^j} \beta& \\
\text{s.t.}& \\
&\sum_{j=1}^{J} \lambda^j y_m^j \geq y_m^0 + \beta(y_m^* - y_m^0) && m = 1, \ldots, M \\
&\sum_{j=1}^{J} \lambda^j z_r^j \geq z_r^0 + \beta(z_r^* - z_r^0) && r = 1, \ldots, R \\
&\sum_{j=1}^{J} \lambda^j x_n^j \leq x_n^0 - \beta(x_n^0 - x_n^*) && n = 1, \ldots, N \\
&\sum_{j=1}^{J} \lambda^j (I_f^j - \delta^j k_f^j) \geq I_f^0 - \delta^j k_f^0 + \beta(I_f^* - I_f^0) && f = 1, \ldots, F \\
&\sum_{j=1}^{J} \lambda^j = 1 \\
&\lambda^j \geq 0 && j = 1, \ldots, J
\end{aligned}
\tag{6}
$$

In the above model β shows by how much variable inputs can be contracted, and investments, marketable output and CSR output can be expanded. Note that the model computes the potential for contraction as a proportion of the gap between the ideal point and the actual observed quantity of inputs, outputs, or investment. In the case of marketable output, Eq. (6) shows that the potential for expansion is a proportion of the gap between the ideal point y_m^* and the observed marketable output quantity, y_m^0.

Using the outcome of model (6), the input-specific (IEV_n), marketable output-specific (IEY_m), CSR output-specific (IEZ_r) and investment-specific dynamic MEA inefficiency scores (IEI_f)for DMU^0 are calculated as:

$$IEV_n = \frac{\beta^*(x_n^0 - x_n^*)}{x_n^0}, \qquad n = 1, \ldots, N \tag{7}$$

$$IEY_m = \frac{\beta^*(y_m^* - y_m^0)}{y_m^0}, \qquad m = 1, \ldots, M \tag{8}$$

$$IEZ_r = \frac{\beta^*(z_r^* - z_r^0)}{z_r^0}, \qquad r = 1, \ldots, R \tag{9}$$

$$IEI_f = \frac{\beta^*\left(I_f^* - I_f^0\right)}{I_f^0}, \qquad f = 1, \ldots, F \tag{10}$$

All specific inefficiency measures ($IEV_n, IEY_m, IEZ_r, IEI_f$) take values larger than 0, where a value of 0 indicates full efficiency with regard to the specific output, input or investment.

3 Dataset

In our empirical application we used the CSR and financial data on firms in Europe for the years 2010–2017. Following Auer and Schuhmacher (2016), we grouped firms into three industries: capital (representing building products, construction and engineering, electrical equipment, industrial conglomerates, machinery, auto components, automobiles, technology hardware, and semiconductors), consumption (comprising of consumer durables, textiles and apparel, media, food retailing, food products, and household products), and other (including oil and gas, chemicals, diversified metals, precious metals, steel, paper and forestry, and pharmaceuticals). The CSR data came from Sustainalytics, which is a global leader in environmental, social, and governance research and ratings, while the source of financial data was ORBIS database prepared by Bureau van Dijk.

Sustainalytics allowed us the measure CSR output. Following prior research (e.g., Auer & Schuhmacher, 2016; Kapelko et al., 2021; Puggioni & Stefanou, 2019) we used a weighted score of indicators within environmental, social and governance domains of CSR, as reported by Sustainalytics.[2] The higher this score, the better the CSR performance of firms. The environmental CSR includes such indicators as firms' involvement in recycling or waste reduction, social CSR relates to employment diversity or community involvement, and governance CSR concerns auditor independence or voting rights.

ORBIS was a source of data to assess marketable output, investments as well variable inputs. Marketable output was proxied by firms' revenues. Gross investments in capital (fixed assets) in year t were calculated as the beginning value of fixed assets in year $t + 1$, minus the beginning value of fixed assets in year t, plus the value of depreciation in year t. Two variable inputs were distinguished: material costs and labor costs. All variables were deflated by country-specific price indices as reported by Eurostat (2022). Revenues were deflated by the producer price index for output, gross investments by the price index for capital goods, material costs by the price index for intermediate goods and labor costs by the price index for labor. As data were downloaded from ORBIS in local currencies, we also adjusted the variables by the purchasing power parity (PPP) of the local currency to the US dollar.

The data obtained in Sustainalytics and ORBIS were further merged. Then firms with at least one missing value and with negative or zero values for inputs and outputs were deleted; also, outliers were removed as detected by the method proposed by Simar (2003). The resulting sample was an unbalanced panel of 201 European firms for the period 2010–2017 (643 observations). The majority of firms represented capital industries (260 observations), followed by other industries (210 observations), and consumption industries (173 observations). The sample consisted of the largest European firms with regard to market capitalization, and represented Austria (9 observations), Belgium (38 observations), Finland (26 observations), France (58 observations), Germany (231 observations), Hungary (1 observation), Italy (46 observations), Luxembourg (12 observations), Netherlands (2 observations), Norway (15 observations), Spain (54 observations), Sweden (3 observations), and Switzerland (148 observations).

Averages and standard deviations for the variables described above are reported in Table 1. The standard deviations relative to their respective averages indicate that the sample is quite homogenous with regard to CSR variable, while a large variation is presented for marketable output, variable inputs, and investments. Looking at industries as a whole, investments present the largest variation, followed by materials, marketable output, labor and finally CSR. Furthermore, other industries have the largest values for all variables, on average, except for materials for which capital industries score the largest value. On the opposite end of the spectrum are capital industries with the considerably lower values for all variables than capital and other

[2] Weighted score was created through the system of weights that represent the importance of each indicator in a specific sector. Sustainalytics reports also raw scores for indicators that range from zero to 100. We find weighed scores more appropriate in our context since we focus on different sectors.

Table 1 Averages and standard deviations of the input–output variables, 2010–2017

Variable and Industry	Capital	Consumption	Other	All
Marketable output (million USD)	5128.232 (18,072.335)	1937.337 (1901.449)	5944.575 (11,230.029)	4336.715 (10,401.271)
CSR output (score)	60.207 (10.164)	59.434 (11.058)	63.811 (11.016)	61.151 (10.746)
Materials (million USD)	3038.085 (11,728.134)	805.186 (1182.613)	2937.029 (5637.078)	2260.100 (6182.608)
Labor (million USD)	730.792 (2062.228)	292.690 (322.949)	797.109 (1954.154)	606.864 (1446.444)
Investments (million USD)	413.265 (992.603)	775.727 (4995.212)	1213.306 (2853.719)	800.776 (2947.178)

Note Standard deviations are in parentheses

industries, apart from investments as capital firms are these that invest least, on average.

4 Results

The estimations of MEA dynamic inefficiency measures were done for all European countries together due to the limited sample size per each country. The aggregation of firms into regions is a common strategy in the studies based on Sustainalytics data (e.g., Auer & Schuhmacher, 2016; Engida et al., 2020). We partitioned the sample by year and estimated by each year accordingly, relying therefore on the contemporaneous frontier (Tulkens & Van den Eeckaut, 1995). Furthermore, in the calculations we pooled all three industries since we wanted to assess the differences between industries. It is motivated by the fact that it is well known that inefficiencies estimated with regard to one frontier (e.g., one industry) cannot be compared with inefficiency levels measured relative to another frontier (e.g., second industry) (O'Donnell et al., 2008).[3]

Table 2 presents the average values of MEA dynamic technical inefficiency estimates for each variable of the production set and sector as well as for all industries, for the entire period 2010–2017. The table also contains the outcomes of the test proposed by Simar and Zelenyuk (2006) to assess the significance of the differences in the inefficiencies between sectors and in the inefficiencies specific for each input, output, and investment. Based on these results, the following remarks can be made.

The average values of dynamic inefficiency for all sectors over the 2010–2017 indicate that the greatest inefficiency encountered was for investments (150.061),

[3] Alternatively, we could rely on the frameworks for the comparison of inefficiencies between groups of firms such as, for example, metafrontier approach (Battese et al., 2004; O'Donnell et al., 2008). Nevertheless it was recently shown that such a framework might be biased in the case of nonradial model and estimation of input- and output-specific inefficiencies (Yu et al., 2021).

Table 2 Averages and standard deviations of dynamic inefficiencies, 2010–2017

Variable and industry	Capital	Consumption	Other	All	Differences between industries
Marketable output	0.933 (1.126)	0.906 (2.002)	0.780 (2.085)	0.873 (1.738)	a b c
CSR output	0.127 (0.117)	0.132 (0.114)	0.087 (0.095)	0.115 (0.109)	b c
Materials	0.315 (0.207)	0.300 (0.199)	0.257 (0.208)	0.291 (0.205)	a b c
Labor	0.280 (0.183)	0.266 (0.166)	0.229 (0.180)	0.258 (0.176)	a b c
Investments	262.816 (791.275)	119.541 (516.995)	67.827 (183.520)	150.061 (497.263)	a b c
Significant differences between inefficiencies at the critical 5 percent level?	Yes	Yes	Yes	Yes	

Note Standard deviations are in parentheses
a denotes significant differences between capital and consumption at the critical 5% level
b denotes significant differences between capital and other at the critical 5% level
c denotes significant differences between consumption and other at the critical 5% level

followed by marketable output (0.873), materials (0.291), labor (0.258), while CSR was used most efficiently among all variables (only 0.115 of inefficiency). Hence, the largest adjustments in proportion to the improvement potential defined by the ideal point were necessary in the case of investments. The average investment inefficiency of 150.061 implies that firms in our sample could expand their investment by as much as 15,006% with reference to an ideal point. Despite the suggested average improvement is extremely large, the high inefficiency of investments is found more often in dynamic inefficiency studies (e.g., Dakpo & Oude Lansink, 2019; Engida et al., 2020; Kapelko et al., 2021). Among the possible explanations, for example, high heterogeneity in investments could drive such finding. Nevertheless, the results in Table 2 indicate also the large standard deviation for investments' inefficiency. Hence, the sample was very diverse in terms of the usage of investments, and still some firms used their investment potential relatively efficiently. On the opposite end of the spectrum is CSR which was used most efficiently in the analyzed sample. The average CSR inefficiency in the period 2010–2017 for the aggregated industries of 0.115 implies that firms in our sample could expand their CSR engagement by almost 12%. The differences between variable-specific inefficiencies are all significant at the critical 5% level, as confirmed by the outcomes of the Simar and Zelenyuk (2006) test.

The second main remark regarding the findings in Table 2 is that when comparing the dynamic inefficiencies between industries, we can observe that firms in the category of the other industry obtain the best efficiency results for all input–output-investment variables considered. Other industry firms, on average, produce the largest values of marketable output, engage mostly in CSR, and make the largest investments in capital, while at the same time they engage the largest values of labor and moderate values of materials (Table 1), which could be one of the explanations for this finding. Conversely, firms in capital industry have the largest inefficiencies for all variables, except for CSR for which consumption firms obtain the worst results. Focusing specifically on the CSR dimension of dynamic performance, the pressure of stakeholders of both internal nature (such as employees) and external one (such as customers, competitors, and regulators) is regarded as one of the factors that gives organizations a significant motivation for CSR adoption (Clarkson, 1995). In this sense, consumption firms in the sample offering such products as textiles, food, beverage, and household products seem to have less pressure for different CSR strategies than firms in other and capital industries, which could be related to their lower dynamically efficient outcomes regarding this variable. Firms in other industries represent such sectors as chemicals, oils and gas, and metals, which offer products that have a direct and negative impact on the environment or use a production process that contributes to environmental pollution and other harmful issues and are subject to high pressure from the public on their CSR engagement leading to substantial increment in CSR (Sun & Stuebs, 2013), and hence obtain a higher dynamic efficiency for CSR. Similarly, the capital industry—which consists of such sectors as building and construction that often exploit natural resources— is claimed to be under strong institutional pressure to adopt CSR strategies, mainly related to environmental protection and fair competition (Li et al., 2019). However, additional analysis would be necessary to provide a further explanation of the differences in dynamic inefficiencies between industries. Furthermore, the results for the Simar and Zelenyuk (2006) test for the differences between inefficiency indicators indicate these differences are significant at the critical 5% level, except for the differences between capital and consumption for CSR.

Thirdly, the results in Table 2 indicate that firms' managers' efforts could mainly focus on improvements in using the potential for increasing investments and marketable output. For example, firms could change the type of investments by switching to retooling investments that focus on the implementation of technological innovations. The improvement in efficiency could be expected after such investment takes place (e.g., Licandro et al., 2003). Also, firms could reduce a possible overproduction that might cause the large inefficiencies in marketable output.

Next we analyze the distributions of inefficiencies in more detail. Figure 1 depicts the kernel distributions by industry and for each of the input-, output- and investment-specific inefficiency scores.[4] The visual inspection of plots indicates that distributions of dynamic inefficiency scores are substantially different between industries and variables. The most pronounced differences in inefficiencies between industries

[4] We followed procedure by Simar and Zelenyuk (2006) to plot kernels.

can be observed in the case of CSR, materials, and labor. Therefore, these variables inefficiencies seem to be mostly impacted by the industry-embedded characteristics. Furthermore, the kernel plots of output-specific inefficiencies reveal that, regardless the industry analyzed, marketable output-specific inefficiencies are concentrated mostly around values between 0 and 2, while inefficiencies related with CSR are concentrated mostly around very small, close to 0 values. However, a part of the sample's inefficiencies is also evenly distributed between 0.1 and 0.4. Input-specific inefficiencies seem to be mostly concentrated around 0 as well around 0.4 for materials (with values slightly larger for other industries) and 0.3 for labor (with values slightly larger for capital industries). Finally, the distributions of investment-specific inefficiencies provide an indication of the very large values of these inefficiencies regardless of the industry studied.

The last part of the analysis of the results concerns the changes in average MEA dynamic inefficiencies over the period 2010–2017 (Fig. 2). Having a look at marketable output-specific inefficiencies, its evolution for consumption and other industries does not have a clear trend, while these inefficiencies clearly increased for the capital industries until 2015. Overall, over the entire period capital and consumption industries' marketable output inefficiencies dominated those of other industries. The evolution of CSR-specific inefficiencies reveals from the findings that consumption industries were the most CSR-inefficient, followed by capital and other industries, are persistent over time. Also, CSR inefficiencies are stable over the entire period as compared to inefficiencies related with other variables of the production set. Materials- and labor-specific inefficiencies tend to change dramatically during the first years of the analyzed period (2010–2012), while they fluctuated only slightly in the period 2013–2017. The considerable changes in these inefficiencies early years could be still related with the developments of 2008 economic crisis, which forced firms to tighten their activities that could lead to underutilization of the existing capacities. Investment-specific inefficiency increased dramatically in the later years of the analyzed period, that is 2015 and 2016. Overall, the inefficiencies dropped in 2017 for all variables and industries (apart from other industries for marketable output) that could indicate some learning process regarding the management of inputs, outputs and investments at firm-level was taking place.

5 Conclusions

This chapter introduced a method for the measurement of inefficiency accounting for firms' CSR engagement. This development was made through a dynamic MEA method that is based on the potential improvements approach in inputs, outputs and investments and offers an advantage in separating the benchmark selection from the inefficiency measurement. The proposed method allowed to extend the production set and analyze the inefficiency related with marketable output, CSR output, inputs, and investments. The empirical application considered a dataset of CSR activities of

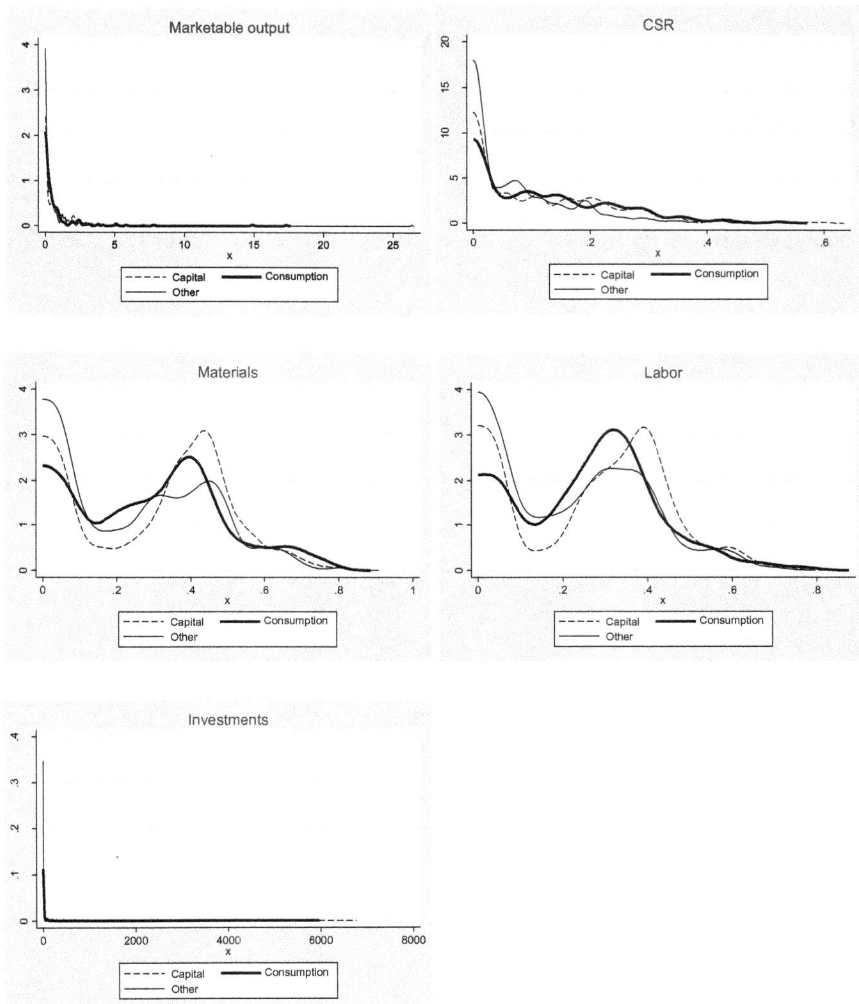

Fig. 1 Technical inefficiencies kernel distributions by the type of inefficiency and industry, 2010–2017

European firms for three industries (capital, consumption, and other) over the period 2010–2017.

The results suggested a considerable scope for reducing dynamic inefficiency across all variables and industries. A more detailed analysis revealed that the largest source of inefficiency, regardless industry studied, were investments, followed by marketable output, materials, labor and finally CSR. This result suggests that there is really a quite a potential for increasing investments in the model developed. This indicates also that analyzed firms could increase inefficiencies mainly through a better management of their investments. Refining the processes, a firm uses to leverage

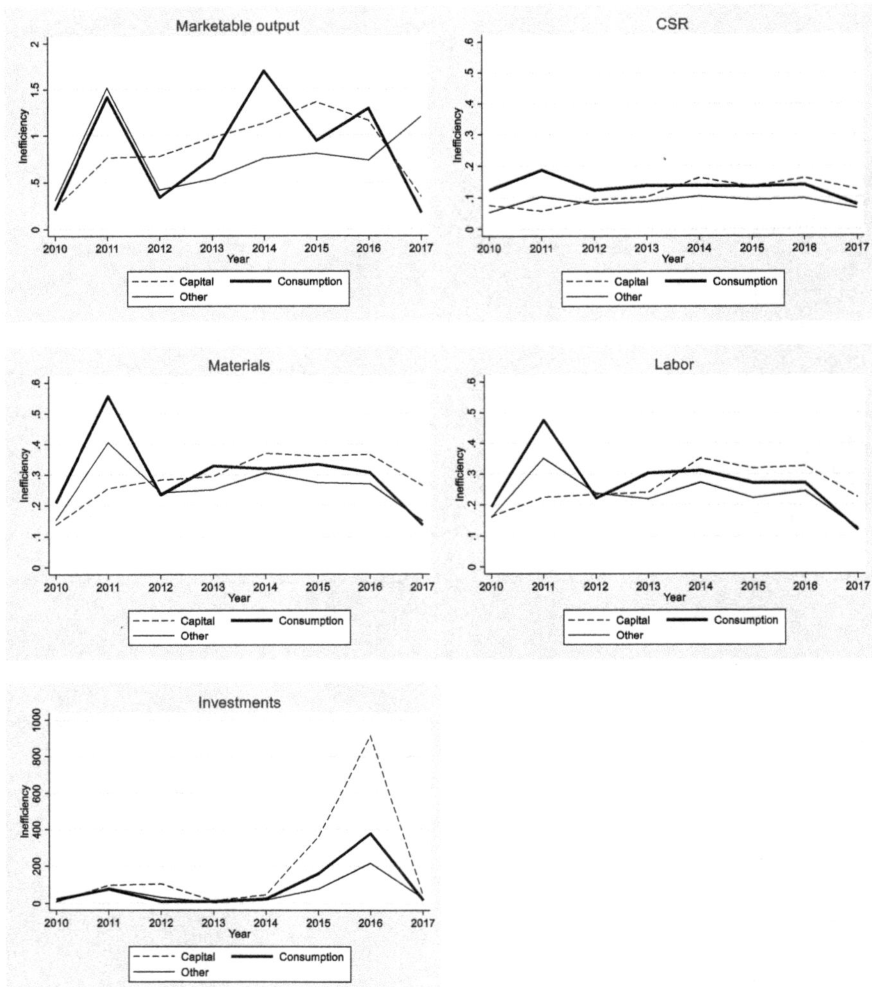

Fig. 2 Evolution of the average dynamic MEA inefficiencies from 2010 till 2017

investment into improved performance is a time-honored focus of management guidance. This is not a surprising result given the greater complexity of the investment input. The alternative to refining existing processes is to invest in new technologies that could provide efficiency benefits in the future. For example, new technologies can have more effective tracking and monitoring systems in place that can signal emergent inefficiencies. Furthermore, this study found that the best dynamic efficient performance was obtained for the other industry, followed by consumption and the capital industry. One of the reasons for these differences in inefficiencies between industries could be different pressures put by the firms' stakeholders on CSR engagement.

Future research efforts could focus on the extensions of the dynamic MEA method developed in this chapter towards the inclusion of undesirable (bad) outputs. Also, the development of dynamic productivity change measures based on the model proposed in this chapter is another promising future research area. Finally, the analysis of CSR inefficiency of firms in the times of COVID-19 pandemic could be of interest too.

Acknowledgements Financial support for this chapter from the National Science Centre in Poland (Grant No. 2016/23/B/HS4/03398) is gratefully acknowledged. The calculations for the adapted Li test were made at the Wroclaw Centre for Networking and Supercomputing (www.wcss.wroc.pl), Grant No. 286. The views expressed in this chapter are of the authors and do not necessarily reflect the official views of the Economic Research Service and U.S. Department of Agriculture.

References

Ait Sidhoum, A., Serra, T., & Latruffe, L. (2020). Measuring sustainability efficiency at farm level: A data envelopment analysis approach. *European Review of Agricultural Economics, 47*(1), 200–225.

Asmild, M., & Matthews, K. (2012). Multi-directional efficiency analysis of efficiency patterns in Chinese banks 1997–2008. *European Journal of Operational Research, 219*, 434–441.

Asmild, M., Hougaard, J. L., Kronborg, D., & Kvist, H. K. (2003). Measuring inefficiency via potential improvements. *Journal of Productivity Analysis, 19*, 59–76.

Asmild, M., & Pastor, J. T. (2010). Slack free MEA and RDM with comprehensive efficiency measures. *Omega, 38*(6), 475–483.

Asmild, M., Baležentis, T., & Hougaard, J. L. (2016). Multi-directional productivity change: MEA-Malmquist. *Journal of Productivity Analysis, 46*, 109–119.

Auer, B. R., & Schuhmacher, F. (2016). Do socially (ir) responsible investments pay? New evidence from international ESG data. *The Quarterly Review of Economics and Finance, 59*, 51–62.

Baležentis, T., & De Witte, K. (2015). One- and multi-directional conditional efficiency measurement—efficiency of Lithuanian family farms. *European Journal of Operational Research, 245*, 612–622.

Battese, G. E., Prasada Rao, D. S., & O'Donnell, Ch. J. (2004). A metafrontier production function for estimatio.n of technical efficiencies and technology gaps for firms operating under different technologies. *Journal of Productivity Analysis, 21*, 91–103.

Bogetoft, P., & Hougaard, J. L. (1999). Efficiency evaluations based on potential (non-proportional) improvements. *Journal of Productivity Analysis, 12*(3), 233–247.

Bogetoft, P., & Hougaard, J. L. (2004). Super efficiency evaluations based on potential slack. *European Journal of Operational Research, 152*, 14–21.

Chambers, R. G., & Serra, T. (2018). The social dimension of firm performance: A data envelopment approach. *Empirical Economics, 54*(1), 189–206.

Clarkson, M. E. (1995). A stakeholder framework for analyzing and evaluating corporate social performance. *Academy of Management Review, 20*(1), 92–117.

Dakpo, K. H., & Oude Lansink, A. (2019). Dynamic pollution-adjusted inefficiency under the by-production of bad outputs. *European Journal of Operational Research, 276*, 202–211.

Engida, T. G., Rao, X., & Oude Lansink, A. (2020). A dynamic by-production framework to examine inefficiency specific to corporate social responsibility. *European Journal of Operational Research, 287*(3), 1170–1179.

Epstein, L. G. (1981). Duality theory and functional forms for dynamic factor demands. *Review of Economic Studies, 48*, 81–95.

Eurostat (2022). *Short-term business statistics*. http://ec.europa.eu/eurostat/web/short-term-business-statistics/data/database. Accessed Jan 2022.

Färe, R., & Grosskopf, S. (2010). Directional distance functions and slacks-based measures of efficiency. *European Journal of Operational Research, 200*(1), 320–322.

Färe, R., & Lovell, C. A. K. (1978). Measuring the technical efficiency of production. *Journal of Economic Theory, 19*, 150–162.

Färe, R., Grosskopf, S., & Lovell, C. A. K. (1994). *Production frontiers*. Cambridge.

Fukuyama, H., & Weber, W. L. (2009). A directional slacks-based measure of technical inefficiency. *Socio-Economic Planning Sciences, 43*(4), 274–287.

Holvad, T., Hougaard, J. L., Kronborg, D., & Kvist, H. K. (2004). Measuring inefficiency in the Norwegian bus industry using multi-directional efficiency analysis. *Transportation, 31*(3), 349–369.

Kapelko, M., & Oude Lansink, A. (2017). Dynamic multi-directional inefficiency analysis of European dairy manufacturing firms. *European Journal of Operational Research, 257*(1), 338–344.

Kapelko, M., & Oude Lansink, A. (2018). Managerial and program inefficiency for European meat manufacturing firms: A dynamic multidirectional inefficiency analysis approach. *Journal of Productivity Analysis, 49*(1), 25–36.

Kapelko, M., Oude Lansink, A., & Stefanou, S. E. (2014). Assessing dynamic inefficiency of the Spanish construction sector pre- and post-financial crisis. *European Journal of Operational Research, 237*, 349–357.

Kapelko, M., Lansink, A. O., & Stefanou, S. E. (2021). Measuring dynamic inefficiency in the presence of corporate social responsibility and input indivisibilities. *Expert Systems with Applications, 176*, 114849.

Li, X., Gao-Zeller, X., Rizzuto, T. E., & Yang, F. (2019). Institutional pressures on corporate social responsibility strategy in construction corporations: The role of internal motivations. *Corporate Social Responsibility and Environmental Management, 26*(4), 721–740.

Licandro, O., Maroto, R., & Puch, L.A. (2003). Innovation, investment and productivity: Evidence from Spanish firms. *FEDEA Working paper*.

Manevska-Tasevska, G., Hansson, H., Asmild, M., & Surry, Y. (2021). Exploring the regional efficiency of the Swedish agricultural sector during the CAP reforms-multi-directional efficiency analysis approach. *Land Use Policy, 100*, 104897.

Murty, S., Russell, R. R., & Levkoff, S. B. (2012). On modeling pollution-generating technologies. *Journal of Environmental Economics and Management, 64*, 117–135.

O'Donnell, C. H. J., Prasada Rao, D. S., & Battese, G. E. (2008). Metafrontier frameworks for the study of firm-level efficiencies and technology ratios. *Empirical Economics, 34*, 231–255.

Puggioni, D., & Stefanou, S. E. (2019). The value of being socially responsible: A primal-dual approach. *European Journal of Operational Research, 276*(3), 1090–1103.

Silva, E., & Stefanou, S. E. (2003). Nonparametric dynamic production analysis and the theory of cost. *Journal of Productivity Analysis, 19*, 5–32.

Silva, E., & Stefanou, S. E. (2007). Dynamic efficiency measurement: Theory and application. *American Journal of Agricultural Economics, 89*, 398–419.

Silva, E., Oude Lansink, A., & Stefanou, S. E. (2015). The adjustment-cost model of the firm: Duality and productive efficiency. *International Journal of Production Economics, 168*, 245–256.

Silva, E., Oude Lansink, A., & Stefanou, S. E. (2020). *Dynamic efficiency and productivity measurement*. Oxford University Press.

Simar, L. (2003). Detecting outliers in frontier models: A simple approach. *Journal of Productivity Analysis, 20*(3), 391–424.

Simar, L., & Zelenyuk, V. (2006). On testing equality of distributions of technical efficiency scores. *Econometric Reviews, 25*(4), 497–522.

Soltani, N., Yang, Z., & Lozano, S. (2021). Ranking decision making units based on the multi-directional efficiency measure. *Journal of the Operational Research Society*, in press.

Stefanou, S. E. (2022). Dynamic analysis of production. In S. C. Ray, R. G. Chambers & S. C. Kumbhakar (Eds.), *Handbook of production economics*. Springer, in press.

Sun, L., & Stuebs, M. (2013). Corporate social responsibility and firm productivity: Evidence from the chemical industry in the United States. *Journal of Business Ethics, 118*(2), 251–263.

Tone, K. (2001). A slacks-based measure of efficiency in data envelopment analysis. *European Journal of Operational Research, 130*, 498–509.

Tulkens, H., & Van den Eeckaut, P. (1995). Non-parametric efficiency, progress, and regress measures for panel data: Methodological aspects. *European Journal of Operational Research, 80*(3), 474–499.

Yu, M. M., See, K. F., & Hsiao, B. (2021). Integrating group frontier and metafrontier directional distance functions to evaluate the efficiency of production units. *European Journal of Operational Research*, forthcoming.

Stochastic DEA

Samah Jradi and John Ruggiero

1 Introduction

Data Envelopment Analysis (DEA) was introduced as a linear programming model by Charnes et al. (1978) and Banker et al. (1984) as a nonparametric model to estimate frontier production and technical efficiency using linear programming. Using only axioms of economic production, DEA is able to estimate the production frontier without specifying a priori the underlying true but unknown frontier production. The model is deterministic in the sense that all deviations from the true production frontier arise from a one-sided noise term that represents only inefficiency. The model proved valuable not only because it was nonparametric, but also because it allowed multiple inputs and multiple outputs.

A competing estimation of the production frontier begins with the work of Aigner et al. (1976) which provided a maximum likelihood estimator for a frontier model where the production frontier was not equivalent to the average production function. In their modeling, they defined the overall error term as using a discontinuous density function defined as two truncated normal distributions with different variances for the positive and negative residuals. The authors show that OLS provided an inconsistent estimate of the intercept term but provide alternative ways to estimate the production frontier assuming a parametric function form and further prove consistency. Notably, however, the authors did not provide a behavioral motivation for the underlying error function.

Aigner et al. (1977) extended Aigner et al. (1976) by providing a behavioral model assuming a composed error stochastic model consisting of noise (traditionally

S. Jradi
EM Normandie University, Le Havre, France

J. Ruggiero (✉)
University of Dayton, Dayton, OH, USA
e-mail: jruggiero1@udayton.edu

P. Macedo et al. (eds.), *Advanced Mathematical Methods for Economic Efficiency Analysis*, Lecture Notes in Economics and Mathematical Systems 692,
https://doi.org/10.1007/978-3-031-29583-6_8

assumed to be normally distributed with mean zero) and a one-sided error term (usually assumed to be half-normal) representing efficiency. As discussed in Aigner et al. (1977), one can obtain unbiased and consistent estimates of the slope parameters by using OLS. However, the intercept would be biased given that the expected value of the composed error is no longer zero. Greene (1980) showed how one could obtain consistent estimates of the intercept from the skewness of the OLS residuals. The model, referred to as COLS provides an alternative to maximum likelihood estimation.

Banker (1988) provided the first stochastic DEA (SDEA) model by introducing a linear programming model that produced a DEA-like piecewise linear frontier through the middle of the data. Banker's model minimized absolute deviations from a nonparametric frontier by constraining production to satisfy the celebrated Afriat constraints (see Afriat, 1972). Banker et al. (1991) applied the SDEA model to analyze the importance of contextual variables and compared the results to the stochastic frontier model. Banker and Maindiratta (1992) showed how one could estimate the model using maximum likelihood with assumed functional forms. In fact, Maindiratta (1984) solved the maximum likelihood version with a small data set.

Banker (1988) further discussed the links of stochastic DEA to quantile regression. In particular, one could generate any empirical quantile of the underlying probability distribution by choosing the appropriate weights on the positive and negative residuals. In the case of equal weights, stochastic DEA estimates the median regression model without an a priori assumption on the production function. In the case that the underlying error structure is normally distributed, then one could of course use least squares instead of least absolute deviation to obtain maximum likelihood estimates. This extension was provided by Kuosmanen (2008); see also Kuosmanen and Johnson (2010). Wang et al. (2014) presented the stochastic DEA model and further discussed the links to the nonparametric quantile regression. In addition, the authors developed a simultaneous quantile function estimator to prevent crossing quantiles and discussed estimation in a stochastic frontier setting.

While Banker (1988) provided the link between the stochastic DEA model and quantile regression, the paper did not further explore the determination of the optimal quantile. One could apply the median regression and correct for the intercept as in COLS; see Kuosmanen and Kortelainen (2012). Alternatively, with additional structure placed on the composed error model, it is possible to estimate the most likely quantile consistent with the underlying production frontier. Jradi and Ruggiero (2019) provided a likelihood based approach to identify which quantile was consistent with maximum likelihood using properties of a normal/half-normal composed error model.[1]

The results from the normal/half normal relied on searching over quantiles to identify the most likely quantile. A search is required because a closed form solution for the composed error CDF does not exist. If, however, the true underlying error has a normal component for measurement error and other statistical noise and an

[1] See also Jradi et al. (2019) for an extension.

exponential component for technical efficiency, one can instead directly estimate the underlying CDF. This approach was derived by Jradi et al. (2021).

In this chapter, we present our work on SDEA and present the approaches for handling both the normal/half-normal and normal/exponential models using SDEA. First, we present the SDEA model introduced by Banker (1988). We then consider additional structure to allow us to estimate the most likely quantile consistent with the production frontier under both distributional assumptions for technical efficiency. Additionally, we provide an alternative measure of firm-level technical efficiency. Ondrich and Ruggiero (2001) prove that the much used conditional expectation measure developed by Jondrow et al. (1982) does not work.[2] In particular, the resulting estimates of inefficiency are perfectly correlated in rank to the observed overall error. As such, the Jondrow et al. estimator is contaminated by measurement error and other statistical noise. Instead, we introduce a measure of individual firm efficiency relative to the median. This measure provides consistent measures of technical efficiency across all estimators including DEA. The drawback, of course, is that the measure will also be contaminated by statistical noise.

2 Stochastic DEA

Following Jradi and Ruggiero (2019) we assume that there are N firms that produce one output y using a vector of M inputs $X \equiv (x_1, ..., x_M)$. The output and inputs of firm j are given by y_j and $X_j \equiv (x_{1j}, ..., x_{Mj})$, respectively. Frontier production and observed production are given by

$$y^f = f(x_1, ..., x_M) \tag{1}$$

and

$$y = f(x_1, ..., x_M) + \varepsilon. \tag{2}$$

The assumptions placed on the error term ε determines which production model is most appropriate. If ε is a standard symmetric error term centered at zero then (2) can be estimated using either least squares or least absolute deviations leading to estimates of the traditional production function.

With the assumption $\varepsilon \leq 0$ represents only technical efficiency, we are interested in estimating the production frontier using either DEA or COLS. The distance between observed production and frontier production provides a measure of firm technical efficiency. In most real world applications, a more appropriate assumption would be that $\varepsilon = v - u$ is a composed error where v measures traditional statistical noise that is symmetric around 0 and $u \leq 0$ represents technical efficiency. The typical econometric approach for estimating the production frontier is with maximum

[2] See also Ruggiero (1999).

likelihood using a priori specified distributional assumptions on the composed error. More recently, there have been alternative approaches to estimating the production function based on quantile regressions.

In addition to the distributional assumptions placed on ε model (2) can be estimated parametrically or nonparametrically. Banker (1988) showed that the median regression could be extended nonparametrically by replacing an assumed functional form with a piecewise linear approximation using the celebrated Afriat constraints. The SDEA model to estimate a production function was first introduced by Banker (1988) with the following linear program:

$$
\begin{aligned}
& Min \sum_{i=1}^{N} (\tau e_{1i} + (1-\tau) e_{2i}) \\
& s.t. \\
& y_i = \alpha_i + \sum_{k=1}^{m} \beta_{ki} x_{ki} + e_{1i} - e_{2i} \qquad \forall i = 1, ..., N \\
& \alpha_i + \sum_{k=1}^{m} \beta_{ki} x_{ki} \leq \alpha_j + \sum_{k=1}^{m} \beta_{kj} x_{kj} \qquad \forall i, j = 1, ..., N \\
& \beta_{mi} \geq 0 \qquad \forall m = 1, ..., M;\ i = 1, ..., N. \\
& e_{1i}, e_{2i} \geq 0 \qquad \forall i = 1, ..., N.
\end{aligned}
\tag{3}
$$

Like the econometric models, this model allows only one output. As discussed in Banker (1988), this model is equivalent to a quantile regression where convexity and monotonicity are assumed without having to specify the functional form for production a priori. The first set of constraints approximates the production quantile piecewise linearly for each DMU. The second set of constraints is the celebrated Afriat conditions that impose concavity of the estimated piecewise linear production function; the predicted value for a given DMU j using its own parameters has to be less than or equal to the predicted output $\alpha_i + \beta_{1i} x_{1i} + ... + \beta_{mi1} x_{mi}$ using the parameters for every other DMU $i, i = 1, ..., N. i, i = 1, ..., N.$

We note that the SDEA model requires a priori specification of the parameter of τ. Given a random variable X with cdf $F_X(x) = P(X \leq x)$ the τth quantile of X is given by $Q_X(\tau) = F_X^{-1}(\tau) = \inf\{x : F_X(x) \geq \tau\}$. As a result, the solution of (3) places the estimated production surface so that (approximately) 100τ percent of the data appear below the production function/frontier.

3 Distributional Assumptions on ε

The Banker (1988) SDEA model provided a useful starting point for extending DEA to allow measurement error. Unfortunately, while τ provides a framework to analyze quantiles of the underlying empirical error, the parameter must be specified a priori. For example, specifying $\tau = 0.5$ leads to a median DEA model where approximately half of the observations appear below the piecewise linear frontier. Notably Banker (1988) does not provide additional guidance.

Jradi and Ruggiero (2019), Jradi et al. (2019, 2021) extended Banker's SDEA model by placing more structure on the error structure with additional assumptions on the composed error model. Jradi and Ruggiero (2019) introduced the concept of the most likely quantile consistent with the true production function/frontier and showed how to estimate the stochastic frontier model using the SDEA model when the underlying error is composed of a normal distribution reflecting statistical noise and a half normal distribution reflecting inefficiency. Jradi et al. (2021) further extended SDEA modeling under the assumption of a normal and exponential composition.

We first consider the case of a normal half-normal composition. In particular, we assume $\varepsilon = v - u$ with $v \sim N(0, \sigma_v)$ and $u \sim |N(0, \sigma_u)|$. The results that follow are presented in Jradi et al. (2021). Given the normal/half normal assumption, the composed error ε has the following theoretical PDF:

$$f(\varepsilon) = \frac{2}{\sigma}\phi\left(\frac{\varepsilon}{\sigma}\right)\Phi\left(\frac{-\varepsilon\lambda}{\sigma}\right) \tag{4}$$

where $\sigma = \sqrt{\sigma_u^2 + \sigma_v^2}$ and $\lambda = \frac{\sigma_u}{\sigma_v}$.

Hence,

$$\begin{aligned}P[\varepsilon \leq 0] &= \int_{-\infty}^{0} \frac{2}{\sigma}\phi\left(\frac{\varepsilon}{\sigma}\right)\Phi\left(\frac{-\varepsilon\lambda}{\sigma}\right)d\varepsilon \\ &= \int_{-\infty}^{0} \frac{1}{\sqrt{2\pi}\sigma}e^{\frac{-\varepsilon^2}{2\sigma^2}}d\varepsilon + \frac{1}{\sqrt{2\pi}\sigma}\int_{-\infty}^{0} e^{\frac{-\varepsilon^2}{2\sigma^2}} erf\left(\frac{-\varepsilon\lambda}{\sqrt{2}\sigma}\right)d\varepsilon.\end{aligned}$$

Let $u = \frac{\varepsilon}{\sqrt{2}\sigma}$ then $du = \frac{1}{\sqrt{2}\sigma}d\varepsilon$. We note that

$$\int_{-\infty}^{0} \frac{1}{\sqrt{2\pi}\sigma}e^{\frac{-\varepsilon^2}{2\sigma^2}}d\varepsilon = \sqrt{2}\sigma\int_{-\infty}^{0} \frac{1}{\sqrt{2\pi}\sigma}e^{-u^2}du = 0.5.$$

Hence, $P[\varepsilon \leq 0] = 0.5 + \frac{1}{\sqrt{2\pi}\sigma}\int_{-\infty}^{0} e^{\frac{-\varepsilon^2}{2\sigma^2}}\left[1 + erf\left(\frac{-\varepsilon\lambda}{\sqrt{2}\sigma}\right)\right]d\varepsilon.$

Given $erf\left(\frac{-\varepsilon\lambda}{\sqrt{2}\sigma}\right) = \frac{2}{\sqrt{\pi}} \sum_{n=0}^{\infty} \frac{(-1)^n(-\varepsilon\lambda)^{2n+1}}{n!(2n+1)(\sqrt{2}\sigma)^{2n+1}}$ we obtain

$$\begin{aligned}
P[\varepsilon \leq 0] &= 0.5 + \frac{1}{\sqrt{2\pi}\sigma} \int_{-\infty}^{0} e^{\frac{-\varepsilon^2}{2\sigma^2}} \frac{2}{\sqrt{\pi}} \sum_{n=0}^{\infty} \frac{(-1)^n(-\varepsilon\lambda)^{2n+1}}{n!(2n+1)(\sqrt{2}\sigma)^{2n+1}} d\varepsilon \\
&= 0.5 + \frac{\sqrt{2}}{\pi\sigma} \sum_{n=0}^{\infty} \frac{(-1)^n(\lambda)^{2n+1}}{n!(2n+1)(\sqrt{2}\sigma)^{2n+1}} \int_{-\infty}^{0} e^{\frac{-\varepsilon^2}{2\sigma^2}} x^{2n+1} d\varepsilon \\
&= 0.5 - \frac{\sqrt{2}}{\pi\sigma} \sum_{n=0}^{\infty} \frac{(-1)^n(\lambda)^{2n+1}}{n!(2n+1)(\sqrt{2}\sigma)^{2n+1}} \frac{-n!}{2\left(\frac{1}{2\sigma^2}\right)^{n+1}} \\
&= 0.5 + \frac{\sqrt{2}}{\pi} \sum_{n=0}^{\infty} \frac{(-1)^n(\lambda)^{2n+1}2^n}{(2n+1)2^{n+0.5}} \\
&= 0.5 + \frac{1}{\pi} \arctan(\lambda).
\end{aligned} \tag{5}$$

This result, first presented inAzzalini (2014),[3] provides the theoretical quantile τ^* associated with the production frontier under the assumption that the underlying composition of the error term is normal/half-normal:

$$\tau^* = 0.5 + \frac{1}{\pi} \arctan(\lambda). \tag{6}$$

Empirically, the signal to noise ratio is not observed and has to be estimated. Unfortunately, a closed form solution for the CDF of the normal/half-normal composed error does not exist. To estimate the most likely quantile, the empirical counterpart of the theoretical quantile τ^* one can search over possible quantiles and choose the most likely quantile that maximizes the likelihood profile.

We illustrate the estimation using simulated data for 300 observations. True production is given by the frontier function $y^f = x^{0.5}$. The single input is generated from a uniform distribution: $x \sim U(1, 10)$. We further assume that $v \sim N(0, 0.1)$ and $u \sim |N(0, 0.2)|$ leading to observed production $y = \exp(v - u)y^f$. Finally, we solve the SDEA model for each τ in the range (0.5, 0.51,…,0.98).[4] After solving for each τ we evaluate the likelihood function based on (4). The particular steps in evaluating the likelihood function are given in Jradi and Ruggiero (2019).

From (6), we know the true but unknown $\tau^* \approx 0.852$. Evaluating the likelihood function for each τ leads to an estimate of the most likely quantile $\hat{\tau}^* = 0.87$. We illustrate the resulting estimated frontier in the following figure. The true but unknown frontier is shown in blue; the SDEA estimated piecewise linear frontier is shown in red.

---

[3] See Proposition 2.7 on page 34–35.

[4] Depending on the sample size, we could fine tune the approach to consider more values of τ.

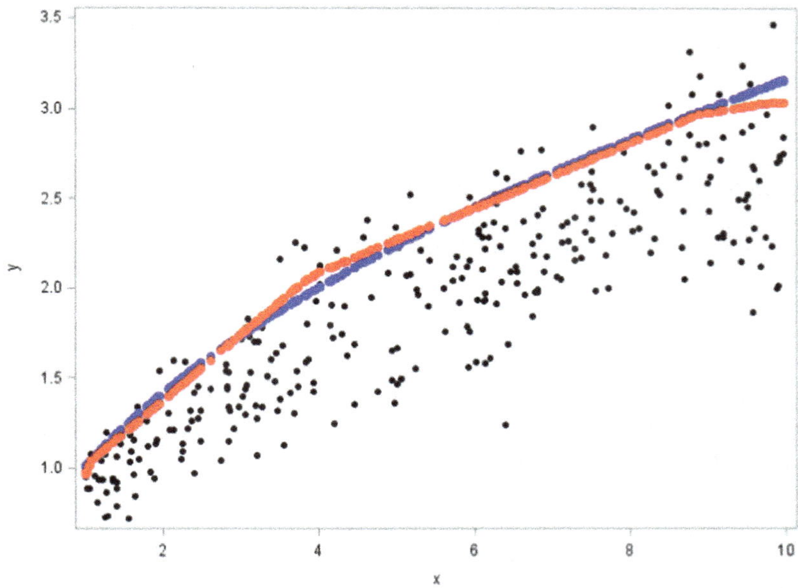

Jradi and Ruggiero (2019) compared the SDEA approach to the maximum likelihood approach using both the correct and a mis-specified functional form. Their results revealed that the SDEA model is modestly outperformed if one chooses the correct functional form. However, the SDEA approach provided much better results than the mis-specified functional form.

Next, we consider an error composed of a normal distribution capturing statistical noise and an exponential distribution measuring technical efficiency. The probability density function of $\varepsilon = v - u$ following normal-exponential distribution with $v \sim N(0, \sigma_v)$ and $u \sim \exp(0, 1/\sigma_u)$ is given by:

$$f(\varepsilon) = \frac{1}{\sigma_u}\Phi\left(-\frac{\varepsilon}{\sigma_v} - \frac{\sigma_v}{\sigma_u}\right)e^{\frac{\varepsilon}{\sigma_u}+\frac{\sigma_v^2}{2\sigma_u^2}}, \tag{7}$$

where Φ is the CDF of the standard normal distribution. Jradi et al. (2021) derived the CDF of the Normal/Exponential model as:

$$F(\varepsilon) = e^{\frac{\varepsilon}{\sigma_u}+\frac{\sigma_v^2}{2\sigma_u^2}}\Phi\left(-\frac{\varepsilon}{\sigma_v} - \frac{\sigma_v}{\sigma_u}\right) + \Phi\left(\frac{\varepsilon}{\sigma_v}\right). \tag{8}$$

Setting $\lambda = \frac{\sigma_v}{\sigma_u}$, we obtain a closed form solution for the most likely quantile[5]

$$\tau^* = e^{\frac{1}{2\lambda^2}}\Phi\left(-\lambda^{-1}\right) + 0.5. \tag{9}$$

[5] See Jradi et al. (2021) for the derivation.

Estimation of this model begins by evaluating the CDF of the Normal/Exponential model at $E(-u) = -\sigma_u$. $E(-u) = -\sigma_u$.

$$F(-\sigma_u) = e^{-1+\frac{1}{2\lambda^2}}\Phi(\lambda - \lambda^{-1}) + \Phi(-\lambda). \tag{10}$$

Given that $F(-\sigma_u)$ is the probability that $\varepsilon \leq 0$, we can apply OLS to obtain an estimate of the percentage of points $\hat{p}$ that have negative residuals and solving $F(-\sigma_u) = \hat{p}$. An estimate $\hat{\lambda}$ is obtained via the solution of

$$e^{-1+\frac{1}{2\hat{\lambda}^2}}\Phi\left(\hat{\lambda} - \hat{\lambda}^{-1}\right) + \Phi\left(-\hat{\lambda}\right) - \hat{p} = 0. \tag{11}$$

Given our estimate $\hat{\lambda}$ we obtain our estimate of the most likely quantile $\hat{\tau}^*$. Importantly, because we have a closed form solution for the CDF of the Normal/Exponential model, we do not need to iterate over quantiles. Rather, the OLS residuals provide sufficient information to directly estimate the most likely quantile.

Like our illustration above, we now illustrate our estimation using simulated data for 300 observations. Similar to the above example, we assume that the frontier function $y^f = x^{0.5}$ where input x is generated from a uniform distribution: $x \sim U(1, 10)$. With $v \sim N(0, \sigma_v)$ we now assume that $u \sim \exp(1/\sigma_u)$, with $\sigma_v = 0.1$ and $\sigma_u = 0.2$. Observed production is given by $y = \exp(v - u)y^f$. To derive an estimate $\hat{p}$ we first apply the median SDEA model to obtain an initial set of residuals. We then derive the percentage of points that are below the average residuals.[6]

Based on the data generating process, we know from (9) that the true but unknown $\tau^* \approx 0.850$. In our sample, we found that approximately 44% of the residuals were below the average residual obtained from the median SDEA model. Solving (11), we estimate $\hat{\lambda} = 0.762$, leading to an estimate of the most likely quantile $\hat{\tau}^* = 0.80$. In the following figure we illustrate our estimated frontier using our estimate of the most likely quantile. Similar to above, the true but unknown frontier is shown in blue; the SDEA estimated piecewise linear frontier is shown in red.

Like the estimation used for the normal/half-normal model above, our approach does a nice job recovering the production frontier. In the next section, we consider how we can extend the SDEA model to estimate individual firm-level efficiency. Due to the absence of additional information (for example, from panel data with assumptions on the efficiency term), we are unable to provide a measure that is not contaminated by statistical noise.

[6] We could have alternatively minimized the sum of square residuals.

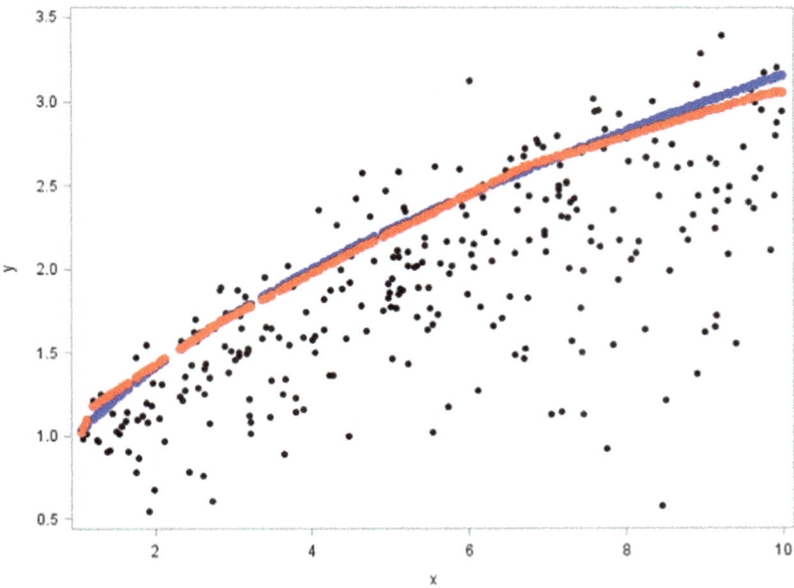

4 Measuring Technical Efficiency

Identifying the true underlying production function/frontier is itself important. The frontier provides information on the maximum output that is achievable for a given input level. In addition, the production function associated with the production frontier provides other useful information including returns to scale, etc. However, in a model that assumes that technical efficiency exists, it is often the case that measuring technical inefficiency is important. The frontier provides the necessary benchmarking information.

By its nature, DEA provides an interpretable measure of technical efficiency as the ratio of observed output to frontier output. Given that the model does not allow measurement error and other statistical noise, the ratio provides the percentage of frontier output that the firm actually produced. Once measurement error is introduced, however, the estimation and interpretation of technical efficiency is not so easy. Jondrow et al. (1982) provided an expected value of efficiency given the overall estimated composed error. However, this measure simply does not work. $E[u|\varepsilon]$ is monotonic in ε. Hence, there is no additional information provided beyond what ε measures. Indeed, two firms with the same ε will achieve the same efficiency score, regardless of the composition of ε. Given that any measure of efficiency can only be relative and not absolute, the Jondrow et al. (1982) measure provides no additional benefit beyond the estimated error term.

Here, we propose evaluating the performance of each firm relative to some benchmark. We could choose any quantile but for this chapter we consider evaluating the performance relative to the median. We solve the SDEA model (3) using $\tau = 0.5$ to obtain residuals defined relative to the median frontier. From the solution of (3) we obtain this estimated residual for each firm i:

$$\hat{e}_i = e_{1i} - e_{2i}. \tag{12}$$

If $\hat{e}_i = 0$ then the observed production coincides with median production. With $\hat{e}_i > (<)0$ the firm is operating above (below) a firm, that is, it has achieved the median level. We can now define performance in output terms relative to the median:

$$\gamma_i^M = \frac{y_i}{y_i - \hat{e}_i}. \tag{13}$$

If $\hat{e}_i = 0$ then $\gamma_i^M = 1$, , the observed unit's performance is consistent with the median: approximately 50% of the sample has a higher (lower) estimated $\hat{e}$. As such, we can consider $\gamma_i^M = 1$ to be median performance. With $\hat{e}_i > 0$ we have $\gamma_i^M > 1$; the observed unit is performing better than the median. As such, $\gamma_i^M - 1$ gives the percentage of output above the median performance level. Finally, for $\hat{e}_i < 0$ and $\gamma_i^M < 1$ gives the observed output as a percentage of the median performance.

Of course, our measure γ^M will only a good measure of individual efficiency if there is little measurement error. As measurement error increases, the correlation between γ^M and the true efficiency will decrease. As such, like the Jondrow et al. (1982) estimator, γ^M captures performance of the output but is biased if one interprets the measure as efficiency. We simulated the normal/half-normal model using the data generating process above. True production is given by the frontier function $y^f = x^{0.5}$. The single input is generated from a uniform distribution: $x \sim U(1, 10)$. We further assume that $v \sim N(0, 0.1)$ and $u \sim |N(0, 0.2)|$ leading to the observed production $y = \exp(v - u)y^f$. . We applied (3) using $\tau = 0.5$ and estimated γ_i^M for each firm using (12) and (13). We also applied the parametric SFA model using the correct production functional form. From the SFA model, we calculated efficiency using the Jondrow et al. (1982) conditional estimator $E[u|\varepsilon]$. . The resulting correlation (rank correlation) between the true efficiency measure and the Jondow et al. estimator was 0.71 (0.67). The correlation (rank correlation) between the true efficiency and γ^M was 0.73 (0.72). We note that the correlation (rank correlation) between the Jondrow et al. estimator and γ^M was 0.93 (0.93).[7] The result of similar overall performance is not surprising given the results of Ondrich and Ruggiero (2001).

[7] We dividing the Jondrow et al. (1982) estimate of efficiency by it's median value and compared it to γ^M. The mean squared difference as 0.004 with a standard deviation of 0.006.

5 Conclusions

In this chapter, we provided an overview of Banker's SDEA model. We also presented recent advances in the literature that show how to select the appropriate τ. In the case of a normal/half-normal composed error, the quantile consistent with the frontier is estimated searching over τ to find the highest value of the likelihood function. In the case of the normal/exponential composed error, we derived the most likely quantile consistent with the true production frontier by solving for the CDF of the composed error. Both approaches were illustrated using simulated data.

We also proposed an alternative measure of performance that evaluates a firm's observed production relative to the performance of the median observation. While this measure does not provide an unbiased measure of efficiency, we feel it provides a more intuitive measure of performance.

References

Afriat, S. (1972). Efficiency estimation of production functions. *International Economic Review, 13*(3), 568–598.

Aigner, D. J., Amemiya, T., & Poirier, D. (1976). On the estimation of production frontiers: Maximum likelihood estimation of the parameters of a discontinuous density function. *International Economic Review, 17*, 377–396.

Aigner, D. J., Lovell, C. A. K., & Schmidt, P. (1977). Formulation and estimation of stochastic frontier production function models. *Journal of Econometrics, 6*, 21–37.

Azzalini, A. (2014). *The skew-normal and related families.* Cambridge University Press.

Banker, R. (1988). Stochastic data envelopment analysis. Working Paper, Carnegie Mellon University.

Banker, R. D., Charnes, A., & Cooper, W. W. (1984). Some models for estimating technical and scale inefficiencies in data envelopment analysis. *Management Science, 30*, 1078–1092.

Banker, R. D., Datar, S. M., & Kemerer, C. F. (1991). A model to evaluate variables impacting the productivity of software maintenance projects. *Management Science, 37*, 1–18.

Banker, R. D., & Maindiratta, A. (1992). Maximum likelihood estimation of monotone and concave production frontiers. *Journal of Productivity Analysis, 3*, 401–415.

Charnes, A., Cooper, W. W., & Rhodes, E. (1978). Measuring the efficiency of decision-making units. *European Journal of Operational Research, 2*, 429–444.

Greene, W. (1980). Maximum likelihood estimation of econometric frontier functions. *Journal of Econometrics, 13*, 27–56.

Jondrow, J., Lovell, C. A. K., Materov, I. S., & Schmidt, P. (1982). On the estimation of technical inefficiency in the stochastic frontier production function model. *Journal of Econometrics, 19*, 233–238.

Jradi, S., Parmeter, C., & Ruggiero, J. (2021). Quantile estimation of stochastic frontiers with the normal-exponential specification. *European Journal of Operational Research, 295*, 475–483.

Jradi, S., Parmeter, C., & Ruggiero, J. (2019). Quantile estimation of the stochastic frontier model. *Economics Letters, 182*, 15–18.

Jradi, S., & Ruggiero, J. (2019). Stochastic data envelopment analysis: A quantile regression approach to estimate the production frontier. *European Journal of Operational Research, 278*, 385–393.

Kuosmanen, T. (2008). Representation theorem for convex nonparametric least squares. *Econometrics Journal, 11*, 308–325.

Kuosmanen, T., & Johnson, A. (2010). Data envelopment analysis as nonparametric least square regression. *Operations Research, 58*, 149–160.

Kuosmanen, T., & Kortelainen, M. (2012). Stochastic non-smooth envelopment of data: Semi-parametric frontier estimation subject to shape constraints. *Journal of Productivity Analysis, 38*, 11–28.

Maindiratta, A. (1984). Studies in the estimation of production frontiers. Ph.D. Thesis, Canegie-Mellon University (pp. 1165). (March 1984).

Ondrich, J., & Ruggiero, J. (2001). Efficiency measurement in the stochastic frontier model. *European Journal of Operational Research, 129*, 434–442.

Ruggiero, J. (1999). Efficiency estimation and error decomposition in the stochastic frontier model: A Monte Carlo analysis. *European Journal of Operational Research, 115*, 555–563.

Wang, Y., Wang, S., Dang, C., & Ge, W. (2014). Nonparametric quantile frontier estimation under shape restriction. *European Journal of Operational Research, 232*, 671–678.

Internal Benchmarking for Efficiency Evaluations Using Data Envelopment Analysis: A Review of Applications and Directions for Future Research

Fabio Sartori Piran, Ana S. Camanho, Maria Conceição Silva, and Daniel Pacheco Lacerda

1 Introduction

Efficiency is a measure often used to evaluate the performance of production units. In general, a production unit can be considered at macro or micro level, such as an industry, a company, and department that uses resources to produce products or services (Park & Cho, 2011). Thus, efficiency evaluation has been a relevant topic for managers and policymakers and has attracted interest from a practical and methodological point of view, mainly in the fields of engineering and economics (Azizi et al., 2016; Aparicio et al., 2017; Førsund, 2018; Kerstens et al., 2019).

Data Envelopment Analysis (DEA) is a widely used technique to conduct efficiency evaluations (Lampe & Hilgers, 2015; Piran et al., 2021). DEA is a non-parametric technique used to evaluate the relative efficiency of production units that are homogeneous and comparable. Such production units are generally called Decision Making Units (DMUs) (Charnes et al., 1978; Banker et al., 1984), and their relative efficiency is evaluated through a benchmarking procedure.

F. S. Piran · D. P. Lacerda
Production Engineering and Systems - PPGEPS/UNISINOS, Research Group on Modeling for Learning - GMAP | UNISINOS, São Leopoldo, Brazil
e-mail: fpiran@unisinos.br

D. P. Lacerda
e-mail: dlacerda@unisinos.br

A. S. Camanho (✉)
Faculdade de Engenharia, Universidade do Porto, Porto, Portugal
e-mail: acamanho@fe.up.pt

M. C. Silva
Católica Porto Business School, CEGE - Centro de Estudos em Gestão e Economia, Lisbon, Portugal
e-mail: csilva@ucp.pt

P. Macedo et al. (eds.), *Advanced Mathematical Methods for Economic Efficiency Analysis*, Lecture Notes in Economics and Mathematical Systems 692,
https://doi.org/10.1007/978-3-031-29583-6_9

However, the benchmarking procedure involved in a DEA application requires relatively large samples of comparable DMUs. The literature (e.g., Kuosmanen et al., 2006) points out that DEA cannot be applied to a specific company in isolation. This belief results from the fact that DEA-based efficiency evaluations are often associated with external benchmarking (Piran et al., 2021). In external benchmarking, a DMU is compared to other similar DMUs (e.g., competitors) at a given point in time or at various points in time (Carpinetti & De Melo, 2002; de Souza et al., 2018a, b). In addition to requiring a relatively large sample of comparable DMUs, external benchmarking requires access to sensitive information from the companies.

However, some organizations have unique characteristics that make it difficult to find appropriate comparators (e.g., NASA, Facebook, Google, Instagram, and others) (Piran et al., 2021). Furthermore, other characteristics (e.g., lack of data access) hinder external benchmarking exercises, making internal benchmarking a promising alternative (Camp & Camp Robert, 1989; Southard & Parente, 2007; Anand & Kodali, 2008; Moriarty & Smallman, 2009). Internal benchmarking considers the comparison between several production units of the same company (Camp & Camp Robert, 1989; Fong et al., 1998). This type of benchmarking can be operationalized through DEA efficiency analysis in many contexts, like the comparison of branches of the same bank, the comparison of stores from the same retailing company, or even the comparison of similar processes within the same company. That is, DMUs are considered multiple subsidiaries of the same firm in a single (cross-sectional data) or multiple (panel data) time periods (Joo et al., 2007, 2009, 2017; Siok & Tian, 2011; Baviera-Puig et al., 2020), to name a few. This type of internal benchmarking is classified by Cox et al. (1997) as collaborative benchmarking, rather than competitive (when external benchmarking is performed).

Nevertheless, there is another possibility of internal benchmarking application, which considers the comparison of the same production unit with itself over time (longitudinal data) (Carpinetti & De Melo, 2002; de Souza et al., 2018a, b). In this case, the DMUs are the single unit of analysis at various points in time, therefore introducing the time dimension into the evaluation of a single company (Piran et al., 2021). In fact, the time dimension seems to be neglected when it comes to internal benchmarking. The classical authors (Camp & Camp Robert, 1989; Cox et al., 1997) and other authors who addressed benchmarking classifications and taxonomies do not consider the possibilities generated through internal benchmarking over time (Fong et al., 1998; Anand & Kodali, 2008; Moriarty & Smallman, 2009).

In this chapter, we focus specifically on DEA-based efficiency analyses, considering internal benchmarking of a single DMU over time, which we will call "internal longitudinal benchmarking". We focus on this type of internal benchmarking that has been neglected in the DEA literature but offers several interesting possibilities for leveraging performance improvements.

This chapter provides a critical revision of the literature that has used DEA in Internal longitudinal benchmarking, pointing out its main contributions and limitations. We exclude from this chapter the literature that performed internal benchmarking of several subsidiaries of the same company. We select articles published in journals indexed in Web of Science and Scopus and conduct a content analysis of 16 papers published from 1978 to 2021. Applications in different industries are explored, and the conditions under which the use of DEA for internal benchmarking is appropriate are analyzed.

The main contribution of the chapter is to show that internal benchmarking with longitudinal data can enhance the performance of companies, and that it should be considered as a valid alternative in two situations: (i) when external benchmarking is not possible; (ii) when external benchmarking is possible and internal benchmarking can be used as a complementary analysis (see Portela et al. (2011)). There are also disadvantages with internal benchmarking, the most important of which is the fact that the best practice of a firm over time may be far from the best possible practice. Analyses considering large samples and including external competitors may lead to the construction of frontiers that are closer to the theoretical optimal performance levels. As a result, with internal benchmarking the improvements that can be sought are always more modest than the ones that could accrue from external benchmarking.

The chapter is organized as follows. The next section describes the methodological procedure adopted for the systematic literature review. Section 3 presents results and the content analysis of the studies that applied the internal benchmarking procedure for efficiency evaluation with DEA in the real world. Section 4 presents an example of empirical application of internal longitudinal benchmarking. Section 5 presents the discussion and future directions. Finally, Sect. 6 presents the conclusion of the study and guidelines for future research.

2 Methodological Procedure

For the systematic literature review, we searched articles by accessing the databases Web of Science and Scopus (Kaffash et al., 2020; Ermel et al., 2021). We considered articles published from 1978 (year of publication of the seminal article on DEA by Charnes et al. (1978)) to 2021. We used the terms "Data Envelopment Analysis" and "Internal Benchmarking" or "Internal Benchmark" or "Internal Analysis" or "Longitudinal Data" or "Time Series". Figure 1 shows the procedure followed. The final scope of the research considered 16 papers. In these papers, a content analysis was performed. The list of 16 papers included in this study is available in Table 1.

Initially, we identified the DEA models used. The papers that perform internal longitudinal benchmarking are characterized by using classical DEA models in its analysis (Charnes et al., 1978; Banker, 1984) or economic analysis (Färe et al., 1985). For this reason, we classify the studies into technical efficiency evaluations (TE) or economic efficiency evaluations (EE) (see Tag in Table 1). In addition, we tried to identify how the DMUs were defined in each study, considering that in internal longitudinal benchmarking this is not a trivial procedure.

The content analysis shows the breakdown of publications considering the four main economic sectors and their respective areas of activities as defined in Kenessey (1987). According to Kenessey (1987), the primary sector corresponds to primary activities, for example, agriculture and commodities. The secondary sector corre-

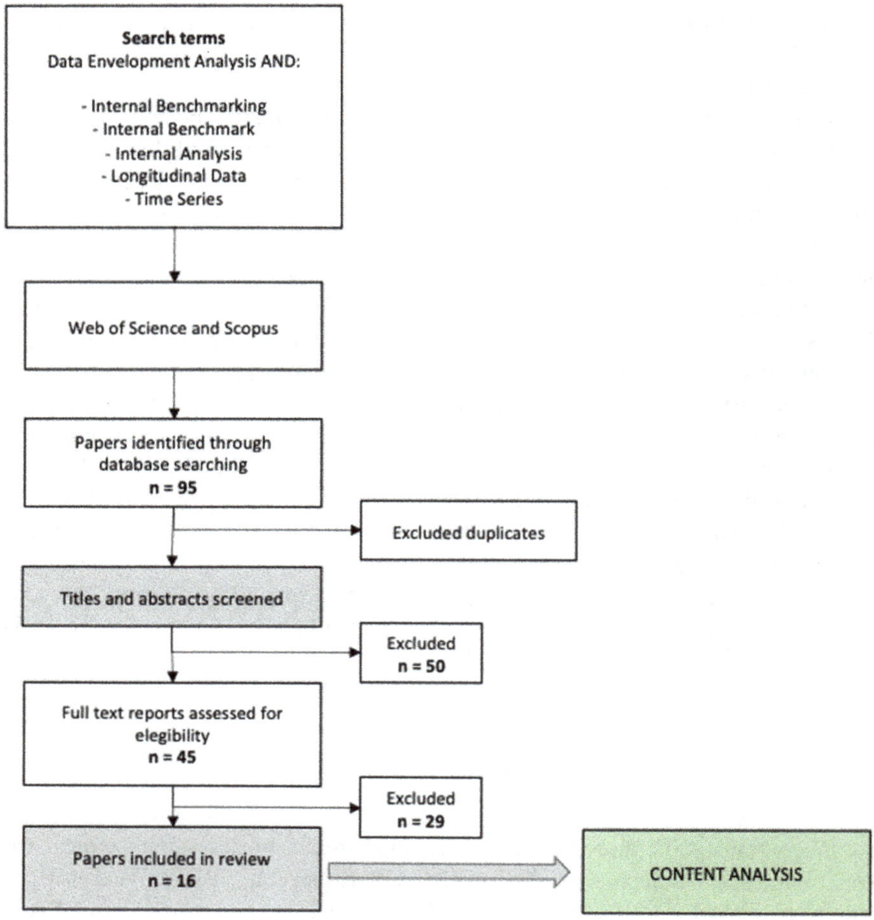

Fig. 1 Flowchart of the results of the systematic literature review

sponds to industrial activities and the tertiary sector to basic services. The quaternary sector considers education, financial, state, and public administration activities.

Subsequently, we identified whether the approach used was a single-stage DEA analysis or a two-stage DEA analysis. In the two-stage DEA analysis, the first stage calculates efficiency using DEA, and the second stage aims to explore contextual factors that affect efficiency (Liu et al., 2016). We classify the two-stage DEA analysis into studies that perform regression analysis, hypothesis testing, and cluster analysis in the second stage. We identified the DEA models and two-stage DEA analysis (when performed) in each of the 16 studies included in this research, and we coded them a posteriori. Finally, to illustrate the concepts discussed we borrow the empirical application performed by Piran et al. (2021) (same authors as in this chapter).

Table 1 List of primary studies composing the analysis

Authors	Tag	Title of the paper
Sueyoshi (1991)	EE	Estimation of stochastic frontier cost function using data envelopment analysis: an application to the AT&T divestiture
Tone and Sahoo (2005)	EE	Evaluating cost efficiency and returns to scale in Life Insurance Corporation of India using data envelopment analysis
Kang (2010)	EE	Liberalization policy, production, and cost efficiency in Taiwan's telecommunications industry
Joo et al. (2013)	TE	Measuring the longitudinal performance of 3PL branch operations
Roets and Christiaens (2015)	TE	Evaluation of railway traffic control efficiency and its determinants
Piran et al. (2016)	TE	Product modularization and effects on efficiency: an analysis of a bus manufacturer using data envelopment analysis (DEA)
Gilsa et al. (2017)	TE	Longitudinal evaluation of efficiency in a petrochemical company
Barbosa et al. (2017)	TE	Exploratory analysis of the variables prevailing on the effects of product modularization on production volume and efficiency
de Souza et al. (2018a)	TE	Efficiency and internal benchmark on an armament company
de Souza et al. (2018b)	TE	Do the improvement programs really matter? An analysis using data envelopment analysis
Gong et al. (2019)	TE	Multi-level and multi-granularity energy efficiency diagnosis scheme for ethylene production process
Telles et al. (2020)	TE	Drum-buffer-rope in an engineering-to-order system: An analysis of an aerospace manufacturer using data envelopment analysis (DEA)
Piran et al. (2020a)	TE	Effects of product modularity on productivity: an analysis using data envelopment analysis and Malmquist index
Piran et al. (2020b)	TE	Overall Equipment Effectiveness: Required but not Enough—An Analysis Integrating Overall Equipment Effect and Data Envelopment Analysis
O'Neal et al. (2020)	TE	Benchmarking aircraft maintenance performances using data envelopment analysis
Piran et al. (2021)	EE	Internal benchmarking to assess the cost efficiency of a broiler production system combining data envelopment analysis and throughput accounting

3 Results

The list of papers included in the analysis is shown in Table 1. We identified that 12 papers performed the technical efficiency (TE) analysis, while 4 papers performed an economic efficiency (EE) analysis. Economic efficiency is in all cases a model of cost functions. So, no revenue or profit functions have been applied yet to the context of internal longitudinal benchmarking.

The first study to perform internal longitudinal benchmarking was Sueyoshi (1991). The author used DEA economic (cost) perspective and analyzed data from AT&T over 31 years (1947–1977). A few years later, Tone and Sahoo (2005) analyzed data from Life Insurance Corporation (LIC) in India over 19 years and Kang (2010) analyzed data from Chunghwa Telecom Company (CHT) over 39 years (1966–2004), both also using an economic perspective. Recently, the internal longitudinal benchmarking using DEA economic (cost) perspective was taken up by Piran et al. (2021). Piran et al. (2021) analyzed data from a broiler production system over 6 years (2014–2019). It is interesting to note that in the studies by Sueyoshi (1991), Tone and Sahoo (2005), and Kang (2010), the DMU was the company itself in annual periods. Piran et al. (2021) adopted another approach by considering as DMU each production batch of the company over the years.

Starting in 2013, we identified a concentration of studies that perform internal longitudinal benchmarking using a technical efficiency perspective. Joo et al. (2013) analyzed the longitudinal efficiency of operations of a subsidiary of a third-party logistics company (3PL) over 36 months (2005–2007). Roets and Christiaens (2015) develop a two-stage internal benchmarking that evaluates and explains the efficiency of the Belgian rail traffic control over 18 months. Piran et al. (2016) analyzed the effects of product modularization on the efficiency of Product Engineering and the Production Process of a bus manufacturer. Barbosa et al. (2017) evaluated the variables influencing the effects of product modularization on the production volume and efficiency of a bus manufacturer. A longitudinal analysis of Product Engineering and Production Process indicators was performed considering the products before and after modularization. The studies by Piran et al. (2016) and Barbosa et al. (2017) analyzed 42 months of company data.

We observe in these studies the trend of defining DMUs in monthly periods, that is, the DMU is the company or department itself in monthly periods over time. The authors argue that this practice facilitates data collection, since most of the companies' indicators (variables) are controlled and measured monthly, facilitating their collection and processing for subsequent analysis.

This practice for defining DMUs considering the same company or department in monthly periods is used by other studies. For example, de Souza et al. (2018a) conducted a longitudinal case study evaluating the efficiency of the production system in an arms manufacturing company over 72 months (6 years). In another paper, de Souza et al. (2018b) analyzed the impact of continuous improvement and learning processes on efficiency and production volume in the same arms manufacturer over the same period (72 months). Telles et al. (2020) analyzed the effects of Drum-buffer-rope (DBR) implementation on the efficiency in production lines of an aerospace manufacturer over 144 months. Piran et al. (2020a) used DEA and the Malmquist index to measure the effects of product modularity on the efficiency and productivity of the product engineering area of a bus manufacturer over 36 months. O'Neal et al. (2020) evaluated the efficiency of maintaining aircraft in the United States Air Force for 41 months.

However, some studies go deeper into the granularity of the DMU, considering periods of the company, department, or process itself in daily periods. For example, Gong et al. (2019) developed a multi-level DEA model to analyze the efficiency of a set of processes and equipment in an ethylene production plant. Piran et al. (2020b) presented the efficiency analysis of a production system using DEA and OEE (Overall Equipment Effectiveness) in an integrated way (DEA/OEE). In the study by Gong et al. (2019), the DMU is the firm in daily periods and in Piran et al. (2020b) the DMU is a single production process in daily periods.

Finally, Gilsa et al. (2017) chose to define the DMU as the production batch over time. Gilsa et al. (2017) longitudinally evaluated efficiency, considering investment projects and technological change, in a second-generation petrochemical company using DEA.

The use of the internal longitudinal benchmarking procedure to evaluate the efficiency of companies is not recent (Sueyoshi, 1991). The practice has been little adopted over the years with few works carried out. However, it seems that the literature has realized the potential of using internal longitudinal benchmarking in recent years (2020–2021). In this regard, 5 papers were published at this time (Telles et al., 2020; Piran et al., 2020a, b; O'Neal et al., 2020; Piran et al., 2021).

Content Analysis—Studies by Economic Sector and Two-Stage DEA

Figure 2 shows the distribution of the revised papers per economic sector, and Table 2 details these numbers per activity (application area) according to Kenessey (1987)'s classification. Of all articles, the secondary sector shows a greater volume of publications (56%), followed by the tertiary sector (25%). The quaternary sector has 13% and the primary sector 6%.

Fig. 2 Number of papers by economic sector

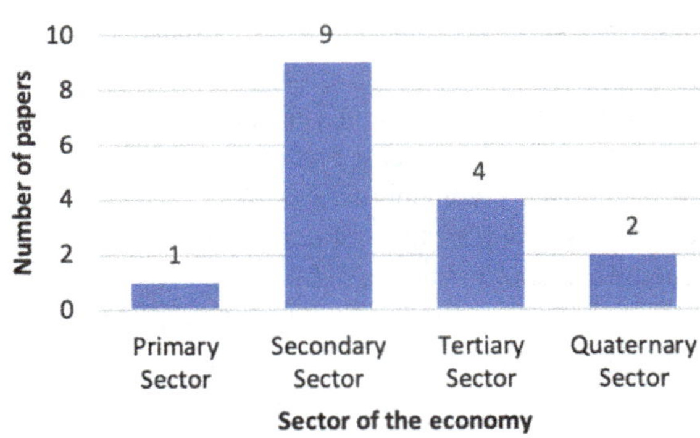

Table 2 Number of papers by economic sector and area of application

Sector of the economy	Application area (activities)	Number of papers	Authors
Primary sector	Agricultural	1	Piran et al. (2021)
Secondary sector	Manufacturing	8	Piran et al. (2016), Gilsa et al. (2017), Barbosa et al. (2017), de Souza et al. (2018a, b), Telles et al. (2020), Piran et al. (2020a), Piran et al. (2020b)
	Supply Chain	1	Joo et al. (2013)
Tertiary sector	Energy	1	Gong et al. (2019)
	Aviation industry	1	O'Neal et al. (2020)
	Telecommunications Industry	2	Sueyoshi (1991), Kang (2010)
Quaternary sector	Insurance Industry	1	Tone and Sahoo (2005)
	Government	1	Roets and Christiaens (2015)

It is possible to notice an emphasis on the manufacturing area (secondary sector) in the internal longitudinal benchmarking. This contrasts with the areas where DEA is typically applied, which are mainly concentrated in the remaining sectors (e.g., banking, insurance, education, health, and agriculture). The manufacturing area has attracted the attention of researchers who use the internal longitudinal benchmarking procedure, probably because manufacturing managers resist releasing sensitive data (of their operations) for comparison with other competing companies, for example the quantity of raw materials and the number of people used in the production processes. In addition, the existence of similar subsidiaries in manufacturing firms is less common than in service firms (e.g., two retail outlets of a retail firm may be similar, but two factories of the same retail company (e.g., Inditex) may not be comparable, because factories tend to be specialized units).

A few studies reviewed perform a second-stage analysis to relate the efficiency calculated by DEA with contextual variables (Table 3). In general, two-stage DEA analyses seek to capture institutional, demographic, and management factors that affect efficiency. These studies perform analyses with regression/classification-based techniques: Ordinary Least Squares (OLS) (2 papers), Tobit Regression (1 paper), and Artificial Neural Network (ANN) (1 paper) are used. This procedure is not new, considering that it is also commonly used in DEA analyses that undertake external benchmarking (Liu et al., 2016).

In addition, some studies perform hypothesis testing. Hypothesis testing generally seeks to evaluate whether there is a significant difference between the efficiency scores of different periods or branches. The hypothesis tests used are ANOVA, CausalImpact, Wilcoxon test, t test, Kruskal-Wallis, and Mann-Whitney U test. Some

Table 3 Two-stage DEA

Approach	Technique	Total	Authors
Regression	Linear Regression (OLS)	2	Roets and Christiaens (2015), de Souza et al. (2018b)
	Tobit Regression	1	O'Neal et al. (2020)
Deep Learning Algorithms (Regression or Classification)	Artificial Neural Network	1	Barbosa et al. (2017)
Hypothesis Testing	ANOVA	7	Piran et al. (2016), Gilsa et al. (2017), Barbosa et al. (2017), de Souza et al. (2018a, b), Telles et al. (2020), Piran et al. (2020b)
	CausalImpact	1	Piran et al. (2016)
	Wilcoxon test	1	Piran et al. (2021)
	t test	1	Gilsa et al. (2017)
	Kruskal-Wallis test	1	Gilsa et al. (2017)
	Mann-Whitney U test	1	Piran et al. (2020a)

studies use more than one technique for hypothesis testing (Gilsa et al., 2017) or perform regression and hypothesis testing jointly (Barbosa et al., 2017).

We postulate that hypothesis testing is one of the great potentials of analyses using internal longitudinal benchmarking. The effect of managerial actions or intervention on efficiency can be evaluated. In addition, it is possible to explore the effect of contextual variables over time, increasing the explanatory power of DEA models.

4 Internal Longitudinal Benchmarking: An Empirical Application

Piran et al. (2021) conducted an exploratory analysis of economic efficiency (EE), specifically cost efficiency, in a Brazilian broiler production system using internal longitudinal benchmarking. An analysis was conducted over 6 years (2014 to 2019) using DEA. The Critical Incident Technique (CIT) was also used to understand the effects of interventions made over time on the economic efficiency of the company.

The aviary works in a batch production process, with an average of seven to eight batches produced annually. A batch corresponds to a quantity of approximately 25,000 birds that are housed in the aviary for six weeks, on average, corresponding to their growth and fattening period. Each batch production process was considered a DMU, with a total of 45 DMUs. The inputs and outputs (Table 4) were determined with the support of two experts in the broiler production system.

Table 4 List of variables used in the analysis (from Piran et al., 2021)

Variable (Measure)	Role in the Model (notation)	Quantity(x_i) Average	Price(c_i) (R$) Average
Water (thousand litres)	Input (x_1, c_1)	58.84	2.18
Energy_Wood (m^3)	Input (x_2, c_2)	13.34	67.97
Energy_Pellets (kg)	Input (x_3, c_3)	1,397.22	0.57
Electricity (kw)	Input (x_4, c_4)	2,054.56	0.42
Energy_Fuel (l)	Input (x_5, c_5)	11.24	4.20
Energy_Gas (m^3)	Input (x_6, c_6)	142.51	5.27
Mortality (un)	Input (x_7, c_7)	635.16	0.38
Lead time (days)	Input (x_8, c_8)	28.73	243.28
Labor (Number of Full Time Equivalents)	Input (x_9, c_9)	1.00	5,722.49
Wood shaving (m^3)	Input (x_{10}, c_{10})	14.73	29.88
Lime (kg)	Input (x_{11}, c_{11})	137.78	0.59
Amount of meat produced (kg)	Output (y_1)	37,039.56	0.32

The cost efficiency levels were estimated assuming constant return scale (CRS), given the comparability of batches in terms of scale size. Given the input-oriented nature of the cost efficiency assessment, the potential reduction to the quantity and/or balance of resources used in the batches produced is explored. The data analyzed (quantities and prices) were collected directly from the database of the aviary management system software and from electronic spreadsheets used for monitoring the aviary indicators. Observed data were collected for each of the 45 batches (DMUs) analyzed.

Figure 3 shows the evolution of the efficiency scores by reporting the average values for batches' efficiency in each year. Considering the six years analyzed, the results show that the average values of technical, allocative, and cost efficiencies were 85.61, 69.24, and 58.63%, respectively. These values represent the relative distance to the best practices observed in the frontier for the 6-year period analyzed. It should also be highlighted that internal benchmarking studies with longitudinal data also reveal efficiency trends over time. For example, Fig. 3 shows an improvement trend in cost efficiency in the last two years considered in this assessment.

Table 5 shows the average values of input targets and gains per batch in monetary value. If the production department of the aviary adopted the target input levels suggested by the DEA cost efficiency analysis, it would be possible to obtain, on average, a cost reduction of 31.57% (R$ 5,341.07/R$ 16,916.64) in relative terms, and R$ 5,341.07 in absolute values per batch produced. A cost-efficient batch would have an average total cost of R$ 11,575.57 associated with the resources considered in the DEA model.

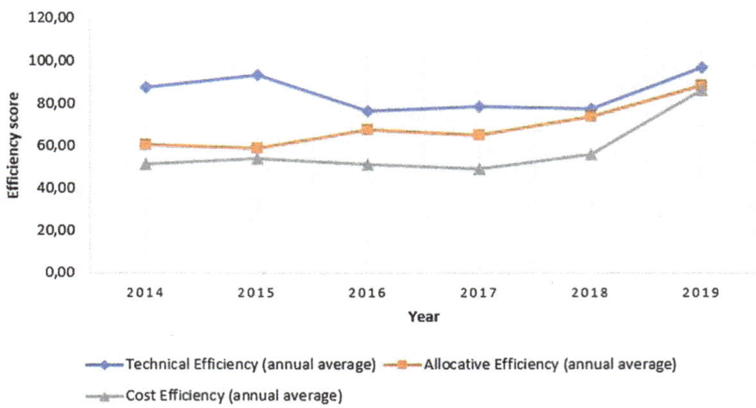

Fig. 3 Annual average efficiency per batch (2014–2019) (from Piran et al., 2021)

Table 5 Average Target and Gain per batch (values in R$) (from Piran et al., 2021)

Input (units of measurement)	Actual	Target	Gain	Gain (%)
Water	128.42	97.58	30.84	24.02
Energy_Wood	788.38	137.06	651.32	82.61
Energy_Pellets	798.14	297.36	500.78	62.74
Electricity	888.94	673.31	215.63	24.26
Energy_Fuel	47.75	25.18	22.58	47.28
Energy_Gas	769.91	279.53	490.38	63.69
Mortality	249.47	186.06	63.42	25.42
Lead time	6,998.16	5,309.40	1,688.75	24.13
Labor	5,722.49	4,383.66	1,338.83	23.40
Wood shaving	443.18	149.79	293.39	66.20
Lime	81.80	36.63	45.16	55.21
Total	**16,916.64**	**11,575.57**	**5,341.07**	**31.57**

To understand the interventions or events that affected cost efficiency, critical incidents (CIs) were mapped over time (Fig. 4). Two classifications were used: (i) Internal Critical Incidents (ICI), which are controllable actions performed by managers to improve the efficiency of the aviary, and (ii) External Critical Incidents (ECI), which are non-controllable events that can affect the productive system. We sought to evaluate the effects provided by these CIs. A CI is expected to include information considering (i) the background (how the cost efficiency was before the CI), (ii) the experienced event (intervention), and (iii) the consequences (the effect of the CI on cost efficiency). Eight CIs (two external and six internal) were identified.

In the production of batches 1 to 4 in 2014, the company used the industrial gas system as a source of heat generation for birds. In April 2014 (batch 5), the

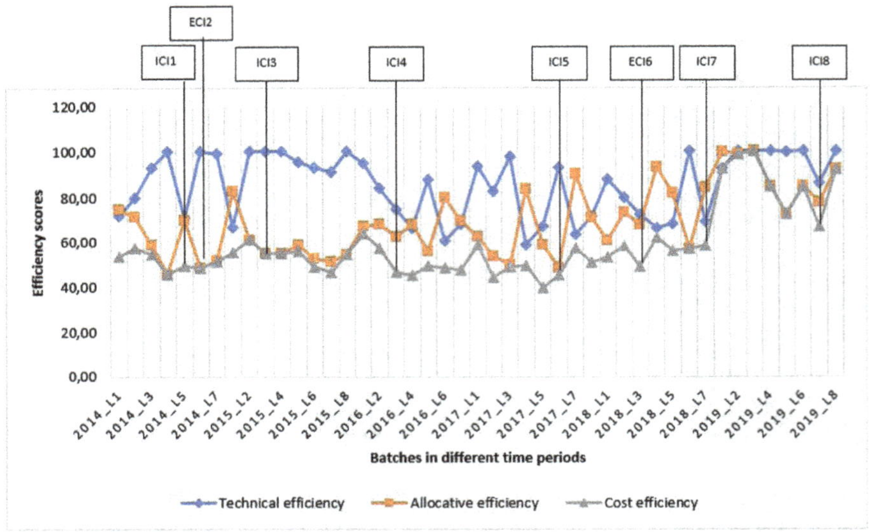

Fig. 4 Evolution of efficiency over time (2014–2019) (from Piran et al., 2021)

company purchased a small furnace to assist in heat generation for birds by burning firewood (ICI1). There were no effects of this intervention on economic efficiency, but there was a decrease in technical efficiency, since in the transition period of the heat generation source it was necessary to use firewood and gas simultaneously. However, this decrease in technical efficiency was offset in terms of allocative efficiency, since the cost of firewood is lower than the cost of gas.

In 2014, specifically in the production of batch 6, the bacterium Salmonella was identified in the aviary (ECI2). The need to interrupt production for cleaning and disinfecting the aviary led to a reduction of 1.11% in the cost efficiency of this period.

The broiler production is specialized in the production of batches with female birds. In 2015 (batch 3), an experimental batch of male birds (ICI3) was introduced in the aviary. The production of male birds requires different production management techniques, thus requiring the specialization of labor. The aviary had difficulties with this process, resulting in a reduction of 6.16% in cost efficiency in this period.

In 2016, the company replaced the small furnace burning firewood with a larger furnace (ICI4) to generate heat. The use of a larger furnace increased the consumption of firewood, causing a reduction in technical efficiency that had a negative impact on cost efficiency (6.27% reduction). In 2017 (batch 6), the company acquired a new furnace that, in addition to burning firewood, could burn pellets for heat generation (ICI5). This generated a 6.50% improvement in cost efficiency. In terms of cost efficiency, the best decision would be to purchase the furnace that generates heat from burning firewood and pellets and not purchase furnaces with different sizes, as was done in previous periods.

In 2018 (batch 3), there was a general strike of truck drivers in Brazil (ECI6), which affected the transportation of bird feed. This implied a reduction in weight and a reduction in the system's output (kg of meat), consequently affecting cost efficiency negatively, with an observed reduction of 9.29%.

In 2018 (from batch 7), there was an agreement between the producer and the integrating company to increase the number of birds per m^2 (ICI7). This enabled an increase in the volume of meat produced per batch. This management action (ICI7) enabled a 30.93% improvement in the cost efficiency level of the aviary.

In 2019 (batch 7), another male poultry production experiment (ICI8) was conducted. It resulted in a 17.93% reduction in the cost efficiency level. After these experiments (ICI3 and ICI8), the managers chose to work only with batches of female birds, which provide higher cost efficiency levels for the aviary.

5 Discussion

The present chapter seeks to show that internal longitudinal benchmarking should be considered a valid alternative especially when external benchmarking is not possible. Indeed, there may be cases where organizations are unique and have no competitors in the market with whom they can be compared. Clearly, to overcome these difficulties, internal longitudinal benchmarking is a promising alternative (Piran et al., 2021). Additionally, performance evaluations based on internal longitudinal benchmarking present various contributions to the DEA efficiency analysis. In what follows, we will mention each of the contributions with examples borrowed from the literature.

Best Practices and Improvements are Easy to Emulate and Implement

In internal longitudinal benchmarking, the identified best practices should be easy to emulate. Most of the times, it regards the performance of the unit in a different period of time and therefore the unit shall be able to replicate, under the same circumstances, the identified performance.

It follows from the above that one may seek to understand why in a specific time period the DMU presents better results than in another time period. Once identified, the causes of best performance and the easiness of adoption of best practices are evident because the unit has been able to do that in the past, so in principle can do that in the present/future (Southard & Parente, 2007; Piran et al., 2021). Thus, internal longitudinal benchmarking is more realistic and accessible to managers, facilitating the implementation of improvements. In this sense, there is a significant probability that the changes arising from this learning will be readily implemented in the organization (Hyland & Beckett, 2002).

For example, de Souza et al. (2018a) identified the best practices performed over time in a production line of an arms manufacturing company. From this, the company developed plans to increase employee dwell time on the production line and to improve the maintenance status of equipment. These two practices were replicated for other production lines and have contributed to increasing the company's production efficiency. Piran et al. (2020b) shows that the steel plate cutting operation of a bus manufacturing company can reduce equipment maintenance time by 16.00% and raw material consumption by 5.00%. The managers agreed on the possibilities of these reductions and make plans to realize them in all steel plate cutting operations in the company. O'Neal et al. (2020) directs managers to reduce the practice of "cannibalization" to improve efficiency in maintaining aircraft in the US Air Force.

Detailed Production Data may be Used

Since only one company is being used in the benchmarking exercise, the access to company's detailed data that is not generally available to the public is much easier (Joo et al., 2009; Claro & Kamakura, 2017). Piran et al. (2021) argue that internal longitudinal benchmarking can help overcome the main limitations of DEA-based evaluations (e.g., unwillingness to share information between firms, or lack of similar firms in the market) and indeed raise the volume of analyzed data. This characteristic is explicit in the number of inputs and outputs used generally in the internal longitudinal DEA benchmarking models.

For example, Piran et al. (2020b) use 13 input and 2 output. The inputs are number of cutting planes, number of different programmed items, number of production orders, number of sheets, volume of raw material, kw consumed, labor time, preventive maintenance time, corrective maintenance time, machine hours, equipment supply time, total cutting time, and total idle hours. The outputs are number of pieces produced and volume of scrap. de Souza et al. (2018b) use 6 inputs (manufacturing time, case, lead, gunpowder, tool, and fuse) and 2 outputs (loaded cartridge and empty cartridge). Among the variables were measured the consumption of the main raw materials and the exact manufacturing time of the products. The example of Piran et al. (2021) reported in "Internal longitudinal benchmarking: an empirical application" shows the use of 11 inputs and 1 output.

This sort of detailed information would rarely be available in external benchmarking exercises. These studies show that researchers have easier access to data, allowing conducting analyses with different types of granularities (e.g., daily, monthly, and yearly). Thus, DEA models can be considered more complete and more focused on the actual processes that happen within the firm, increasing their usefulness for managers.

Information on Evolution and Disruptions of Efficiency Became Evident

Since a time series of efficiency is obtained for the same company, it is possible to evaluate the trend of efficiency over time, e.g., if the efficiency of the unit of analysis is improving, getting worse, or stagnating (see Fig. 5). Furthermore, it allows to evaluate the effects of management actions over time (causal effect) and to evaluate the effect of interventions made in the unit of analysis (e.g., the implementation of a new technology) (see Fig. 5). Finally, it allows observing effects that have a delay in efficiency. These possibilities help managers to make plans for improving efficiency.

For example, Joo et al. (2013) identified that the branch efficiency of a third-party logistics (3PL) company had been deteriorating over time (worsening trend). The authors found that the rate of expense spending outpaces the rate of revenue increase

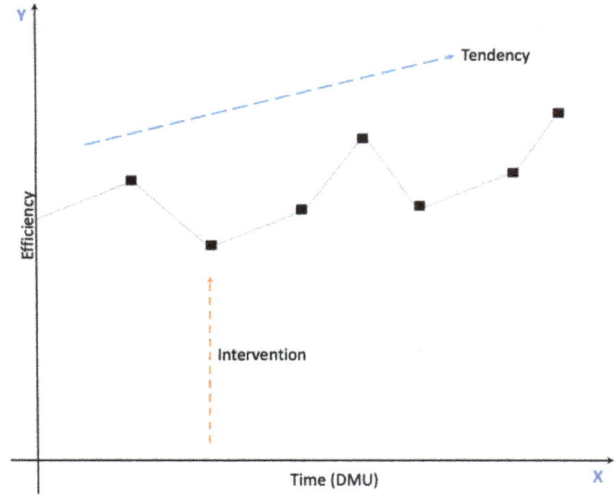

Fig. 5 Longitudinal efficiency analysis

and worsens the recent efficiency of the firm. To be efficient, the company needs cost containment efforts, especially in other expenses.

In another example, Piran et al. (2016) shows that the implementation of product modularity improved the efficiency of a bus manufacturing company. The efficiency of Product Engineering and Production Process shows an improvement trend after the intervention (implementation of product modularity) that was carried out by the managers. Thus, internal longitudinal benchmarking through DEA was used to evaluate the effect of an intervention in the company. Piran et al. (2021) performed a similar analysis and identified which management actions provided improvements in the efficiency of a broiler production system (see Fig. 4).

Internal Longitudinal Benchmarking can be an Important Complement of External Benchmarking

In many situations, researchers use panel data—a number of firms repeatedly assessed in a number of time periods. When such data are used, methods employed are fit to handle "time"; for example, Malmquist indices or similar indices, where for each company one assesses the extent to which it moved toward or away from the frontier and whether the frontier of each year actually moved upwards or downwards. Interestingly, these studies tend to analyze in detail the time effects, but not many studies paid attention to the other component of the panel which is the "firm" and its effects. When stochastic methods are used, most of the times there are parameters reflecting the effects of the firm and the effects of the time period. But somehow in non-parametric techniques, the effect of the firm is not generally assessed.

One exception is found in Portela et al. (2011) that not only assessed a number of water companies in the UK computing Malmquist indices to understand the efficiency of each unit and the progression of the frontier over time, but also computed firm frontiers (frontiers constituted just by the units of the same firm observed in different periods of time). Each of these firm frontiers corresponds to what we call in this paper internal longitudinal benchmarking. Clearly, the existence of more than one firm frontier allows a further analysis (as shown in Portela et al. (2011)) which is the computation of frontier gaps between companies (comparing the best performance of each company observed in a fixed time range). As a result, internal longitudinal benchmarking may prove of utility to be used alone, or complementarily to external benchmarking—something that the literature has not explored much.

Limitations

There are also limitations with internal benchmarking. The first limitation is that best practices are restricted to the observed unit of analysis. Thus, the possibilities for improvement should be smaller than those that could come from external benchmarking. The second limitation is the increased likelihood of lack of discrimination occurring in the efficiency results. In internal benchmarking, the difference between the DMUs' efficiency scores is usually small, and in many cases very close to 100%, making it difficult to detect differences between DMUs' efficiency, reducing the usefulness of the analysis.

The third limitation is the lack of certainty that both data sourcing and system characteristics are maintained over time. This limitation can also be attributed to external benchmarking; however, it is more evident in internal longitudinal benchmarking. Internal longitudinal benchmarking analyses typically assess the efficiency of several years or months (Kang, 2010; Piran et al., 2016, 2021)). This characteristic increases the possibility that the evaluated system will undergo substantive changes

Table 6 Contributions and limitations of internal longitudinal benchmarking

Contributions	Limitations
• Allows benchmarking when external data is not available	• Best practices are restricted to the observed unit of analysis
• Applicable when units of analysis are unique and not comparable	• Increased likelihood of lack of discrimination problems occurring in model efficiency scores
• Best practices and improvements are easy to emulate and implement (benchmarking that is more realistic and accessible to managers, facilitating the implementation of improvements)	• Lack of certainty that both data sourcing and system characteristics are maintained over time
• Detailed production data may be used (ease of access to data, ease of conducting analyses with different types of granularities)	
• Information on evolution and disruptions of efficiency became evident (ability to evaluate the effects of management actions over time (causal effect), ability to observe effects that have a delay on efficiency)	
• Internal longitudinal benchmarking can be an important complement of external benchmarking	

undermining the comparability that is inherent and necessary for efficiency analyses by benchmarking with DEA. In addition, companies may have difficulty providing structured data for the more distant past (past) reducing the time span that is possible to be evaluated.

Table 6 summarizes the main contributions and limitations of internal longitudinal benchmarking as just described.

6 Conclusions

This chapter provides a critical revision of the literature that has used DEA in Internal longitudinal benchmarking, pointing out its main contributions and limitations. We select articles published in journals indexed in Web of Science and Scopus and conduct a content analysis of 16 papers published from 1978 to 2021. We identified that DEA-based efficiency analysis using the internal longitudinal benchmarking procedure is still not well explored in the literature, despite showing some growth in recent years. In this regard, studies in manufacturing have been prominent.

We conclude that internal benchmarking can be a powerful mechanism to drive continuous improvement for companies in which external benchmarking is difficult or even impractical. We postulate that two-stage DEA analyses are especially important in internal longitudinal benchmarking studies that focus on providing insights for managerial decision-making. Thus, evaluations incorporating contextual variables or hypothesis testing can guide the improvement plans drawn up by managers. The two-stage model analysis offers realistic guidelines for a critical management problem from internal benchmarking.

Regarding the limitations of the work, although we used appropriate keywords and selected the appropriate databases for this study, using other keywords in other databases may yield different results. Thus, there may be works not identified in the search we performed.

Our main suggestion from this review is that there is unexplored scope for developing empirical studies on internal longitudinal benchmarking in all four sectors of the economy. In the primary sector, for example, only one study was identified. Regarding economic efficiency, revenue and profit efficiency studies can be introduced in the literature that so far has focused only on cost efficiency. We understand that internal longitudinal benchmarking provides high potential for leveraging economic efficiency analyses due to the ease of access to data (quantities and prices of inputs and outputs) from companies. In addition, we suggest conducting a systematic literature review to identify works that use other techniques besides DEA (e.g., SFA) to measure efficiency through internal longitudinal benchmarking.

References

Anand, G., & Kodali, R. (2008). Benchmarking the benchmarking models. *Benchmarking: An International Journal*.

Aparicio, J., Pastor, J. T., Vidal, F., & Zoffío, J. L. (2017). Evaluating productive performance: A new approach based on the product-mix problem consistent with data envelopment analysis. *Omega, 67*, 134–144.

Azizi, R., Matin, R. K., & Amin, G. R. (2016). A ratio-based method for ranking production units in profit efficiency measurement. *Mathematical Sciences, 10*(4), 211–217.

Banker, R. D. (1984). Estimating most productive scale size using data envelopment analysis. *European Journal of Operational Research, 17*(1), 35–44.

Banker, R. D., Charnes, A., & Cooper, W. W. (1984). Some models for estimating technical and scale inefficiencies in data envelopment analysis. *Management Science, 30*(9), 1078–1092.

Barbosa, L. M., Lacerda, D. P., Piran, F. A. S., & Dresch, A. (2017). Exploratory analysis of the variables prevailing on the effects of product modularization on production volume and efficiency. *International Journal of Production Economics, 193*, 677–690. https://doi.org/10.1016/j.ijpe.2017.08.028.

Baviera-Puig, A., Baviera, T., Buitrago-Vera, J., & Escribá-Pérez, C. (2020). Internal benchmarking in retailing with DEA and GIS: The case of a loyalty-oriented supermarket chain. *Journal of Business Economics and Management, 21*(4), 1035–1057. https://doi.org/10.3846/jbem.2020.12393.

Camp, R. C., & Camp Robert, C. (1989). *Benchmarking: The search for industry best practices that lead to superior performance*. WI: Quality Press Milwaukee.

Carpinetti, L. C., & De Melo, A. M. (2002). What to benchmark? A systematic approach and cases. *Benchmarking: An International Journal*.

Charnes, A., Cooper, W. W., & Rhodes, E. (1978). Measuring the efficiency of decision making units. *European Journal of Operational Research, 2*(6), 429–444.

Claro, D. P., & Kamakura, W. A. (2017). Identifying sales performance gaps with internal benchmarking. *Journal of Retailing, 93*(4), 401–419. https://doi.org/10.1016/j.jretai.2017.08.001.

Cox, J. R. W., Mann, L., & Samson, D. (1997). Benchmarking as a mixed metaphor: Disentangling assumptions of competition and collaboration. *Journal of Management Studies, 34*(2), 285–314. https://doi.org/10.1111/1467-6486.00052.

de Souza, I. G., Lacerda, D. P., Camargo, L. F. R., Dresch, A., & Piran, F. A. S. (2018a). Efficiency and internal benchmark on an armament company. *Benchmarking: An International Journal, 25*(7), 2018–2039. https://doi.org/10.1108/bij-08-2016-0128, https://doi.org/10.1108%2Fbij-08-2016-0128.

de Souza, I. G., Lacerda, D. P., Camargo, L. F. R., Dresch, A., & Piran, F. S. (2018). Do the improvement programs really matter? An analysis using data envelopment analysis. *BRQ Business Research Quarterly, 21*(4), 225–237. https://doi.org/10.1016/j.brq.2018.08.002, https://doi.org/10.1016%2Fj.brq.2018.08.002.

Ermel, A. P. C., Lacerda, D. P., Morandi, M. I. W. M., & Gauss, L. (2021). Systematic literature review. In: *Literature Reviews* (pp. 19–30). Springer International Publishing. https://doi.org/10.1007/978-3-030-75722-9_3.

Fong, S. W., Cheng, E. W., & Ho, D. C. (1998). Benchmarking: A general reading for management practitioners. *Management Decision, 36*(6), 407–418. https://doi.org/10.1108/00251749810223646.

Førsund, F. R. (2018). Economic interpretations of DEA. *Socio-Economic Planning Sciences, 61*, 9–15.

Färe, R., Grosskopf, S., & Lovell, C. A. K. (1985). *The Measurement of Efficiency of Production* (Vol. 6). Netherlands: Springer. https://doi.org/10.1007/978-94-015-7721-2.

Gilsa, C. V., Lacerda, D. P., Camargo, L. F. R., Souza, I. G., & Cassel, R. A. (2017). Longitudinal evaluation of efficiency in a petrochemical company. *Benchmarking: An International Journal, 24*(7), 1786–1813. https://doi.org/10.1108/bij-03-2016-0044

Gong, S., Shao, C., & Zhu, L. (2019). Multi-level and multi-granularity energy efficiency diagnosis scheme for ethylene production process. *Energy, 170*, 1151–1169. https://doi.org/10.1016/j.energy.2018.12.203.

Hyland, P., & Beckett, R. (2002). Learning to compete: The value of internal benchmarking. *Benchmarking: An International Journal, 9*(3), 293–304. https://doi.org/10.1108/14635770210429036.

Joo, S. J., Stoeberl, P. A., Kwon, I. W. G. (2007). Benchmarking efficiencies and strategies for resale operations of a charity organization. *Benchmarking: An International Journal, 14*(4), 455–464. https://doi.org/10.1108/14635770710761861.

Joo, S. J., Stoeberl, P. A., & Fitzer, K. (2009). Measuring and benchmarking the performance of coffee stores for retail operations. *Benchmarking: An International Journal, 16*(6), 741–753. https://doi.org/10.1108/14635770911000088, https://doi.org/10.1108%2F14635770911000088

Joo, S. J., Keebler, J. S., & Hanks, S. (2013). Measuring the longitudinal performance of 3pl branch operations. *Benchmarking: An International Journal, 20*(2), 251–262. https://doi.org/10.1108/14635771311307704, https://doi.org/10.1108%2F14635771311307704.

Joo, S. J., Stoeberl, P. A., Liao, K.,& Ke, K. (2017). Measuring the comparative performance of branches of a credit union for internal benchmarking. *Benchmarking: An International Journal, 24*(6):1663–1674, https://doi.org/10.1108/bij-03-2016-0029, https://doi.org/10.1108%2Fbij-03-2016-0029.

Kaffash, S., Azizi, R., Huang, Y., & Zhu, J. (2020). A survey of data envelopment analysis applications in the insurance industry 1993–2018. *European Journal of Operational Research, 284*(3), 801–813. https://doi.org/10.1016/j.ejor.2019.07.034.

Kang, C. C. (2010). Liberalization policy, production and cost efficiency in Taiwan's telecommunications industry. *Telematics and Informatics, 27*(1), 79–89. https://doi.org/10.1016/j.tele.2009.05.004.

Kenessey, Z. (1987). The primary, secondary, tertiary and quaternary sectors of the economy. *Review of Income and Wealth, 33*(4), 359–385. https://doi.org/10.1111/j.1475-4991.1987.tb00680.x, https://doi.org/10.1111%2Fj.1475-4991.1987.tb00680.x.

Kerstens, K., Sadeghi, J., & Van de Woestyne, I. (2019). Convex and nonconvex input-oriented technical and economic capacity measures: An empirical comparison. *European Journal of Operational Research, 276*(2), 699–709.

Kuosmanen, T., Cherchye, L., & Sipiläinen, T. (2006). The law of one price in data envelopment analysis: Restricting weight flexibility across firms. *European Journal of Operational Research, 170*(3), 735–757.

Lampe, H. W., & Hilgers, D. (2015). Trajectories of efficiency measurement: A bibliometric analysis of DEA and SFA. *European Journal of Operational Research, 240*(1), 1–21.

Liu, J. S., Lu, L. Y., & Lu, W. M. (2016). Research fronts in data envelopment analysis. *Omega, 58*, 33–45.

Moriarty, J. P., & Smallman, C. (2009). En route to a theory of benchmarking. *Benchmarking: An International Journal.*

O'Neal, T., Min, H., Cherobini, D., & Joo, S. J. (2020). Benchmarking aircraft maintenance performances using data envelopment analysis. *International Journal of Quality & Reliability Management, 38*(6), 1328–1341. https://doi.org/10.1108/ijqrm-05-2020-0157. https://doi.org/10.1108%2Fijqrm-05-2020-0157

Park, K. S., & Cho, J. W. (2011). Pro-efficiency: Data speak more than technical efficiency. *European Journal of Operational Research, 215*(1), 301–308.

Piran, F. A. S., Lacerda, D. P., Camargo, L. F. R., Viero, C. F., Dresch, A., & Cauchick-Miguel, P. A. (2016). Product modularization and effects on efficiency: An analysis of a bus manufacturer using data envelopment analysis (DEA). *International Journal of Production Economics, 182*, 1–13, https://doi.org/10.1016/j.ijpe.2016.08.008, https://doi.org/10.1016%2Fj.ijpe.2016.08.008.

Piran, F. A. S., Lacerda, D. P., Camargo, L. F. R., & Dresch, A. (2020). Effects of product modularity on productivity: An analysis using data envelopment analysis and malmquist index. *Research in Engineering Design, 31*(2), 143–156, https://doi.org/10.1007/s00163-019-00327-3, https://doi.org/10.1007%2Fs00163-019-00327-3.

Piran, F. A. S., Paris, A. D., Lacerda, D. P., Camargo, L. F. R., Serrano, R., & Cassel, R. A. (2020). Overall equipment effectiveness: Required but not enough—An analysis integrating overall equipment effect and data envelopment analysis. *Global Journal of Flexible Systems Management, 21*(2), 191–206. https://doi.org/10.1007/s40171-020-00238-6, https://doi.org/10.1007%2Fs40171-020-00238-6.

Piran, F. S., Lacerda, D. P., Camanho, A. S., & Silva, M. C. (2021). Internal benchmarking to assess the cost efficiency of a broiler production system combining data envelopment analysis and throughput accounting. *International Journal of Production Economics, 238*, 108173.

Portela, M. C. A. S., Thanassoulis, E., Horncastle, A., & Maugg, T. (2011). Productivity change in the water industry in england and wales: Application of the meta-malmquist index. *Journal of the Operational Research Society, 62*(12), 2173–2188. https://doi.org/10.1057/jors.2011.17, https://doi.org/10.1057%2Fjors.2011.17.

Roets, B., & Christiaens, J. (2015). Evaluation of railway traffic control efficiency and its determinants. *European Journal of Transport and Infrastructure Research, 15*(4).

Siok, M. F., & Tian, J. (2011). Benchmarking embedded software development project performance. In *IEEE 13th International Symposium on High-Assurance Systems Engineering*. IEEE. https://doi.org/10.1109/hase.2011.59, https://doi.org/10.1109%2Fhase.2011.59.

Southard, P. B., & Parente, D. H. (2007). A model for internal benchmarking: When and how? *Benchmarking: An International Journal, 14*(2), 161–171. https://doi.org/10.1108/14635770710740369, https://doi.org/10.1108%2F14635770710740369

Sueyoshi, T. (1991). Estimation of stochastic frontier cost function using data envelopment analysis: An application to the AT&t divestiture. *Journal of the Operational Research Society, 42*(6), 463–477. https://doi.org/10.1057/jors.1991.95, https://doi.org/10.1057%2Fjors.1991.95.

Telles, E. S., Lacerda, D. P., Morandi, M. I. W. M., & Piran, F. A. S. (2020). Drum-buffer-rope in an engineering-to-order system: An analysis of an aerospace manufacturer using data envelopment analysis (DEA). *International Journal of Production Economics, 222*, 107500. https://doi.org/10.1016/j.ijpe.2019.09.021, https://doi.org/10.1016%2Fj.ijpe.2019.09.021.

Tone, K., & Sahoo, B. K. (2005). Evaluating cost efficiency and returns to scale in the life insurance corporation of India using data envelopment analysis. *Socio-Economic Planning Sciences, 39*(4), 261–285. https://doi.org/10.1016/j.seps.2004.06.001, https://doi.org/10.1016%2Fj.seps.2004.06.001.

Part 3

Recent Advances in the Construction of Nonparametric Stochastic Frontier Models

Christopher F. Parmeter and Subal C. Kumbhakar

1 Introduction

Efficiency has been, and most likely will remain a dominant research agenda within economics. The classic production possibilities frontier, while simple in idea, is profound in transmitting key economic ideas. To that end, statistical methods that allow economists to study efficiency are of utmost importance. A variety of methods exist that achieve exactly such a goal, yet two main approaches receive the lion's share of attention: data envelopment analysis (DEA) and stochastic frontier analysis (SFA). These two methods have the same aim, to construct a boundary of the economic agent's potential and measure distance to it, but do so with different constructs on the problem.

Since the inception of both methods, there has been a lively debate over the modeling advantages of each estimator. DEA, relying on shape restrictions on the frontier and ignoring the presence of stochastic noise, constructs a fully nonparametric boundary, while SFA, relying on a series of parametric functional forms and distributional assumptions while also incorporating stochastic noise, constructs a stochastic parametric boundary. Both estimators clearly have benefits and costs and there has been a great deal of attention that has been paid to 'merging' or 'blending' these two distinct estimation frameworks into a singular entity that contains many of

All errors are ours alone.

C. F. Parmeter (✉) · S. C. Kumbhakar
University of Miami, Miami, USA
e-mail: cparmeter@bus.miami.edu

S. C. Kumbhakar
e-mail: kkar@binghamton.edu

Binghamton University, Binghamton, USA

S. C. Kumbhakar
Department of Economics, Binghamton University, Binghamton, USA

C. F. Parmeter
Department of Economics, University of Miami, Miami, USA

P. Macedo et al. (eds.), *Advanced Mathematical Methods for Economic Efficiency Analysis*, Lecture Notes in Economics and Mathematical Systems 692,
https://doi.org/10.1007/978-3-031-29583-6_10

the virtues of each. Parmeter and Zelenyuk (2019) detail much of the progress that has been made in this area.

The benchmark stochastic frontier model considers maximal output in a composed error setting, which is mostly formulated as

$$y_i = m(\boldsymbol{x}_i; \beta) + v_i - u_i = \boldsymbol{x}_i'\beta + v_i - u_i = \boldsymbol{x}_i'\beta + \varepsilon_i, \tag{1}$$

where $\boldsymbol{x}_i$ is a vector of inputs, v_i is the traditional noise while $u_i \geq 0$ acknowledges that output, y_i, is a maximum (conditional on $\boldsymbol{x}_i$ and v_i). u is assumed to have distribution $f_u(u)$, and v is assumed to have distribution $f_v(v)$. This model has seen wide application and has been reviewed extensively Greene (2008), Parmeter and Kumbhakar (2014), Kumbhakar et al. (2015), Kumbhakar et al. (2021a). Statistically, this is just a shifted conditional mean model given that $E[u_i] \neq 0$. A routine criticism of this model is the reliance on parametric assumptions for which there are three bottlenecks that the modeler must confront:

(1) Specification of the structure of u_i;
(2) Specification of the structure of v_i;
(3) Specification of the functional form $m(\cdot; \beta)$.

How these three assumptions are approached is summarized succinctly in Aragon et al. (2005): "These methods may lack robustness if the assumed distributional form does not hold. In particular, outliers in the data may unduly affect the estimate of the frontier function, or, it may be biased if the error structure is not correctly specified." The DEA estimator avoids parametric structure on the distribution of u and the frontier, but precludes the existence of v in the model. With these assumptions, coupled with monotonicity (and concavity) restrictions on the shape of the frontier, the frontier can be estimated nonparametrically.

Assumptions aside, estimation of the stochastic frontier model in Eq. (1) is straightforward with maximum likelihood (if the error components produce a density with closed form solution), or simulated maximum likelihood Greene (2003) when the error components have a complicated density. Even least squares approaches exist (corrected ordinary least squares) where moment matching can be used to estimate the parameters of the distributions of the error components. At issue is how one goes about estimating the frontier without making, or minimizing, parametric assumptions.

It is common to think of starting from the stochastic frontier model and trying to insulate it from the variety of distributional assumptions that are needed to identify the objects of interest (the frontier and distance to it). Early approaches were successful at eliminating some, but not all, of these assumptions. Fan et al. (1996) estimated the frontier (up to location) fully nonparametrically, but required distributional assumptions on both u and v to estimate inefficiency. The constrained estimators of Parmeter and Racine (2012) and Kuosmanen and Kortelainen (2012), which mimic the approach of Fan et al. (1996), not only suffer the same drawback but also enforce axioms of production which aligns them more with DEA. A similar approach to Fan et al. (1996) is that of Kumbhakar et al. (2007) who also estimate the frontier, up to location, fully nonparametrically, but require distributional assumptions to engage in local maximum likelihood. This estimator, while computationally more demanding than that of Fan et al. (1996), allows the distributional parameters of both noise and inefficiency to depend on covariates and to be estimated locally. A similar fate is shared by the pseudo-likelihood approach of Martins-Filho and Yao (2015) and the local-least squares approach of Simar et al. (2017). These approaches allow the frontier to be estimated nonparametrically, as with DEA, but still need parametric distributional assumptions, albeit some are only enforced locally.

One known practical issue with these approaches is that because they rely to varying degrees on distributional assumptions, they may be subject to the wrong skew, or locally wrong skew Waldman (1982). Wrong skew arises when the skewness of the residuals inside the likelihood function for a specific pair of distributional assumptions is of the incorrect sign (hence wrong skew). This issue not only is commonly encountered in applied settings with fully parametric stochastic frontier analysis but can also impact the more advanced semi-/nonparametric approaches that rely on the same types of distributional assumptions. That is, even though these more advanced approaches rely on a first-stage nonparametric regression, the identification of the variance parameters depends on their skewness, locally.

The second approach has been to focus attention on the parametric assumptions surrounding the distributional assumptions of the error components at the expense of more structure placed on the production frontier. Both Tran and Tsionas (2009) and Parmeter et al. (2017) deploy Robinson (1988)'s partly linear estimator to nonparametrically model the impact that a set of determinants have on inefficiency at the expense of a fully parametric production frontier. This approach fixes one of the main criticisms of SFA, the reliance on virtually unjustifiable distributional assumptions for the error components. This also allows the researcher the ability to explain inefficiency in a manner that is conducive to policy prescription.

Notice that the two main veins that currently exist, and that are detailed in Parmeter and Zelenyuk (2019), focus attention on one part of the frontier model: either the frontier or the error components. The reason for this is simple; it is difficult to identify both the production frontier and the nature of inefficiency (when noise is present) in a fully nonparametric setup. Hall and Simar (2002) proposed the first estimator in this direction, but they obtained a nonvanishing bias that was linked to the size of the noise variance. Horrace and Parmeter (2011) tried to augment the estimator of Hall and Simar (2002) but they relied on deconvolution techniques which still required full distributional specification of the noise component, which, under an assumption of Normality, resulted in exceedingly slow rates of convergence. It has only been recently that more flexible and deployable methods have entered the arena for applied efficiency analysis.

The remainder of the chapter is focused on discussing some of the newest results that presently exist in this exciting area.

2 Identification

If one ignores the structure of the error term, standard regression methods can be used to estimate the conditional expectation of y. Additionally, if there is no connection between the covariates in the frontier and the error components, then the shape of the frontier can be estimated consistently with nonparametric methods. This might lead one to believe that there really is no issue at hand. However, to say something about inefficiency requires more structure on the model. The stochastic frontier model at its core is a deconvolution exercise. Typically, parametric assumptions guarantee that the model is fully identified, but this has the requisite shortcomings. Moreover, even if one makes parametric assumptions, the model may be only weakly identified. For example, Greene (1990) provided a simple approach to estimate the stochastic frontier model assuming u was distributed Gamma and v was distributed Normal. However, Ritter and Simar (1994, 1997) demonstrated effectively that quite large sample sizes are needed before the distributional parameters can be reliably estimated. Another example of this is when u is assumed to be distributed Truncated Normal. The pretruncation mean parameter, when it diverges to $-\infty$, effectively makes the Truncated Normal behave as an Exponential. These types of identification issues arise due to a distinction between local and global identification, which has been discussed generally in Rothenberg (1971) and in the specific context of the stochastic frontier model by Das and Bandyopadhyay (2008). More specifically, Das and Bandyopadhyay (2008) use a concept known as *near identifiability* Prakasa Rao (1992) to investigate potential identification issues of the stochastic frontier model (specifically the Normal-Gamma pairing). As these approaches are heavily embedded in the parametric framework, we refrain from further discussion.

The main issue we simply wish to highlight is that even in a world where parametric assumptions are made and are correct, it may be hard to identify the stochastic frontier model fully because deconvolution is a difficult statistical problem. A clever attempt to nonparametrically identify the two pieces of the deconvolution was made in Hall and Simar (2002). They used the fact that u was one-sided to persuasively argue that there should be a large jump discontinuity at $u = 0$. One can estimate the stochastic frontier model using standard nonparametric methods and then recover $-E[u]$ by searching for this large jump discontinuity. However, as Hall and Simar (2002) showed, the estimator itself depends on the unknown variance of v. To produce a consistent estimator then assumes that $\sigma_v^2 \to 0$, which is not an appealing assumption to make in empirical work.

Kneip et al. (2015) remedy this issue by relying on a result of Schwarz and Van Bellegem (2010) who show that in a deconvolution, it is possible to identify both components provided one of the components has a probability distribution P_u belonging to

$$(0, \infty) \times \{P \in \mathcal{P} | \exists A \in \mathcal{B}(\mathbb{R}) : |A| > 0 \text{ and } P(A) = 0\}, \quad (2)$$

where $\mathcal{P}$ is the set of all probability distributions on R, $\mathcal{B}(\mathbb{R})$ is the set of Borel sets in $\mathbb{R}$ and $|A|$ is the Lebesgue measure of the set A. This set is the set of all probability distributions that contain an interval that has a positive Lebesgue measure but no probability mass on that interval. In essence, Schwarz and Van Bellegem (2010) use the fact that one of the distributions has limited support to achieve identification. This is similar to Hall and Simar (2002) insight, but more formal.

Another avenue to achieve identification in the stochastic frontier model is to assume that u depends on some set of characteristics that can be measured and then introduce them into the model through the conditional mean of u. This is an approach that has been undertaken by a great many authors, originally suggested by Simar et al. (1994). However, one typically has to make some type of assumption on the overlap, or lack thereof, between those variables that influence output directly (through the frontier) and those that influence output indirectly (through inefficiency) when eschewing parametric functional form assumptions. This is sometimes known as separability. In the full separability case, Parmeter et al. (2017) have shown how to nonparametrically identify both the frontier and the conditional mean of inefficiency, which we will detail below. This type of identification has also been exploited by Centorrino and Parmeter (2022) (which we will also detail below), but with different smoothness assumptions so that full separability can be dispensed with.

We end here by noting that it is impossible to nonparametrically identify the two pieces of deconvolution without some types of assumptions. One route is full parametric identification, another is to place various smoothness assumptions on the distribution based on available determinants of inefficiency while a third route is to rely on the statistical features of u and v to achieve identification. A researcher will need to assess which of these types of assumptions they are comfortable making as this will ultimately dictate which approach to use to estimate the stochastic frontier model.

3 Estimation

We focus our attention here on the estimation of the more recent nonparametric stochastic frontier estimators. Earlier semi-/nonparametric approaches are detailed in Parmeter and Kumbhakar (2014), Parmeter and Zelenyuk (2019) and Kumbhakar et al. (2021b).

To begin, we note that many of the semi-/nonparametric approaches to estimating the stochastic frontier model do so in a variety of steps. This is mainly done to decouple estimation of the frontier with the estimation of the structure of the error term. To see how this materializes, consider that due to the one-sided nature of inefficiency $E[\varepsilon|\boldsymbol{x}] = \gamma \neq 0$. Thus, the level of the production frontier cannot be identified in a standard regression setup as it is the case that

$$y_i = m(\boldsymbol{x}_i) + \varepsilon_i = m(\boldsymbol{x}_i) + \gamma + (\varepsilon_i - \gamma) \equiv m^*(\boldsymbol{x}_i) + \varepsilon_i^*. \tag{3}$$

Typical approaches try to estimate the shifted frontier, $m^*(\boldsymbol{x}_i)$, recover γ from the residuals from Eq. (3) and then 'un-shift' the frontier. This is the approach followed by both Fan et al. (1996) and Simar et al. (2017). Under fairly weak conditions, Fan et al. (1996) show that the parameters of the composed error distribution can be consistently estimated at the parametric $\sqrt{n}$ rate, though Martins-Filho and Yao (2015) later showed that these parameter estimators are biased and inefficient.

3.1 Full Separability

As noted in the discussion on identification, when there exist variables that influence inefficiency, this can aid identification. Assume the production frontier only depends on $\boldsymbol{x}$ while the moments of the distribution of inefficiency depend only on z. This is a variation of the so-called separability assumption Simar and Wilson (2007). With such a model structure, all relevant features of both the production frontier and inefficiency can be identified nonparametrically. To begin, start with Eq. (3)

$$y_i = m(\boldsymbol{x}_i) + \varepsilon_i$$

and re-center ε by $E[u_i|z_i]$ to produce

$$y_i = m(\boldsymbol{x}_i) + v_i - u_i + E[u_i|z_i] - E[u_i|z_i] = c + m(\boldsymbol{x}_i) + g(z_i) + \varepsilon_i^*, \tag{4}$$

where $\varepsilon_i^* = v_i - u_i - g(z_i) - c$ and $g(z_i) + c = E[u_i|z_i]$. The generic constant c is included here to indicate that with this nonparametric setup, the function $g(\cdot)$ is only identified up to location.

This is a generalized-additive model, and it is possible to estimate both $m(\cdot)$ and $g(\cdot)$ without requiring distributional assumptions on v or u (or any notion of independence for that matter since expectation is a linear operator). For identification purposes, it is typically assumed that $E\left[m(\boldsymbol{x})\right] = E\left[g(z)\right] = 0$ and hence $E\left(y\right) = c$, which is the intercept. We note that this is not a restrictive assumption as the shifting of any of the functions does not change their shape, which is important for measuring marginal changes and elasticities. However, in the context of estimating a production frontier it is somewhat awkward since it holds that $m(\boldsymbol{x}) \geq 0\ \forall \boldsymbol{x}$ and $g(\boldsymbol{x}) \leq 0\ \forall z$. In this case, centering y prior to estimation ensures that $c = 0$.

This model can be estimated in a variety of ways. In a kernel smoothing approach, Kim et al. (1999) have proposed a two-step estimator that is quite simple to implement in existing statistical software. Begin by assuming that y is centered. Then, conditional on $\boldsymbol{x}$:

$$E\left(y|\boldsymbol{x}\right) = m\left(\boldsymbol{x}\right) + E\left[g\left(z\right)|\boldsymbol{x}\right]. \tag{5}$$

Estimation of the mean of y conditional on $\boldsymbol{x}$ would introduce a bias due to ignorance of z (presuming that $\boldsymbol{x}$ and z are not independent of one another). This bias can be eliminated by finding a scaling function, $\varpi\left(\boldsymbol{w}\right)$ which satisfies Kim et al. (1999),

$$E\left[\varpi\left(\boldsymbol{w}\right)|\boldsymbol{x}\right]=1$$
$$E\left[\varpi\left(\boldsymbol{w}\right)g\left(z\right)|\boldsymbol{x}\right]=0.$$

The vector $\boldsymbol{w}$ may or may not contain the elements $\boldsymbol{x}$ and z. To see that this works, observe that

$$\begin{aligned}E\left[\varpi\left(\boldsymbol{x},z\right)y|\boldsymbol{x}\right]&=E\left[\varpi\left(\boldsymbol{x},z\right)m(\boldsymbol{x})|\boldsymbol{x}\right]+E\left[\varpi\left(\boldsymbol{x},z\right)g(\boldsymbol{x})|\boldsymbol{x}\right]\\&=E\left[\varpi\left(\boldsymbol{x},z\right)|\boldsymbol{x}\right]m(\boldsymbol{x})+E\left[\varpi\left(\boldsymbol{x},z\right)g(\boldsymbol{x})|\boldsymbol{x}\right]\\&=m(\boldsymbol{x}).\end{aligned}\tag{6}$$

The function $\varpi\left(\boldsymbol{w}\right)$ counteracts the presence of z in Eq. (5), essentially unwrapping any dependence that might exist between $\boldsymbol{x}$ and z so that, after rescaling, the presence of $g(z)$ can be ignored when estimating conditional on $\boldsymbol{x}$. This is similar in spirit to the orthogonalization that OLS engages in when constructing estimates of the slope coefficients embodied in the 'Frisch-Waugh-Lovell' Theorem.

Kim et al. (1999) proposed

$$\varpi\left(\boldsymbol{w}\right)=\frac{f\left(\boldsymbol{x}\right)f\left(z\right)}{f\left(\boldsymbol{w}\right)},\tag{7}$$

where $f\left(\boldsymbol{x}\right)$, $f\left(z\right)$ and $f\left(\boldsymbol{w}\right)$ are the marginal and joint probability densities of $\boldsymbol{x}$, z and $\boldsymbol{w}=\left(\boldsymbol{x},z\right)$, respectively. In this case, $\varpi\left(\boldsymbol{w}\right)$ represents a scaled version of the inverse of the copula function. This specific form of $\varpi\left(\boldsymbol{w}\right)$ is a measure of the degree of independence between $\boldsymbol{x}$ and z.

To construct $\varpi\left(\boldsymbol{w}\right)$, kernel density estimators can be deployed:

$$\widehat{f}\left(\boldsymbol{x}\right)=n^{-1}\sum_{i=1}^{n}K_{i\boldsymbol{x}},\quad\widehat{f}\left(z\right)=n^{-1}\sum_{i=1}^{n}K_{iz},$$

and

$$\widehat{f}(\boldsymbol{w})=\widehat{f}\left(\boldsymbol{x},z\right)=n^{-1}\sum_{i=1}^{n}K_{i\boldsymbol{x}}K_{iz},\tag{8}$$

where $K_{i\boldsymbol{w}}=\prod_{s=1}^{q_w}h_{w_s}^{-1}k\left(\frac{w_{is}-w_s}{h_{ws}}\right)$ is the product kernel with bandwidth h_s for each element of the vector $\boldsymbol{w}$ (of length q_w and individual kernel function $k(\cdot)$) Li and Racine (2007), Henderson and Parmeter (2015). For a given set of bandwidths, these densities can easily be constructed in existing software. For example, in the `R` programming environment, one can call `npudens()` in the `np` package Hayfield and Racine (2008) to calculate each of the three densities just described.

From here, the estimator of $\nu_m\left(\boldsymbol{x}\right)=E\left[\varpi\left(\boldsymbol{x},z\right)y|\boldsymbol{x}\right]$ is given by

$$\widehat{\nu}_m(\boldsymbol{x}) = n^{-1}\sum_{j=1}^{n} K_{jx}\frac{\widehat{f}(z_j)}{\widehat{f}(\boldsymbol{x}, z_j)}y_j = n^{-1}\sum_{j=1}^{n}\frac{K_{jx}}{\widehat{f}(\boldsymbol{x})}\frac{\widehat{f}(z_j)\widehat{f}(\boldsymbol{x})}{\widehat{f}(\boldsymbol{x}, z_j)}y_j = n^{-1}\sum_{j=1}^{n}\frac{K_{jx}}{\widehat{f}(\boldsymbol{x}_j)}y_j^{*}. \quad (9)$$

This estimator looks akin to the local-constant least-squares estimator except that y is transformed by $\frac{\widehat{f}(z_j)\widehat{f}(x)}{\widehat{f}(x,z_j)}$ prior to estimation. Again, in the `R` programming environment one can call `npreg()` in the `np` package after y has been transformed to estimate this conditional mean. Note the subtlety here that the weights have a fixed $\boldsymbol{x}$ but variable z. Thus, for each point of evaluation for the frontier $m(\cdot)$, the weights have to be recalculated.

Alternatively, a simple matrix implementation for this estimator exists. Define

$$S_x = n^{-1}\left[K_{x_i x_j}\right]_{i,j}; \quad S_z = n^{-1}\left[K_{z_i z_j}\right]_{i,j},$$

to be the $n \times n$ matrices of kernel weights for $\boldsymbol{x}$ and z. Then,

$$\widehat{\nu}_m(\boldsymbol{x}) = S_x\left(y \odot (S_z \boldsymbol{i}) \oslash (n(S_x \odot S_z)\boldsymbol{i})\right)$$

where $\boldsymbol{i}$ is the $n \times 1$ vector of ones, $\odot$ represents Hadamard multiplication and $\oslash$ represents Hadamard division. The matrices S_x and S_z can be calculated quite quickly through the use of the command `npksum()` in the `np` package.

This process can be replicated to estimate $E[\varpi(\boldsymbol{x}, z)\, y|z]$. Following the logic from above, this leads to the estimator of $\nu_g(z) = E[\varpi(\boldsymbol{x}, z)\, y|z]$, given by

$$\widehat{\nu}_g(z) = n^{-1}\sum_{j=1}^{n}\frac{K_{jz}}{\widehat{f}(z)}y_j^{*}, \quad (10)$$

where $y_j^{*} = \frac{\widehat{f}(z)\widehat{f}(\boldsymbol{x}_j)}{\widehat{f}(\boldsymbol{x}_j, z)}y_j$. Again, note that for each point of evaluation, z the weights used to transform y_j need to be recalculated over $\boldsymbol{x}$. In matrix notation, the equivalent setup is

$$\widehat{\nu}_g(z) = S_z\left(y \odot (S_x \boldsymbol{i}) \oslash (n(S_x \odot S_z)\boldsymbol{i})\right).$$

While both $\widehat{\nu}_m(\boldsymbol{x})$ and $\widehat{\nu}_g(\boldsymbol{x})$ are consistent estimators, they are inefficient given that they transform the model to ignore the presence of the other set of covariates. It is possible to use the second step adjustment (backfitting) to produce estimators of $m(\boldsymbol{x})$ and $g(z)$ which are both consistent and efficient Kim et al. (1999). This approach is described in detail for the stochastic frontier model in Parmeter and Zelenyuk (2019).

While the approach described here was proposed in passing in Parmeter et al. (2017), and elaborated on in Parmeter and Zelenyuk (2019), to our knowledge it has yet to be applied in the literature and would mark a fruitful area of application. This additive estimator is attractive when the separability assumption is valid. However, where and when this assumption is likely to hold remains an open question. Justification for separability is that there exist variables/factors that only influence output directly through the technology ($m(\cdot)$) or indirectly through inefficiency ($g(\cdot)$). One may think of the degree of workers who are unionized as such a variable that influences output through inefficiency, but not through technology. A simple test of this assumption can be predicated on testing the full nonparametric model $y_i = m(\boldsymbol{x}_i, z_i) + \varepsilon_i$ against $y_i = m(\boldsymbol{x}_i) + g(z_i) + \varepsilon_i$, however, a rejection of the additive model precludes identification of the production technology and technical efficiency.

A practical issue is that of the selection of the bandwidths. There are two stages and multiple objects for which kernel smoothing is engaged. Kim et al. (1999) suggested using the same bandwidths in all stages for $\boldsymbol{x}$ and z, where the rule-of-thumb bandwidth deployed is $0.5\widehat{\sigma}_{\boldsymbol{x}} n^{-1/(4+q_x)}$ and $0.5\widehat{\sigma}_{z} n^{-1/(4+q_z)}$ with $\sigma_{\boldsymbol{x}}$ the vector of standard deviations for $\boldsymbol{x}$ and σ_z the vector of standard deviations for z. This is a relatively simple approach to get started, but ideally would receive pushback over concerns that optimal smoothness was not imparted on the model. An alternative would be least-squares cross-validation (LSCV), which determines the optimal bandwidths based on

$$LSCV = \min_{h_x, h_z} \sum_{i=1}^{n} \left(y_i - \widehat{\nu}_{m,-i}(\boldsymbol{x}_i) + \widehat{\nu}_{g,-i}(z_i) \right)^2, \tag{11}$$

where $\widehat{\nu}_{m,-i}(\boldsymbol{x}_i)$ and $\widehat{\nu}_{g,-i}(z_i)$ are the leave-one-out estimators of $m(\cdot)$ and $g(\cdot)$, respectively. These can be constructed by setting the corresponding kernel weights to 0 inside the appropriate matrices defined above. To our knowledge, at present no implementation of LSCV is available in statistical software.

3.2 *Weak Separability*

One practical limitation of this additive estimator is its reliance on the separability assumption. The intent of the assumption is to avoid making parametric assumptions that typically encapsulate the benchmark stochastic frontier model. However, it can be hard to justify in specific empirical settings. Recently, Centorrino and Parmeter (2022) considered a slightly more general setup based on weak separability, where some of the variables are allowed to overlap.

Centorrino and Parmeter (2022) consider the model where, defining $\boldsymbol{w}_i = (z_i, \boldsymbol{x}_i) = (z_{1,i}, z_{-1,i}, \boldsymbol{x}_i) = (z_i, \boldsymbol{\omega}_i)$,

$$y_i = m(\boldsymbol{w}_i) - g(\boldsymbol{w}_i) + \varepsilon_i^*, \tag{12}$$

with $g(\boldsymbol{w}_i) = g_1(z_i) g_2(\boldsymbol{\omega}_i) = E[u|\boldsymbol{w}_i] \geq 0$ such that $g_1(\cdot)$ and $g_2(\cdot)$ are strictly monotone in all their arguments and $\frac{\partial m}{\partial z}(\boldsymbol{w}) = 0$. Centorrino and Parmeter (2022) model takes the additive frontier model and makes several shape restrictions on it. First, they assume that the frontier function does not depend on z. Second, they assume that the inefficiency function is monotonic in all of its arguments. And third, they assume that the inefficiency function can be multiplicatively decomposed into two distinct pieces. Centorrino and Parmeter (2022) also need several additional high-level statistical assumptions, but these are less impactful from the standpoint of the stochastic frontier model.

More intuitively, the identification that Centorrino and Parmeter (2022) provide simply requires that there exists one variable that influences output **only** through inefficiency. Technically, their identification proof could be reconfigured so that identification only requires one variable to influence output solely through technology. Thus, what is needed is (at least) a single variable that affects output through one, and only one, component of the model (either technology or inefficiency). They refer to this condition as weak separability, which is in line with the full separability assumption made in the additive stochastic frontier model.

Unfortunately, the existence of this special covariate by itself is not enough to identify the two functions of the model. This is because there is some inversion required, hence the additional assumptions of monotonicity. However, these monotonicity assumptions are not restrictive if one considers that the vast majority of applied work commonly operates with assumptions such as $u = u_0 \exp(z\delta_1 + \boldsymbol{\omega}'\delta_2)$, and u_0 is taken to be independent of z Wang and Schmidt (2002).[1] In this case,

$$g(\boldsymbol{w}) = E\left[u_0\right] e^{z\delta_1+\boldsymbol{\omega}'\delta_2} = E\left[u_0\right] e^{z\delta_1} e^{\boldsymbol{\omega}'\delta_2}, \tag{13}$$

where the definitions of $g_1(\cdot)$ and $g_2(\cdot)$ should be apparent with both satisfying strict monotonicity. Perhaps the most restrictive assumption is the multiplicative separability enforced between $g_1(\cdot)$ and $g_2(\cdot)$.

To see how the model is identified, let $r(\boldsymbol{w}) = m(\boldsymbol{w}) - g(\boldsymbol{w})$. Next, differentiating $r(\boldsymbol{w})$ with respect to z leads to

$$\partial r(\boldsymbol{w})/\partial z = \partial m(\boldsymbol{w})/\partial z - \partial g(\boldsymbol{w})/\partial z = -\left(dg_1(z)/dz\right) g_2(\boldsymbol{\omega}).$$

z can be integrated out from here to identify $g_2(\boldsymbol{\omega})$:

$$-\int_{\mathcal{Z}} \left(\partial r(\boldsymbol{w})/\partial z\right) dz = -g_2(\boldsymbol{\omega}) \int_{\mathcal{Z}} \left(dg_1(z)/dz\right) dz = -Rg_2(\boldsymbol{\omega}),$$

with R a constant. Here, the monotonicity of $g_1(z)$ ensures that $R \neq 0$. Next, note that for each value of z, we have $\boldsymbol{\omega}$ such that $z = g_1^{-1}\left(C/Rg_2(\boldsymbol{\omega})\right)$, where C is a constant. Here is where some of the high-level conditions from Centorrino and Parmeter (2022) come into play. We need enough variation in $\boldsymbol{\omega}$ to ensure that, given the monotonicity of $g_1(\cdot)$, we can invert the function to recover z.

[1] Moreover, these monotonicity restrictions can be tested (see, e.g., Du et al. 2013).

Next, it holds

$$
\begin{aligned}
E\left[y|\omega, \frac{C}{Rg_2(\omega)}=t\right] &= m(\omega) - E\left[g(\boldsymbol{w})|\omega, \frac{C}{Rg_2(\omega)}=t\right] \\
&= m(\omega) - E\left[g_1(z)g_2(\omega)|\omega, \frac{C}{Rg_2(\omega)}=t\right] \\
&= m(\omega) - \frac{C}{Rt}E\left[g_1(z)|\omega, \frac{C}{Rg_2(\omega)}=t\right] \\
&= m(\omega) - \frac{C}{Rt}E\left[g_1(z)|\omega, z=g_1^{-1}(t)\right] \\
&= m(\omega) - \frac{C}{Rt}g_1\left(g_1^{-1}(t)\right) \\
&= m(\omega) - \frac{C}{R}. \qquad (14)
\end{aligned}
$$

The first equality holds by conditioning on ω, the second by definition, the third by conditioning on the value for $g_2(\cdot)$, the fourth by the inversion of $g_1(\cdot)$ to recover z and the fifth by definition. This shows that the frontier is identified up to location if we nonparametrically regress y on ω where the value of ω ensures that we can recover z. Once the frontier is identified, the inefficiency function can be identified by subtracting the frontier from $r(\boldsymbol{w})$: $g(\boldsymbol{w}) = -r(\boldsymbol{w}) + m(\omega) - (C/R)$.

This identification strategy leads Centorrino and Parmeter (2022) to produce a three-step estimator for the stochastic frontier model:

(1) Estimate a local polynomial regression of even order ρ_1 with bandwidth vector $\boldsymbol{h}_1$ of y on $\boldsymbol{w}$, and obtain a consistent estimator of the first partial derivative of $r(\boldsymbol{w})$ with respect to z (call this $\widehat{r}_z(\boldsymbol{w})$. Local polynomial estimation can be undertaken in R using the `npglpreg()` command available in the `crs` package Racine and Nie (2021). For any local polynomial estimator, the derivatives are automatically calculated provided the order of the polynomial $\rho_1 > 0$. Next, marginally integrate z from $\widehat{r}_z(\boldsymbol{w})$ to obtain an estimator of $g_2(\omega)$ up to scale, denoted by $\widehat{g}_2$.
(2) Estimate the second local polynomial regression of odd order $\rho_2 < \rho_1$ with bandwidth vector $\boldsymbol{h}_2$ of y on $(\omega, 1/\hat{g}_2(\omega))$ and obtain the estimator of the production frontier $m(\cdot)$, denoted by $\hat{m}$.
(3) Finally, obtain a consistent estimator of the function r by local ρ_2-th order smoothing with bandwidth h_2, denoted by $\hat{r}$, and write $\hat{g}(\boldsymbol{w}) = \hat{m}(\omega) - \hat{r}(\boldsymbol{w})$.

Notice that the estimation strategy is relatively straightforward and only involves the estimation of conditional expectations. Compared to Simar et al. (2017), for example, the proposed approach does not require higher moments of the residuals or distributional assumptions, which may be estimated imprecisely and, as detailed earlier, may lead to issues of locally wrong skew.

The reason to allow for polynomial regressions of different orders is two-fold. From existing theoretical results, it is known that local polynomial regression estimators have better properties at the boundaries of the support when the order of the polynomial minus the order of the derivative to be estimated is odd Ruppert and Wand (1994). Since Centorrino and Parmeter (2022) need the first derivative of $r(\cdot)$, they require ρ_1 to be even. ρ_2 is set to be an odd number since both the second and third steps of the estimation routine are focused on the conditional expectation itself and not a derivative. Additionally, as the second step involves a nonparametrically generated regressor Mammen et al. (2012), depending on the choice of the bandwidth vectors, $(\boldsymbol{h}_1, \boldsymbol{h}_2)$, and the orders of the polynomial, (ρ_1, ρ_2), the first step estimator may or may not affect the asymptotic properties of the second step estimator.

3.3 No Separability

Both of the estimators just described for the stochastic frontier model have involved some sort of restriction on where a covariate can appear in the model and types of separability. While there exist empirical settings where this type of argument can be justified, it may still be viewed as restrictive. To that end, (Kneip et al. 2015, KSVK hereafter) have proposed an estimator focused more around the statistical structure of v and u than on which covariates can or cannot enter specific places in the model. In this case, concerns over separability can be dispensed with.

KSVK's estimator of the stochastic frontier model can best be understood by first focusing on the setting where the frontier is constant, i.e., ignoring the shape of the frontier and focusing only on its fixed location. KSVK denotes the fixed/constant frontier as τ_0. In this setup, the density of output is $g_0(\cdot)$ and the density of the frontier times inefficiency is $f_0(\cdot)$. Lastly, v is assumed to be Normally distributed with unknown variance σ_0^2. Following KSVK, the true density of y is

$$g_0(y) = \frac{1}{\sigma_0 y} \int_0^1 h_0(t)\phi\left(\sigma_0^{-1} \ln \frac{y}{t\tau_0}\right) dt, \tag{15}$$

where $h_0(t) = \tau_0 f_0(t\tau_0)$. KSVK suggested estimation of $h_0(\cdot)$ through estimation of $g_0(\cdot)$ with $g_{h,\tau,\sigma}(\cdot)$ such that h is a probability density with support $[0, 1]$,[2] $\tau > 0$ and $\sigma > 0$.

Clearly, $h_0(\cdot)$ is unknown and as such KSVK suggested approximation of h with a histogram type estimator:

$$h_\gamma(t) = \gamma_1 \boldsymbol{I}\{t = 0\} + \sum_{k=1}^{M} \gamma_k \boldsymbol{I}\{q_{k-1} < t \le q_k\}$$

[2] KSVK's model assumes that $h_0(\cdot)$ is the density of τe^{-u} and since u is assumed to have support on $[0, \infty)$, this leads to $h_0(\cdot)$ having support on $[0, 1]$.

such that $\gamma = (\gamma_1, \gamma_2, \ldots, \gamma_M)$ with $\gamma_k > 0\ \forall k$ (k =0, 1, ...,M) and $M^{-1}\sum_{k=1}^{M}\gamma_k = 1$ with $q_k = k/M$ and M a prespecified number. This set is referred to as Γ. The final density estimator is

$$g_{h_\gamma,\tau,\sigma}(y) = \frac{1}{\sigma y}\sum_{k=1}^{M}\gamma_k \int_{q_{k-1}}^{q_k} \phi\left(\sigma^{-1}\ln\frac{y}{t\tau}\right)dt$$

$$= \frac{1}{\sigma y}\sum_{k=1}^{M}\gamma_k\left(\Phi\left(\sigma^{-1}\ln\frac{y}{q_k\tau}\right) - \Phi\left(\sigma^{-1}\ln\frac{y}{q_{k-1}\tau}\right)\right) \quad (16)$$

and estimates can be found by maximizing over τ, σ and γ of the averaged log density of $g_{h_\gamma,\tau,\sigma}(\cdot)$ over the observations $y_1, y_2, \ldots, y_n$. KSVK also suggests penalizing the estimator to account for how smooth (or unsmooth) the density of $g(\cdot)$ is. The final optimization is

$$(\widehat{\tau}, \widehat{\sigma}, \widehat{\gamma}) = \max_{\tau>0,\sigma>0,\gamma\in\Gamma}\left\{n^{-1}\sum_{i=1}^{n}\ln g_{h_\gamma,\tau,\sigma}(y_i) - \lambda\text{pen}\left(g_{h_\gamma,\tau,\sigma}(y)\right)\right\}.$$

Here, the penalty term $\lambda \geq 0$ controls the overall smoothness of the estimated density. KSVK suggests $\text{pen}\left(g_{h_\gamma,\tau,\sigma}(y)\right) = \max_{3\leq j\leq M}\left|\gamma_j - 2\gamma_{j-1} + \gamma_{j-2}\right|$ as the penalty function. Note that as $\lambda \to 0$, the estimated density becomes rougher as there is less penalization and as $\lambda \to \infty$, the estimated density becomes smoother. Clearly, selection of the parameters here is key for the performance of the estimator. As M increases a less smooth (more noisy) estimate of the density is produced and as M decreases a smoother (less noisy) estimate of the unknown density is produced.

With this penalized estimator in hand, incorporating covariates in a nonparametric framework follows accordingly. First, for a given bandwidth vector $\boldsymbol{b}$, focus attention on only those observations that are within distance $||\boldsymbol{b}||$ of a given point $\boldsymbol{w}_0$. In this case, one uses the same estimator, only with fewer observations:

$$(\widehat{\tau}(\boldsymbol{w}_0), \widehat{\sigma}(\boldsymbol{w}_0), \widehat{\gamma}(\boldsymbol{w}_0)) = \max_{\tau>0,\sigma>0,\gamma\in\Gamma}\left\{n_{\boldsymbol{b}}^{-1}\sum_{i:||\boldsymbol{w}_i-\boldsymbol{w}_0||\leq||\boldsymbol{b}||}\ln g_{h_\gamma,\tau,\sigma}(y_i) - \lambda\text{pen}\left(g_{h_\gamma,\tau,\sigma}(y)\right)\right\},$$

where $n_{\boldsymbol{b}}$ is the number of observations that fall within a 'bandwidth' of the point $\boldsymbol{w}_0$. An alternative would be to introduce kernel weights through a product kernel. In this case, smoothing takes place over all the observations. KSVK did not study the impact that this modification would have on their proposed estimator.

Aside from the additional notation, this estimator is identical to the covariate-free estimator. Implementation requires the user to select a bandwidth vector (which should ideally depend on the sample size) and then for each point of interest, calculate the penalized likelihood estimator. That is, rather than estimating the density once, now the density is estimated for each point of interest for the covariates. It is also important to highlight that KSVK keep λ fixed when introducing covariates. This penalization parameter is designed to ensure the same level of smoothness penalization on the density regardless of location in the support of $\boldsymbol{w}$. Thus, it only needs to be set once, not changed for each $\boldsymbol{w}_0$ under consideration.

Note that in this approach of KSVK, the estimation of conditional efficiency is obscured. Rather than the penalized likelihood approach over a specified set of observations, a different framework would be to estimate a nonparametric regression (some type of local-polynomial say) first, using the same level of smoothness, and then implement the penalized likelihood approach over the adjusted output variables. To be precise, for a given bandwidth vector $\boldsymbol{b}$, one can approximate a function $m(\cdot)$ around $\boldsymbol{w}_0$ as

$$m(\boldsymbol{w}_i) \approx \sum_{0\leq|\mathbf{j}|\leq\rho_1} \frac{1}{\mathbf{j}!}(D^{(\mathbf{j})}m)(\boldsymbol{w}_0)(\boldsymbol{w}_i - \boldsymbol{w}_0)^{\mathbf{j}}, \tag{17}$$

where

$$\mathbf{j} = (j_1, \ldots, j_d), \qquad \mathbf{j}! = j_1! \times \cdots \times j_d!, \qquad |\mathbf{j}| = \sum_{k=1}^{d} j_k,$$

$$(\boldsymbol{w}_i - \boldsymbol{w}_0)^{\mathbf{j}} = (w_{1i} - w_{10})^{j_1} \times \cdots \times (w_{di} - w_{d0})^{j_d},$$

$$\sum_{0\leq|\mathbf{j}|\leq\rho} = \sum_{k=0}^{\rho} \sum_{\substack{j_1=0 \\ j_1+\cdots+j_d=|\mathbf{j}|}}^{k} \cdots \sum_{j_d=0}^{k},$$

and

$$(D^{(\mathbf{j})}m)(\boldsymbol{w}_0) = \frac{\partial^{\mathbf{j}} m(\boldsymbol{w}_0)}{\partial w_{10}^{j_1} \ldots \partial w_{d0}^{j_d}}.$$

Finally, the local ρth order polynomial estimator of y is

$$\sum_{i:||\boldsymbol{w}_i - \boldsymbol{w}_0|| \leq ||\boldsymbol{b}||} \left[y_i - \sum_{0\leq|\mathbf{j}|\leq\rho} \beta_{\mathbf{j}}(\boldsymbol{w}_i - \boldsymbol{w}_0)^{\mathbf{j}} \right]^2. \tag{18}$$

Next, construct $\widetilde{\widetilde{y}}_i = y_i - \sum_{1\leq|\mathbf{j}|\leq\rho} \beta_{\mathbf{j}}(\boldsymbol{w}_i - \boldsymbol{w}_0)^{\mathbf{j}}$. Note here that we have held out $\widehat{\beta}_{\mathbf{j}}$ for $\mathbf{j} = (0, \ldots, 0)$ as this is the constant term from the frontier model. We then use $\widetilde{\widetilde{y}}_i$ instead of y_i in the penalized log-likelihood setup to determine the frontier. Lastly, conditional inefficiency ($E[u|\boldsymbol{w} = \boldsymbol{w}_0]$) is estimated as $\widehat{\tau}(\boldsymbol{w}_0) - \widehat{\beta}_{\mathbf{0}}$ where $\widehat{\beta}_{\mathbf{0}}$ is the intercept from the local ρth order polynomial estimator.

Variations of this approach exist. For example, Noh and Van Keilegom (2020) describe estimation in this framework but with a parametric frontier structure, a partially linear structure and an additive structure. Note that this estimator is predicated on an assumption of Normality for v but otherwise does not hinge on any distributional assumptions for u or parametric functional form assumptions on the frontier.

Again, this is linked to the deconvolution identification result in Schwarz and Van Bellegem (2010).

4 Concluding Remarks

Recent advances in nearly nonparametric estimation of the stochastic frontier model offer an exciting avenue for applications. The newest set of methods can be easily deployed and the hope is that this chapter has removed some of the barriers to implementation for practitioners. Naturally, one might ask what is left to accomplish. Plenty. Perhaps the greatest barrier that remains is conducting inference in these nonparametric settings, specifically of issues like symmetry of the error term and independence between u and v (which are needed for KSVK's estimator to be identified) or multiplicative separability and weak separability (which are needed for CP's estimator to be identified). At present robust inferential techniques for these estimators do not exist, or are in their infancy.

We end here by noting that this review was intended to highlight several of the most recent proposals for estimation of the stochastic frontier model in a nearly nonparametric framework and to draw attention to the types of assumptions (both statistical and economic) that can be used to help in identification. There are undoubtedly other approaches as well that have been put forth and their omission here is more of an indictment on us as authors rather that the research. This area remains an important and crucial aspect of the frontier literature.

References

Aragon, Y., Daouia, A., & Thomas-Agnan, C. (2005). Nonparametric frontier estimation: A conditional quantile-based approach. *Econometric Theory, 21*(2), 358–389.

Centorrino, S., & Parmeter, C. F. (2022) , Nonparametric estimation of stochastic frontier models with weak separability. Working Paper, University of Miami.

Das, A., & Bandyopadhyay, D. (2008). Identifiability of the stochastic frontier models. *Journal of Quantitative Economics, 6*(1&2), 57–70.

Du, P., Parmeter, C. F., & Racine, J. S. (2013). Nonparametric kernel regression with multiple predictors and multiple shape constraints. *Statistica Sinica, 23*(3), 1347–1371.

Fan, Y., Li, Q., & Weersink, A. (1996). Semiparametric estimation of stochastic production frontier models. *Journal of Business & Economic Statistics, 14*(4), 460–468.

Greene, W. H. (1990). A gamma-distributed stochastic frontier model. *Journal of Econometrics, 46*(1–2), 141–164.

Greene, W. H. (2003). Simulated likelihood estimation of the normal-gamma stochastic frontier function. *Journal of Productivity Analysis, 19*(2), 179–190.

Greene, W. H. (2008). The econometric approach to efficiency analysis. In H. O. Fried, C. A. K. Lovell & S. S. Schmidt (Eds.), *The Measurement of Productive Efficiency and Productivity Change*. Oxford, UK: Oxford University Press, Chapter 2.

Hall, P., & Simar, L. (2002). Estimating a changepoint, boundary or frontier in the presence of observation error. *Journal of the American Statistical Association, 97*, 523–534.

Hayfield, T., & Racine, J. S. (2008). Nonparametric econometrics: The np package. *Journal of Statistical Software,* **27**(5). http://www.jstatsoft.org/v27/i05/.

Henderson, D. J., & Parmeter, C. F. (2015). *Applied Nonparametric Econometrics.* Cambridge, Great Britain: Cambridge University Press.

Horrace, W. C., & Parmeter, C. F. (2011). Semiparametric deconvolution with unknown error variance. *Journal of Productivity Analysis, 35*(2), 129–141.

Kim, W., Linton, O. B., & Hengartner, N. W. (1999). A computationally efficient oracle estimator for additive nonparametric regression with bootstrap confidence intervals. *Journal of Computational and Graphical Statistics, 8*(2), 278–297.

Kneip, A., Simar, L., & Van Keilegom, I. (2015). Frontier estimation in the presence of measurement error with unknown variance. *Journal of Econometrics, 184*, 379–393.

Kumbhakar, S. C., Park, B. U., Simar, L., & Tsionas, E. G. (2007). Nonparametric stochastic frontiers: A local maximum likelihood approach. *Journal of Econometrics, 137*(1), 1–27.

Kumbhakar, S. C., Parmeter, C. F., & Zelenyuk, V. (2021a). Stochastic frontier analysis: Foundations and advances I. In: S. Ray, R. Chambers & S. C. Kumbhakar (Eds.), *Handbook of Production Economics* (Vol. 1). Springer.

Kumbhakar, S. C., Parmeter, C. F., & Zelenyuk, V. (2021b). Stochastic frontier analysis: Foundations and advances II. In: S. Ray, R. Chambers & S. C. Kumbhakar (Eds.), *Handbook of Production Economics* (Vol. 1). Springer. https://link.springer.com/referenceworkentry/10.1007/978-981-10-3450-3_11-1.

Kumbhakar, S., Wang, C.H.-J., & Horncastle, A. (2015). *A Practitioners Guide to Stochastic Frontier Analysis Using Stata.* Cambridge, UK: Cambridge University Press.

Kuosmanen, T., & Kortelainen, M. (2012). Stochastic non-smooth envelopment of data: Semi-parametric frontier estimation subject to shape constraints. *Journal of Productivity Analysis, 38*(1), 11–28.

Li, Q., & Racine, J. (2007). *Nonparametric Econometrics: Theory and Practice.* Princeton University Press.

Mammen, E., Rothe, C., & Schienle, M. (2012). Nonparametric Regression with Nonparametrically Generated Covariates. *Annals of Statistics, 40*(2), 1132–1170.

Martins-Filho, C. B., & Yao, F. (2015). Semiparametric stochastic frontier estimation via profile likelihood. *Econometric Reviews, 34*(4), 413–451.

Noh, H., & Van Keilegom, I. (2020). On relaxing the distributional assumption of stochastic frontier models. *Journal of the Korean Statistical Society, 49*, 1–14.

Parmeter, C. F., & Racine, J. S. (2012). Smooth constrained frontier analysis, In: X. Chen & N. Swanson (Eds.), *Recent Advances and Future Directions in Causality, Prediction, and Specification Analysis: Essays in Honor of Halbert L. White Jr.* (pp. 463–489). New York, New York: Springer, chapter 18.

Parmeter, C. F., Wang, H.-J., & Kumbhakar, S. C. (2017). Nonparametric estimation of the determinants of inefficiency. *Journal of Productivity Analysis, 47*, 205–221.

Parmeter, C. F., & Zelenyuk, V. (2019). Combining the virtues of stochastic frontier and data envelopment analysis. *Operations Research, 67*(6), 1628–1658.

Parmeter, C., & Kumbhakar, S. (2014). *Efficiency Analysis: A Primer on Recent Advances.* Now: Foundations and Trends. in Econometrics.

Prakasa Rao, B. L. S. (1992). *Identifiability in Stochastic Models.* New York, NY: Academic Press Inc.

Racine, J. S., & Nie, Z. (2021) , *crs: Categorical Regression Splines.* R package version 0.15-33. https://CRAN.R-project.org/package=crs

Ritter, C., & Simar, L. (1994). Another look at the American electrical utility data. Technical report, Institut de Statistique, Université Catholique de Louvain, B-1348 Louvain-la-Neuve, Belgium.

Ritter, C., & Simar, L. (1997). Pitfalls of normal-gamma stochastic frontier models. *Journal of Productivity Analysis, 8*(2), 167–182.

Robinson, P. M. (1988). Root-n consistent semiparametric regression. *Econometrica, 56*, 931–954.

Rothenberg, T. J. (1971). Identification in parametric models. *Econometrica, 39*(3), 577–591.

Ruppert, D., & Wand, M. P. (1994). Multivariate Locally Weighted Least Squares Regression. *The Annals of Statistics, 22*(3), 1346–1370.

Schwarz, M., & Van Bellegem, S. (2010). Consistent density deconvolution under partially known error distribution. *Statistics & Probability Letters, 80*(3), 236–241.

Simar, L., Lovell, C. A. K., & van den Eeckaut, P. (1994). Stochastic frontiers incorporating exogenous influences on efficiency. Discussion Papers No. 9403, Institut de Statistique, Universite de Louvain.

Simar, L., Van Keilegom, I., & Zelenyuk, V. (2017). Nonparametric least squares methods for stochastic frontier models. *Journal of Productivity Analysis, 47*(3), 189–204.

Simar, L., & Wilson, P. W. (2007). Estimation and inference in two-stage, semi-parametric models of production processes. *Journal of Econometrics, 136*(1), 31–64.

Tran, K. C., & Tsionas, E. G. (2009). Estimation of nonparametric inefficiency effects stochastic frontier models with an application to British manufacturing. *Economic Modelling, 26*, 904–909.

Waldman, D. M. (1982). A stationary point for the stochastic frontier likelihood. *Journal of Econometrics, 18*(1), 275–279.

Wang, H.-J., & Schmidt, P. (2002). One-step and two-step estimation of the effects of exogenous variables on technical efficiency levels. *Journal of Productivity Analysis, 18*, 129–144.

A Hierarchical Panel Data Model for the Estimation of Stochastic Metafrontiers: Computational Issues and an Empirical Application

Christine Amsler, Yi Yi Chen, Peter Schmidt, and Hung Jen Wang

1 Introduction

A recent paper (Amsler et al., 2021, hereafter ACSW) proposes a hierarchical stochastic frontier panel data model for the estimation of metafrontiers. The current paper presents an application of this model, with emphasis on computational issues.

The idea of a metafrontier originated with Hayami and Ruttan (1971, 1985), and was later operationalized in a large number of other papers, including Pitt (1983), Lau and Yotopoulos (1989), Battese and Rao (2002), Battese et al. (2004), O'Donnell et al. (2008), Moreira and Bravo-Ureta (2010), Villano et al. (2015) and Amsler et al. (2017). In this literature firms are assigned to groups defined by technology or geography, and one is interested in the technical inefficiency of a firm relative to its group-specific frontier, but also in the "technology gap" between the group-specific frontier and the overall maximal frontier (the metafrontier). That is, for each group there is a frontier and a firm in that group has an inefficiency relative to that frontier, but there is also a potential inefficiency from being in an inefficient group (e.g. using the wrong technology), reflected in the distance between the group-specific frontier and the metafrontier.

Several recent papers have proposed models to separate long-run and short-run technical inefficiency from each other and from long-run and short-run heterogeneity that is not regarded as inefficiency. In these papers, there are two composed errors, each of which has two parts. There is a long-run (time-invariant) composed error

C. Amsler · P. Schmidt (✉)
Michigan State University, East Lansing, USA
e-mail: schmidtp@msu.edu

Y. Y. Chen
Tamkang University, New Taipei, Taiwan

H. J. Wang
National Taiwan University, New Taipei, Taiwan

P. Macedo et al. (eds.), *Advanced Mathematical Methods for Economic Efficiency Analysis*, Lecture Notes in Economics and Mathematical Systems 692,
https://doi.org/10.1007/978-3-031-29583-6_11

$c_i = c_i^o - c_i^*$, where c_i^o is normal and represents long-run heterogeneity, and $c_i^* \geq 0$ is half-normal and represents long-run inefficiency. There also is a short-run (independent over time) composed error $u_{it} = u_{it}^o - u_{it}^*$, where u_{it}^o is normal and represents short-run heterogeneity, and $u_{it}^* \geq 0$ is half-normal and represents short-run inefficiency. See, e.g., Filippini and Greene (2016) and Lai and Kumbhakar (2018), and the references therein. In this paper we will add a third composed error $w_g = w_g^o - w_g^*$, where $g = g(i)$ represents the "group" for firm i. In our model the technology gap, for firms in group g, is captured by $w_g^* \geq 0$.

The model with two composed errors has been estimated successfully. Adding a third composed error adds an extra level of complexity. In this paper we will discuss computational issues and illustrate them in the context of an empirical application.

2 The Model and Its Estimation

Suppose that we have T time periods, indexed by $t = 1, \ldots, T$, and G groups, indexed by $g = 1, \ldots, G$. We have n_g firms in group g, so that the total number of firms is $N = \sum_g n_g$. We will index firms by $i = 1, \ldots, N$. We assume that each firm is in only one group and that firms do not change groups over time, so that we can represent the group to which firm i belongs as $g(i)$. Sometimes we will simplify notation by just using g in place of $g(i)$. This is a "hierarchical" or "multi-level" data structure. Firms are nested in groups, where "nested" is a term that dates back to the seminal article of Fuller and Battese (1973), because for each firm there is a unique group.

The model of ACSW is similar to the model of Yang and Schmidt (2021), in that it has fixed time effects and random firm and group effects. Specifically, the model is:

$$y_{it} = x_{it}^{'}\beta + d_t^{'}\theta + c_i + w_{g(i)} + u_{it}. \tag{1}$$

Here y_{it} is the output (in logs) of firm i at time t; x_{it} is a vector of measures of inputs; and d_t is a dummy variable for time t, so that the elements of θ are the fixed time effects. The c_i, w_g and u_{it} are the long-run, group-specific and short-run composed errors, respectively. Thus $c_i = c_i^o - c_i^*$, where c_i^o is normal and $c_i^* \geq 0$ is half-normal; similarly, $w_g = w_g^o - w_g^*$ and $u_{it} = u_{it}^o - u_{it}^*$. These composed errors have a skew-normal distribution (Azzalini (1985)). The fixed time effects are not essential to the model and could be omitted.

The interpretation of the model is as follows. The overall frontier (metafrontier) for firm i at time t is $y_{it} = x_{it}^{'}\beta + d_t^{'}\theta + \left(c_i^o + w_{g(i)}^o + u_{it}^o\right)$. Here c_i^o is firm-specific heterogeneity; $w_{g(i)}^o$ captures heterogeneity across groups; and u_{it}^o is idiosyncratic heterogeneity. For firm i at time t, its inefficiency relative to the overall frontier is $(c_i^* + w_{g(i)}^* + u_{it}^*)$; its inefficiency relative to its group-specific frontier is $c_i^* + u_{it}^*$; and $w_{g(i)}^*$ is the inefficiency of its group relative to the metafrontier.

We follow ACSW and estimate the model by maximum likelihood. To do so we require the following assumptions. These are strong assumptions but they mirror the assumptions made by Filippini and Greene (2016) for the model with two composed errors.

Assumptions

A. (i) The c_i^o are iid $N(0, \sigma_{c^o}^2)$. (ii) The w_g^o are iid $N(0, \sigma_{w^o}^2)$. (iii) The u_{it}^o are iid $N(0, \sigma_{u^o}^2)$. (iv) The c_i^* are iid $N^+(0, \sigma_{c^*}^2)$. (v) The w_g^* are iid $N^+(0, \sigma_{w^*}^2)$. (vi) The u_{it}^* are iid $N^+(0, \sigma_{u^*}^2)$. (Here N^+ denotes a half-normal distribution.)

B. $x_{it}, c_j^o, c_k^*, w_g^o, w_h^*, u_{ms}^o, u_{qr}^*$ are mutually independent for all $i, j, k, m, q = 1, \ldots, N$, $g, h = 1, \ldots, G$ and $r, s, t = 1, \ldots, T$.

Define $\varepsilon_{it} = c_i + w_{g(i)} + u_{it}$. We have independence of the ε's across different groups, but within a group we have correlation across individuals and over time because of the group effect. Suppose that within group g we re-index the individuals as $i = 1, \ldots, n_g$ (a separate re-indexing for each group), and we let $\varepsilon_{(g)} = \left(\varepsilon_{11}, \ldots, \varepsilon_{1T}, \ldots, \varepsilon_{n_g1}, \ldots, \varepsilon_{n_gT}\right)'$ be the $Tn_g \times 1$ vector of ε's for group g. Let $f_g(\varepsilon_{(g)})$ be the density of $\varepsilon_{(g)}$. Define $y_{(g)}$and$x_{(g)}$ analogously to $\varepsilon_{(g)}$ and $d_{(g)} = (d_1, \ldots, d_T, \ldots, d_1, \ldots, d_T)'$, so that $\varepsilon_{(g)} = y_{(g)} - x_{(g)}\beta - d_{(g)}\theta$. Then the log-likelihood function is

$$\ln L = \sum_{g=1}^{G} \ln f_g(y_{(g)} - x_{(g)}\beta - d_{(g)}\theta). \tag{2}$$

We will maximize this to calculate the maximum likelihood estimator. The parameters with respect to which the likelihood is maximized are $\beta, \theta, \sigma_{c^o}^2, \sigma_{c^*}^2, \sigma_{w^o}^2, \sigma_{w^*}^2, \sigma_{u^o}^2$ and $\sigma_{u^*}^2$.

The remaining issue is how to calculate the density f_g for each group$g$. The vector of random elements on which $\varepsilon_{(g)}$ depends is$\xi_g = (c_1, \ldots, c_{n_g}, w_g, u_{11}, \ldots, u_{n_gT})$, a vector of $Tn_g + n_g + 1$ independent skew-normal random variables. These have densities that depend on the univariate normal pdf and cdf; for example, $f_c(c_i) = \frac{2}{\sigma_c}\varphi\left(\frac{c_i}{\sigma_c}\right)\Phi\left(-\lambda_c\frac{c_i}{\sigma_c}\right)$, where $\sigma_c^2 = \sigma_{c^o}^2 + \sigma_{c^*}^2$ and$\lambda_c = \sigma_{c^*}/\sigma_{c^o}$, and similarly for w_g andu_{it}. Because the elements of ξ_g are mutually independent, the density of ξ_g is:

$$f_\xi(\xi_g) = f_g(w_g)\prod_{i=1}^{n_g}[f_c(c_i)\prod_{t=1}^{T} f_u(u_{it})]. \tag{3}$$

Then we can obtain the density of $\varepsilon_{(g)}$ by integrating out w_g and the c_i:

$$f_g(\varepsilon_{(g)}) = \int \ldots \int f_g(w_g)\prod_{i=1}^{n_g}[f_c(c_i)\prod_{t=1}^{T} f_u(\varepsilon_{it} - w_g - c_i)]dw_g dc_1 \ldots dc_{n_g}. \tag{4}$$

This integration is of dimension $n_g + 1$ for a given g and is best evaluated numerically using the Monte Carlo integration method. To apply this method, we first transform the integral's domain to (0, 1) in each of the dimensions by a change of variables:

$$f_g(\varepsilon_{(g)}) = \int \cdots \int_{(0,1)^{n_g+1}} \tilde{f}_g(\tilde{w}_g) \prod_{i=1}^{n_g} [\tilde{f}_c(\tilde{c}_i) \prod_{t=1}^{T} \tilde{f}_u(\varepsilon_{it} - \tilde{w}_g - \tilde{c}_i)] d\tilde{w}_g d\tilde{c}_1 \ldots d\tilde{c}_{n_g}. \quad (5)$$

In the expression, $\tilde{f}_g(\tilde{w}_g) = f_g(\rho(\tilde{w}_g))\rho'(\tilde{w}_g)$ where $w_g = \rho(\tilde{w}_g)$ is the transformation equation and $\rho'(\tilde{w}_g)$ is the Jacobian. Other similar notation is similarly defined. Let $\{(r_j^1, r_j^2, \ldots, r_j^{(n_g+1)}), j = 1, \ldots, S\}$ being a $S \times (n_g + 1)$ matrix of random numbers in $(0, 1)^{n_g+1}$, then the integration is approximated by

$$\widehat{f}_g(\varepsilon_{(g)}) = \frac{1}{S} \sum_{j=1}^{S} \tilde{f}_g(r_j^{n_g+1}) \Pi_{i=1}^{n_g} [\tilde{f}_c(r_j^1) \Pi_{t=1}^{T} \tilde{f}_u(\varepsilon_{it} - r_j^{n_g+1} - r_j^i)]. \quad (6)$$

Finally, the simulated log likelihood is $\ln\widehat{L} = \sum_{g=1}^{G} \ln \widehat{f}_g(y_{(g)} - x_{(g)}\beta - d_{(g)}\theta)$. We maximize this function to obtain our estimates of the parameters (β, θ and the parameters in the distributions of u, c and w).

As will be shown in our empirical example, obtaining numerical robustness of the estimation is a challenge even with a modest size of n_g. We discuss some computational issues in Sect. 3.

Inefficiency Index

After we obtain the model estimates, we can compute the group-, firm-, and observation-specific inefficiency indexes. The state-level index may be defined as $E(w_g^*|\varepsilon_{(g)})$ which is derived from the density function

$$f(w_g^*|\varepsilon_{(g)}) = f(\varepsilon_{(g)}, w_g^*)/f(\varepsilon_{(g)}). \quad (7)$$

The denominator is the density of $\varepsilon_{(g)}$ where $\varepsilon_{(g)} = (\varepsilon_{1(g)}, \ldots, \varepsilon_{N_g(g)})'$. The element $\varepsilon_{i(g)}$ is a $T \times 1$ vector of the i th firm in the g th group. Thus, elements in $\varepsilon_{(g)}$ contain random variables from $Z_g = (c_1, \ldots, c_{n_g}, w_g, u_{11}, \ldots, u_{1T}, \ldots, u_{n_g 1}, \ldots, u_{n_g T})'$, and $\varepsilon_{(g)}$ can be represented by linear combinations of Z_g. Note that because c_i, w_g, and u_{it} are all independently distributed closed skew normal (CSN) random variables, Z_g is itself a CSN random variable. As a result, $\varepsilon_{(g)}$, which is a vector of linear combinations of Z_g, would also has a CSN distribution. The density function $f(\varepsilon_{(g)})$ could be derived accordingly after tedious algebra.

The density function on the numerator could be derived similarly with the exception that w_g in Z_g is decomposed into w_g^o and w_g^*. Here, w_g^o is a normal random variable which is CSN with the scale parameter equal to 0. The w_g^* is a half-normal random variable which is also CSN with the scale parameter approaching infinity.

The group-specific inefficiency index is thus

$$E(w_g^*|\varepsilon_{(g)}) = \int_0^\infty w_g^* f(w_g^*|\varepsilon_{(g)}) dw_g^*, \tag{8}$$

which can be evaluated using the quadrature or the Monte Carlo method. The firm- and the observation-specific inefficiency could be similarly derived.

3 Computational Issues

Quadrature and Monte-Carlo methods are commonly used for numerical integrations. In the case of multi-dimensional integration, the former suffers from the curse of dimensionality, making the Monte Carlo method the only feasible choice.

To illustrate, the numerical evaluation of $I = \int\int_{(0,1)^2} f(x, y)dxdy$ is equivalent to computing $I = E[f(x, y)]$. By way of the probability integral transformation, the evaluation may be done by drawing a S × 2 array of random numbers $\{(r_j^1, r_j^2), j = 1, \ldots, S\}$ from uniform distributions in (0, 1) and then $I \approx (1/S)\sum_{j=1}^S f(r_j^1, r_j^2)$. The random-number based Monte Carlo (MC) method has a convergence rate of $1/\sqrt{S}$. The MC method can nevertheless still be slow and thus time-consuming particularly when accuracy and numerical robustness are required. We adopt two strategies to overcome this hurdle.

Quasi-Monte Carlo Integration Using Low Discrepancy Sequences

Instead of using random numbers r_j, a more efficient method is to use quasi-random numbers to perform quasi-Monte Carlo (QMC) integration, which has a convergence rate equal to $1/S$. The quasi-random numbers are also referred to as a low discrepancy sequence (LDS), which has equidistant values in (0, 1). The values are not random; they are generated algorithmically with the purpose of evenly and efficiently covering the unit cube $(0, 1)^d$, where d is the number of dimensions, while avoiding correlations across dimensions. Some of the well-known types of LDS include the Halton sequence, the Sobol sequence, and variations of these sequences. Given a two-dimensional arrays of an LDS $\{(l_j^1, l_j^2), j = 1, \ldots, S\}$, the two-dimensional integration problem is similarly evaluated by $I \approx (1/S)\sum_{j=1}^S f(l_j^1, l_j^2)$.

How large does S have to be? The answer to this question has direct consequences for the run time and the robustness of the estimates. This question could be addressed in (at least) two ways. First, S should be big enough so that increasing it does not materially change the estimates. That is, the estimates converge across S for a given type of LDS (e.g., Halton). Second, S should be big enough so that two or more sets of LDS (e.g., Halton and Sobol) do not yield materially different estimates. That is, the estimates converge across types of LDS.

As a preview, our empirical results indicate that the latter is a stronger requirement in the sense that it accommodates the former. However, the required S in such a case is likely in the order of millions even with the efficient QMC method. Note that if the integration has d dimensions, the required number of function evaluations would be dS and that is just for one (log-likelihood) function value without gradients and

Hessians. Because the maximum-likelihood estimation is iterative such that values of the function have to be calculated repeatedly, it raises significant concerns about the necessary computing power. We turn to parallel computing for the required resources.

Parallel Computing using GPU

GPU computing is a type of parallel computing. A GPU (graphical processing unit) contains a large amount of small cores which individually may not be as powerful as a CPU's core, but together they deliver massive performance. GPU computing is particularly suitable for tasks that can be divided up and allocated across cores. Monte Carlo integration, by which the main task is to evaluate a function using a sequence of independent numbers, is a prime example. Special computer instructions, by way of GPU programming, are required in order to communicate with the GPU for dividing and distributing the tasks and calling back the results.

GPU computing used to require specialized hardware and advanced computer instructions, both of which are likely beyond the reach of most of empirical researchers. However, recent developments in computer software and hardware have made GPU computing accessible for individuals using consumer-grade hardware and high-level programming languages (Owens et al., 2008).

We use a personal computer with an Intel i9 CPU and a Nvidia RTX 3090 GPU. The GPU has 10,496 cores and 24 GB memory. The software is Julia v1.7.2 which is a free, open-source, and high-level programming language with dedicated packages for GPU computing.[1] According to our preliminary investigation, the GPU-based estimation of our empirical model is at least 14 times faster than the CPU-based serial estimation when $S = 2^{23}$. Still, using GPU one iteration of the maximum likelihood estimation takes about 8.65 min to finish when $S = 2^{23}$.

4 Empirical Application

As an illustration, we apply the model to estimate the production efficiency of India's agricultural productions at the district level, where districts are administrative units within states. The annual data is derived from Duflo and Pande (2007) and is prepared in a similar way as in Chen et al. (2020), with two major differences from the latter.

The first is that we do not reduce the data's panel structure to cross-sectional. The second difference, which is more substantial, is that we merge districts in a state with similar elevations together to form a combined district which becomes the base unit of observation in the data.[2] In this way, the model's g ($g = 1, \ldots, G$) denotes

[1] Information can be found in https://juliagpu.org/.

[2] The original data of Duflo and Pande (2007) contains geographical information of districts including the percentages of areas where the elevations are below 250 M (meters, same below), between 250 to 500 M, between 500 to 1000 M, and above. We use the information to calculate the district's average evaluation and use the result to assign a district to one of the four elevation categories defined similarly as above. Then, districts in the same state and of the same elevation category are combined together.

Table 1 Summary statistics of variables

	Mean	Std.dev.	Min	Max
yit	9.667	0.959	5.070	11.835
$fert_{it}$	16.587	1.116	13.661	18.819
gca_{it}	6.188	0.526	3.912	7.077
gia_{it}	4.658	1.068	−1.204	6.637
$rain_{it}$	−0.025	0.180	−0.538	0.449
dam_{it}	0.984	2.030	0	11.857

Note The total number of observations is 408

a state and i ($i = 1 \ldots, n_g$) denotes a combined district in a state. Other variables are measured at the combined-district level by taking the average of the data from the involved districts. Duflo and Pande (2007) and Chen et al. (2020) both identify elevation as an important factor affecting agricultural production in India. Therefore, instead of using the original district units, which are administrative-based, we use the elevation-based districts which have characteristics that bear direct relevance to production efficiency. Furthermore, by merging and combining, we reduce the value of n_g in each state which greatly eases the complexity of estimating the hierarchical model.

The estimation data contains 12 states ($G = 12$) with a total of 34 combined districts ($\sum_{g=1}^{G} n_g = 34$) and 408 observations from the year 1976 to 1987 ($T = 12$). The value of n_g is either 2, 3, or 4 in the states. The dependent variable y_{it} is the log of agricultural production of the combined district i in year t, and the explanatory variables include $fert_{it}$ (the log of the fertilizer), gca_{it} (the log of the gross cultivated area), gia_{it} (the log of the gross irrigated area), $rain_{it}$ (the rainfall measured as the fractional deviation of the combined district's rainfall from the combined district's mean in the period), dam_{it} (the combined district's number of dams), and a constant intercept. All of the variables are the mean values from the combined districts. Table 1 reports the summary statistics of the variables.

4.1 Estimation

To perform the Monte Carlo integration, we transform the integration function's domain to (0, 1) in every dimension using a change of variables, while noting the Jacobian. We include two widely-used versions of LDS in the analysis, namely the Halton sequence (Halton) and the Sobol sequence (Sobol). When applied to multidimensional settings, both the Halton and the Sobol may exhibit strong correlations between sequences used in different dimensions. The correlation causes the quasi-random numbers to distribute in specific patterns in the unit cube, compromising the LDS's goal of evenly filling the space with the points. The problem could be particularly serious when the sequence is short. To avoid this pitfall, the sequence can be scrambled using permutation methods to reduce the correlation. In this study, we also adopt two versions of the scrambled sequences, which we refer to as the scrambled Halton sequence (scramHalton) and the scrambled Sobol sequence (scramSobol). We use Julia's RCall package to import these sequences from R's randtoolbox package. A discussion of these sequences may be found in various texts including Brandimarte (2014).

To gauge the estimates' convergence over S and across different LDS, we use a range of values for S: $S = \{2^n, n = 13, 14, \ldots, 23\} = (8{,}192, 16{,}384, 32{,}768, 65{,}536, 131{,}072, 262{,}144, 524{,}288, 1{,}048{,}576, 2{,}097{,}152, 4{,}194{,}304, 8{,}388{,}608)$. Because the integration has n_g+1 dimensions and each dimension requires a different sequence, so the matrix of the sequences is of dimension $S \times (n_g + 1)$. As suggested by Owen (2022), we do not burn-in the sequences (i.e., remove the first few points in the sequence). Given the large value of S, removing a few points is likely to be inconsequential anyway.

Figure 1 plots log-likelihood values of the model estimated using various LDS and S. The graph shows different convergence paths of models using different LDS, and the paths appear to arrive at similar values ("converged") only when $S = 2^{23}$. We also plot the estimated values of the model's variance parameters ($\sigma^2_{w^o}$, $\sigma^2_{w^*}$, $\sigma^2_{c^o}$, $\sigma^2_{c^*}$, $\sigma^2_{u^o}$, $\sigma^2_{u^*}$) in Fig. 2. If we require the model's estimates and key statistics to be numerically robust to the choice of LDS, these results indicate that a very large amount of quasi-random numbers is needed.

The size of S required to achieve robustness is somewhat surprising and unexpected, and it has important implications for the model's run time. As shown in Fig. 3,

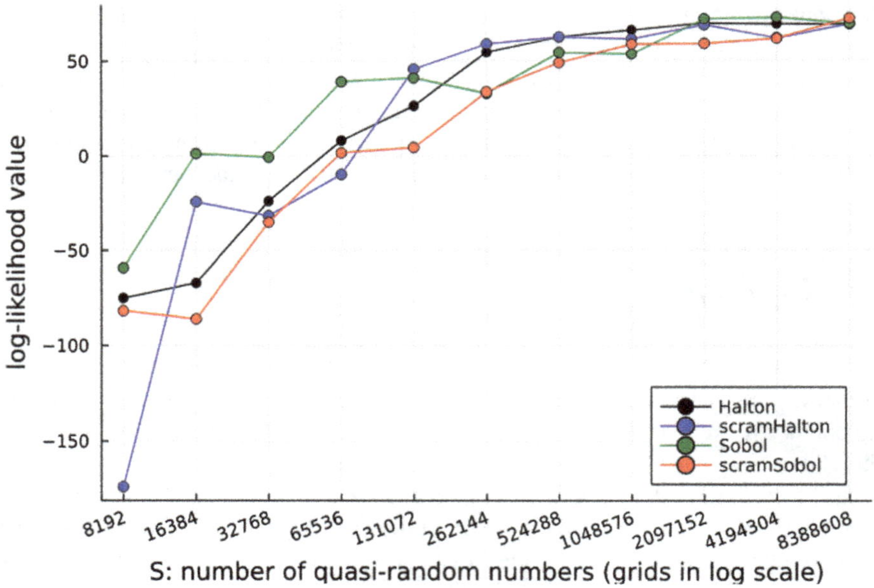

Fig. 1 Log-likelihood values across S with different LDS

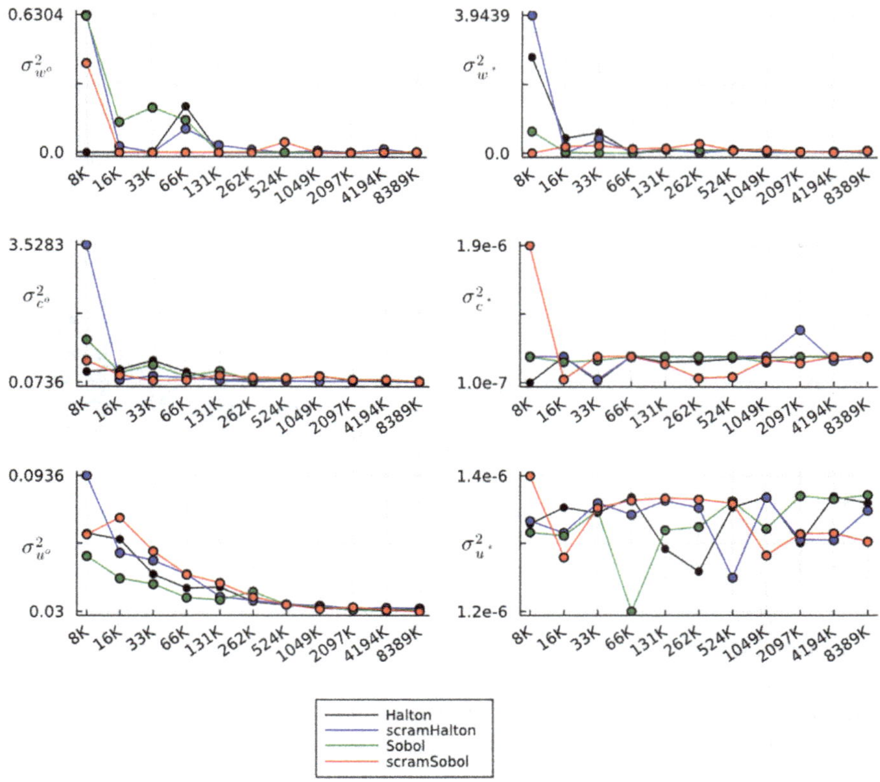

Fig. 2 Estimated variance parameters across S with different LDS

the run time per iteration is roughly proportional to the size of S, which is expected. In the case of $S = 2^{23}$, it takes on average 8.65 min for one iteration. Given that the MLE estimation of the model is likely, according to our experiment, to take about 50 to 70 iterations to finish, the required run time may take up to 10 h per model. Better initial values would cut down the number of iterations, and our best effort shows that 8 to 9 iterations may suffice in the best scenario.[3] Using CPU-based serial computing, the required time is about 14 times larger.

Table 2 shows the estimation results, for the Sobol-based estimation. The slope coefficients of major production inputs, including fertilizer, two types of land, and rain, are all significant and positive. The number of dams, however, has a negative impact on production. We suspect that the dam variable may have been a proxy for local geographical and economic conditions. According to Duflo and Pande (2007),

[3] Using results from a corresponding true random effect stochastic frontier model as initial values for the slope coefficients and $\exp(-2)$ for the other variance parameters, the required number of iterations is typically 50 to 70 using different LDS. Using results from the Sobol-based estimation (with $S = 2^{23}$) as initial values, it takes 8 iterations for the Halton-based estimation (also with $S = 2^{23}$) to complete.

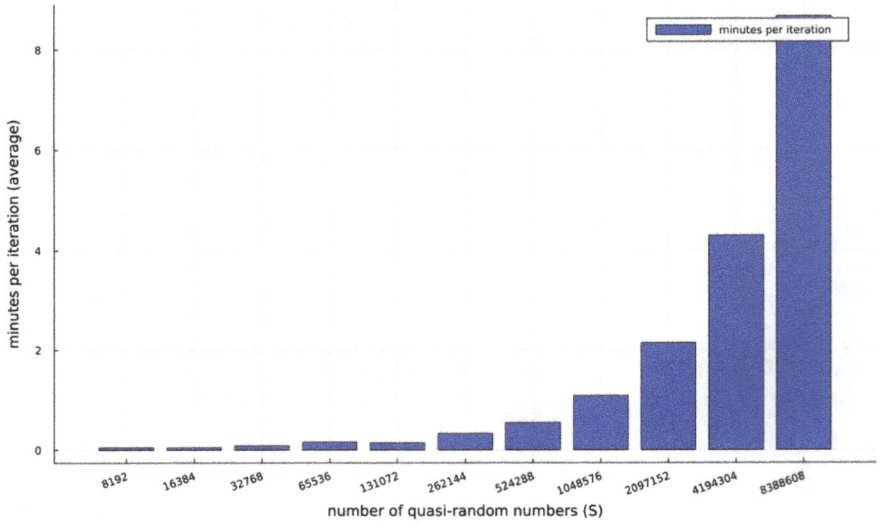

Fig. 3 Time per iteration for different numbers of quasi-random numbers

dams were mostly located in areas with specific geographical properties including rivers exhibiting certain gradients. These properties are usually not ideal for agricultural cultivation. Furthermore, the decision to build dams could be endogenous to local economic conditions. All these factors may imply that districts with more dams are likely to be poorer areas with geographical conditions that are less suitable for agricultural production.

The model's estimated variance parameters vary widely in size. In particular, the estimated value of σ_{w*}^2 is large and significant while the values of σ_{c*}^2 and σ_{u*}^2 are negligible. The results indicate that there is substantial technical inefficiency across states when measured against the metafrontier. Within a state, the district level inefficiency was minimal or could not be identified.

Table 2 Results from the sobol-based estimation

Var	Coeff	Std.err		Param	Value	Std.err
$fert_{it}$	0.226***	0.017		σ_{wo}^2	0.0002	0.001
gca_{it}	0.260***	0.035		σ_{w*}^2	0.079***	0.045
gia_{it}	0.221***	0.028		σ_{co}^2	0.097**	0.026
$rain_{it}$	0.312***	0.050		σ_{c*}^2	0.000001	0.000002
dam_{it}	−0.195***	0.011		σ_{uo}^2	0.031***	0.002
constant	3.647***	0.618		σ_{u*}^2	0.000001	0.0002

Note 1 Significance: ***: 1% level, **: 5% level; *: 10% level
Note 2 Use Sobol sequences with $S = 2^{23}$
Note 3 The ML estimation takes 52 iterations and 7.13 h to complete using GPU computing

Table 3 Summary statistics of state inefficiency index

	Mean	Std.dev	Min	Max
$E(w_g^*\|\varepsilon_{(g)})$	0.197	0.095	0.085	0.420

We can use the results to compute the state-, the combined district- and the observation-specific inefficiency index. Because the inefficiency at the district and the observation levels are negligible, as implied by the small values of $\widehat{\sigma}_{c*}^2$ and $\widehat{\sigma}_{u*}^2$, we only report the state-level inefficiency $E(w_g^*|\varepsilon_{(g)})$. As shown in Table 3, when compared to the metafrontier, the most efficient state in India has the index equal to 0.085, indicating that the production could increase by 8.5% if it becomes fully efficient. The least efficient state, on the other hand, could gain a 42% improvement, which is quite substantial. In sum, India's agricultural production showed great disparity across states, and there is a potential of gaining 19.7% improvement if all the states could catch up to the meta-technical frontier.

5 Conclusions

In this paper we report an empirical application of the hierarchical panel data stochastic frontier model proposed by Amsler et al. (2021). The estimation of this model is numerically challenging, and the main issue in the paper is to see whether it is feasible. We answer this in the context of a data set previously used by Duflo and Pande (2007) and Chen et al. (2020). For our model and data set, estimation is feasible, but it requires a large of amount of computer time and attention to numerical detail.

Evaluating the likelihood function requires numerically integrating with respect to $d \equiv \mathrm{n_g} + 1$ skew normal random variables. We use a quasi-Monte Carlo method, which relies on low discrepancy sequences, such as Halton or Sobol sequences, to evaluate the integral in the $(0, 1)^{\mathrm{d}}$ domain. This method is much more efficient than a pure Monte Carlo method. However, to obtain numerically stable results (robust to the choice of low discrepancy sequence and number of replications of the quasi-Monte Carlo procedure) requires a huge number of replications. In our application where $\mathrm{n_g} + 1$ is 4, the required number of replications is 2^{23}. This is far larger than the number of replications used in previous studies of which we are aware that use Monte Carlo evaluation of a likelihood. The massive amount of computation becomes an obstacle.

We use graphic processing unit (GPU) computing to alleviate the problem. GPU computing is a massively parallel method of computing originally designed for graphical applications (e.g. games). Recent developments in the related software and hardware have made GPU computing feasible for scientific applications by individual researchers with consumer-grade equipment. For our application, GPU computing is about 14 times faster than CPU-based serial computing. Even so, in our application (with 12 parameters and 408 observations) one iteration of the maximization of the likelihood took about 8.65 min. Depending on the quality of the initial starting values, the number of iterations was on the order of 10 to 70, implying a computing time of the order of 1.5 to 10 h for the likelihood maximization.

Despite these considerable numerical challenges, we were able to reliably estimate the model and to obtain apparently reasonable results.

References

Amsler, C., O'Donnell, C. J., & Schmidt, P. (2017). Stochastic metafrontiers. *Econometric Reviews, 36*, 1007–1020.

Amsler, C., Chen, Y.-Y., Schmidt, P., & Wang, H.-J. (2021). A hierarchical panel data stochastic frontier model for the estimation of stochastic metafrontiers. *Empirical Economics, 60*, 353–363.

Azzalini, A. (1985). A class of distributions which includes the normal ones. *Scandanavian Journal of Statistics, 12*, 171–178.

Battese, G. E., & Rao, D. S. P. (2002). Technology gap, efficiency, and a stochastic metafrontier function. *International Journal of Business and Economics, 1*, 87–93.

Battese, G. E., Rao, D. S. P., & O'Donnell, C. J. (2004). A metafrontier production function for estimation of technical efficiencies and technology gaps for firms operating under different technologies. *Journal of Productivity Analysis, 21*, 91–103.

Brandimarte, P. (2014). *Handbook in Monte Carlo Simulation.* Wiley.

Chen, Y. T., Hsu, Y. C., & Wang, H. J. (2020). A stochastic frontier model with endogenous treatment status and mediator. *Journal of Business and Economic Statistics, 38*, 243–256.

Duflo, E., & Pande, R. (2007). Dams. *The Quarterly Journal of Economics, 122*, 601–646.

Filippini, M., & Greene, W. (2016). Persistent and transient productive inefficiency: A maximum simulated likelihood approach. *Journal of Productivity Analysis, 45*, 187–196.

Fuller, W. A., & Battese, G. E. (1973). Transformations for estimation of linear models with nested-error structure. *Journal of the American Statistical Association, 68*, 626–632.

Hayami, Y., & Ruttan, V. W. (1971). *Agricultural development: An international perspective.* Baltimore: The Johns Hopkins University Press.

Hayami, Y., & Ruttan, V. W. (1985). *Agricultural development: An international perspective*, revised and expanded edition, Baltimore: The Johns Hopkins University Press.

Lai, H.-P., & Kumbhakar, S. C. (2018). Panel data stochstic frontier model with determinants of persistent and transient inefficiency. *European Journal of Operational Research, 271*, 746–755.

Lau, L. J., & Yotopoulos, P. A. (1989). The meta-production function approach to technological change in world agriculture. *Journal of Development Economics, 31*, 241–269.

Moreira, V., & Bravo-Ureta, B. (2010). Technical efficiency and metatechnology ratios for dairy farms in three southern cone countries: A stochastic meta-frontier model. *Journal of Productivity Analysis, 33*, 33–45.

O'Donnell, C. J., Rao, D. S. P., & Battese, G. E. (2008). Metafrontier frameworks for the study of firm-level efficiencies and technology ratios. *Empirical Economics, 34*, 231–255.

Owen, A. B. (2022). On dropping the first Sobol's point. In: Keller, A. (Ed.), *Monte Carlo and Quasi-Monte Carlo methods. MCQMC 2020.* Springer Proceedings in Mathematics & Statistics (Vol. 387). Springer.

Owens, J. D., Houston, M., Luebke, D., Green, S., Stone, J. E., & Phillips, J. C. (2008). GPU computing. *Proceedings of the IEEE, 96*(5), 879–899.

Pitt, M. M. (1983). Farm-level fertilizer demand in java: A meta-production function approach. *American Journal of Agricultural Economics, 65*, 502–508.

Villano, R., Bravo-Ureta, B., Solis, D., & Fleming, E. (2015). Modern rice technologies and productivity in the Philippines: Disentangling technology from managerial gaps. *Journal of Agricultural Economics, 66*, 129–154.

Yang, Y., & Schmidt, P. (2021). An econometric approach to the estimation of multi-level models. *Journal of Econometrics, 220*, 532–543.

Robustness in Stochastic Frontier Analysis

Alexander D. Stead, Phill Wheat, and William H. Greene

1 Introduction

Stochastic frontier (SF) analysis involves the estimation of an efficient frontier function—e.g. a production, cost, or revenue frontier. The robustness of our estimators and predictors is critical to accurate prediction of efficiency levels or rankings. The presence of contaminating outliers and other departures from distributional assumptions is problematic, since maximum likelihood estimation (MLE) is the most commonly employed estimation method in the SF literature and it is well known that MLE and other classical estimation methods are usually non-robust, in that they perform poorly in the presence of departures from distributional assumptions.

In SF modelling, robustness has additional relevance since a primary concern, especially in a regulatory context, is the prediction of efficiency levels of individual firms against the estimated frontier function. We therefore have reason to be concerned not only with the robustness of our estimation of the frontier, but the robustness of our efficiency predictions.

Despite this relevance, relatively little attention has been given to the issue of robustness in the SF literature. In recent years, several studies have proposed alternative models or estimators relevant to the discussion of robustness, though with little explicit discussion of what is meant by 'robustness' and how it might be measured

A. D. Stead (✉) · P. Wheat
Institute for Transport Studies, University of Leeds, 34-40 University Rd, Leeds LS2 9JT, UK
e-mail: a.d.stead@leeds.ac.uk

P. Wheat
e-mail: p.e.wheat@its.leeds.ac.uk

W. H. Greene
Department of Economics, Stern School of Business, New York University, 44 West 4th Street, New York NY10012, USA
e-mail: wgreene@stern.nyu.edu

P. Macedo et al. (eds.), *Advanced Mathematical Methods for Economic Efficiency Analysis*, Lecture Notes in Economics and Mathematical Systems 692,
https://doi.org/10.1007/978-3-031-29583-6_12

and compared across specifications and estimation methods. This chapter discusses two different aspects of robustness in stochastic frontier analysis: first, robustness of parameter estimates, by comparing the influence of outlying observations across different specifications—a familiar approach in the wider literature on robust estimation; second, the robustness of efficiency predictions to outliers across different specifications—a consideration unique to the efficiency analysis literature.

The remainder of this chapter is structured as follows. Sections 2 and 3 briefly introduce the SF model and some background and key concepts from the robustness literature, respectively. Section 4 discusses robust estimation of the SF model, drawing on the concepts introduced in Sect. 3 and relevant SF literature. Section 5 discusses the robustness of efficiency prediction. Section 6 summarises and concludes.

2 The Stochastic Frontier Model

Introduced by Aigner et al. (1977) and Meeusen and van Den Broeck (1977), the basic cross-sectional stochastic frontier model is given by

$$y_i = \boldsymbol{x}_i'\boldsymbol{\beta} + \varepsilon_i, \qquad \varepsilon_i = v_i - su_i, \tag{1}$$

where for the ith firm, y_i is the dependent variable, $\boldsymbol{x}_i$ is a vector of independent variables, $\boldsymbol{\beta}$ is a vector of frontier coefficients, and ε_i is an error term. The latter is composed of a two-sided term, v_i, representing random measurement error and other noise factors, and a non-negative term, u_i, representing departure from the frontier function as a result of inefficiency. In the case of a production frontier, firms may only be on or below the frontier, in which case $s = 1$. On the other hand, in the case of a cost frontier, firms may only be on or above the frontier, in which case $s = -1$. However, note that the composed error ε_i may take on either sign owing to the presence of the two-sided noise term v_i. Many extensions of this model, particularly to panel data settings, have been proposed; see reviews of the literature by Kumbhakar and Lovell (2000), Murillo-Zamorano (2004), Coelli et al. (2005), Greene (2008), Parmeter and Kumbhakar (2014), and Stead et al. (2019).

Though alternative parametric and semiparametric approaches to estimation of the model have been explored, some of which are of particular interest from a robustness perspective and discussed further in Sect. 4, MLE is the most common approach in theoretical and applied SF literature. This necessitates specific distributional assumptions regarding the error terms v_i and u_i; Aigner et al. (1977) and Meeusen and van Den Broeck (1977) explored estimation of the model under the assumption that

$$v_i \sim \mathcal{N}(0, \sigma_v^2), \qquad u_i = |w_i|, \qquad w_i \sim \mathcal{N}(0, \sigma_u^2),$$

known as the normal-half normal (N-HN) model, or alternatively that

$$v_i \sim \mathcal{N}(0, \sigma_v^2), \qquad u_i \sim \text{Exponential}(1/\sigma_u),$$

known as the normal-exponential (N-EXP) model. In both cases it is assumed that v_i and u_i are independent. The marginal density of the composed error, ε_i, is then derived via the convolution

$$f_\varepsilon(\varepsilon_i, \boldsymbol{\theta}) = \int_0^\infty f_{v,u}(\varepsilon_i + su_i, u_i, \boldsymbol{\theta})\mathrm{d}u_i, \qquad \boldsymbol{\theta} = (\boldsymbol{\beta}', \boldsymbol{\vartheta}')', \tag{2}$$

where $f_{v,u}$ is the joint density of v_i and u_i—which under the assumption of independence, is just the product of their marginal densities—and $\boldsymbol{\theta}$ is a vector of parameters. This is then used to form the log-likelihood function.

The next step in SF modelling is then to predict observation-specific efficiency, u_i, relative to the estimated frontier. These predictions are based on the conditional distribution

$$f_{u|\varepsilon}(u_i|\varepsilon_i, \boldsymbol{\theta}) = \frac{f_{v,u}(\varepsilon_i + su_i, u_i, \boldsymbol{\theta})}{f_\varepsilon(\varepsilon_i, \boldsymbol{\theta})}, \tag{3}$$

following Jondrow et al. (1982). Both the log-likelihood function and $f_{u|\varepsilon}$ clearly depend on our underlying assumptions about the distribution of v_i and u_i. Many generalisations of and departures from the N-HN and N-EXP cases have been proposed, and a detailed discussion of these is beyond the scope of this chapter, but is available in the aforementioned reviews of the SF literature. From a robustness perspective, the essential point is that distributional assumptions about v_i and u_i are critical to both estimation of the frontier function and the prediction of efficiency relative to it.

3 Definitions and Measures of Robustness

Robustness, broadly defined, means a lack of sensitivity to small departures from our model assumptions. Ideally, we would like a robust estimator, i.e. one that is not unduly sensitive such departures. The main concern in many applications is robustness to 'outliers', i.e. outlying observations drawn from some contaminating distribution. Robustness to outliers will also be the main focus of this chapter, though given the critical role of specific distributional assumptions in SF modelling, it is worth keeping in mind a broader definition, especially in relation to issues from the SF literature such as 'wrong skew'. Two related but distinct concepts in the literature on robust estimation are *robustness*, measured in terms of the *influence* of contaminating observations on an estimator, and *resistance*, measured in terms of the estimator's *breakdown point*. These are discussed below.

3.1 The Influence Function

Analysis of robustness often centres around the influence of contaminating observations on a statistical functional. Introduced by Hampel et al. (1986), the *influence function* measures the effect on a statistical functional T of an infinitesimal perturbation of the distribution of the data F at a point $(y, \boldsymbol{x})$, and is given by

$$L_T(y, \boldsymbol{x}) = \lim_{\epsilon \to 0} \frac{T\big((1-\epsilon)F + \epsilon\delta_y\big) - T(F)}{\epsilon}, \tag{4}$$

where δ_y is a point mass at y. A statistical functional, in turn, is any function that maps a distribution to a real scalar or vector. The greater the magnitude of L_T, the greater the influence of the contaminating observation. The usefulness of this concept is clear, since it can be applied to many different kinds of estimators and predictors, enabling the discussion of the robustness of SF estimation and efficiency prediction in terms of a consistent set of concepts and measures.

The influence function can be derived as a limiting case of the *Gâteaux derivative*, a generalisation of the directional derivative to differentiation in vector spaces. In the current context, we could generalise the influence function given by Eq. 4 to the Gâteaux derivative

$$\mathrm{d}T(F, G) = \lim_{\epsilon \to 0} \frac{T\big((1-\epsilon)F + \epsilon G\big) - T(F)}{\epsilon},$$

where G is potentially any contaminating distribution. This potentially offers a useful way of analysing the influence of other kinds of departures from our distributional assumptions, though such a discussion is beyond the scope of this chapter. However, we will exploit a useful property of the influence function which it owes to this relationship. As with ordinary derivatives, a chain rule exists for Gâteaux derivatives, and therefore for influence functions. If we have a statistical functional which can be expressed in terms of J functionals such that

$$T(F) = T\big(T_1(F), \ldots, T_J(F)\big),$$

then the overall influence on the functional $T(F)$ of an infinitesimal peturbation of the data F at y is given by

$$L_T(y, \boldsymbol{x}) = \sum_{j=1}^{J} \frac{\partial T(F)}{\partial T_j(F)} L_j(y, \boldsymbol{x}), \tag{5}$$

where $T_j(F)$ is the jth functional, and L_j is its corresponding influence function.

Some key measures derived from the influence function include the *rejection point*, the *gross-error sensitivity*, and the *local-shift sensitivity*, discussed below. We also discuss the related concept of *leverage*.

3.1.1 Gross-Error Sensitivity

A key measure of the robustness of a functional is the *gross-error sensitivity*

$$\gamma_T^*(\boldsymbol{x}) = \inf_y \left|L_T(y, \boldsymbol{x})\right| \tag{6}$$

which is the infimum of the magnitude of the influence function over all points for which the influence function exists. If the gross-error sensitivity is finite, that is the influence is bounded, we say that a functional is *bias-robust* or simply robust; the lower the gross-error sensitivity, the more robust the model is to contaminating outliers.

3.1.2 Local-Shift Sensitivity

An alternative metric, the *local-shift sensitivity*

$$\lambda_T^*(\boldsymbol{x}) = \inf_{y \neq z} \frac{\left|L_T(y, \boldsymbol{x}) - L_T(z, \boldsymbol{x})\right|}{|y - z|}, \tag{7}$$

where the infimum is taken over all y, z for which $y \neq z$ and the influence functions exist. The local-shift sensitivity measures the sensitivity of the estimator to a *small* change in observed values.

3.1.3 Rejection Point

Another measure of interest is the *rejection point*

$$\rho_T^*(\boldsymbol{x}) = \inf_{r>0} \left\{r : L_T(y, \boldsymbol{x}) = 0, |y| > r\right\}, \tag{8}$$

which tells us how large an outlier must become before the influence function becomes zero. If the influence of gross outliers becomes zero, the effect is the same as removing them from the sample, hence the name *rejection point*.

3.1.4 Leverage

In regression-type settings which involve linear functions of vectors of covariates, such as our SF model described by Eq. 1, we are interested in the predicted values of the dependent variable for some $\boldsymbol{x}_i$, given by $\hat{y} = \boldsymbol{x}_i'\hat{\boldsymbol{\beta}}$, where $\hat{\boldsymbol{\beta}}$ is our estimator for $\boldsymbol{\beta}$. Predicted values are, of course, of particular interest in SF modelling, given our interest in estimating the distance of the firm from the frontier. The sensitivity of $\hat{y}_i$ to contaminating observations is therefore of particular interest. Given that we can think of $\hat{y}_i$ as a function of functionals as in Eq. 5, it has an influence function. Applying the influence function chain rule as given in Eq. 4, this is simply

$$L_{\hat{y}_i}(y, \boldsymbol{x}) = \boldsymbol{x}_i' \boldsymbol{L}_{\hat{\boldsymbol{\beta}}}(y, \boldsymbol{x}), \tag{9}$$

where $\boldsymbol{L}_{\hat{\boldsymbol{\beta}}}$ is the influence function for $\hat{\boldsymbol{\beta}}$. It is of natural interest here to consider the evaluation of $L_{\hat{y}_i}$ at $(y_i, \boldsymbol{x}_i)$,

$$L_{\hat{y}_i}(y_i, \boldsymbol{x}_i) = \boldsymbol{x}_i' \boldsymbol{L}_{\hat{\boldsymbol{\beta}}}(y_i, \boldsymbol{x}_i), \tag{10}$$

that is the influence on the predicted value for a observation i of a peturbation of the data at $(y_i, \boldsymbol{x}_i)$. This is known as the self-influence or *leverage* of the observation, and depends fundamentally on the on the values of the independent variables. The direct relationship between the leverage, the influence of an observation on $\hat{\boldsymbol{\beta}}$, and the values of the covariates is apparent from Eq. 10.

3.2 Breakdown Point

Aside from influence-based measures and definitions of robustness, there is a related but distinct concept of the *breakdown point*. The concept of the breakdown point, introduced by Hampel (1968, 1971), gives the smallest fraction of sampled observations that would be needed to make an estimator take on an arbitrary value. There is a distinction between the *replacement breakdown point* (the minimum fraction of existing observations one would need to contaminate) and the *addition breakdown point* (the fraction of additional contaminating observations one would need to add to an existing sample). Both asymptotic and finite sample definitions exist—see Donoho and Huber (1983). An estimator with a high breakdown point is said to be *resistant*.

The concept of the breakdown point is easily illustrated by contrasting the sample mean and sample median; the former has a breakdown point of $1/n$ in small samples and zero asymptotically, while the latter has a breakdown point of $1/2$. The sample mean and median therefore represent opposite extremes in terms of breakdown points and resistance to contaminating observations.

In terms of the asymptotic breakdown points of functionals, the asymptotic addition breakdown point is, from Huber (1981)

$$\epsilon_T^* = \inf \left\{ \epsilon > 0 : \sup_G \left| T\big((1-\epsilon)F + \epsilon G\big) - T(F) \right| = \infty \right\}, \tag{11}$$

where the supremum is taken over all possible contaminating distributions.

3.3 Summary of Measures

In this section, we have introduced several related concepts from the literature on robustness. With the exception of the breakdown point these measures relate to the influence of contaminating observations, which can be evaluated using the influence function describing the relationship between the influence of an observation and its value. The gross-error sensitivity is defined as the supremum of the magnitude of the influence function, while the local-shift sensitivity reflects its slope, and the rejection point is how large $|y|$ must be for influence to become zero.

The breakdown point offers an alternative approach. In contrast to the influence function, we are not aware of convenient formulae for the breakdown point of general classes of functionals, and therefore most of the discussion hereafter will focus on influence-based measures. The influence function also offers a more natural way to extend the discussion to the robustness of post-estimation predictions. It will, however, be useful to discuss breakdown points in the context of, e.g. conditional quantile regression estimation approaches that have been proposed in the SF literature.

4 Robustness and Stochastic Frontier Estimation

In this section, we move on to discuss SF estimation in light of the concepts from the robustness literature introduced in the previous Sect. 3. MLE is the workhorse of the SF literature, in terms of both the attention it has received in theoretical literature, and its useage in empirical applications. Other estimation methods we consider are the corrected ordinary least squares (COLS), and quantile regression (QR) approaches, and some generalisations of MLE. We discuss the robustness properties of each of these estimators in the context of SF modelling, and some of the approaches that have been taken to deal with outliers, including both alternative approaches to estimation and alteration of distributional assumptions.

This discussion is aided by the fact that all of the estimation methods under consideration belong to a broader class of *M-estimators*. Following Huber (1964, 1981), an M-estimator is any that can be defined as the solution to the problem

$$\hat{\boldsymbol{\theta}} = \arg\min_{\boldsymbol{\theta}} \sum_{i=1}^{n} \rho(y_i, \boldsymbol{x}_i, \boldsymbol{\theta}). \tag{12}$$

where n is the number of observations and ρ is a loss function. If the loss function has a derivative with respect to $\boldsymbol{\theta}$, denoted $\boldsymbol{\psi}$, we can define the M-estimator as the solution to the equation

$$\sum_{i=1}^{n} \boldsymbol{\psi}(y_i, \boldsymbol{x}_i, \hat{\boldsymbol{\theta}}) = \mathbf{0}, \qquad \boldsymbol{\psi}(y_i, \boldsymbol{x}_i, \hat{\boldsymbol{\theta}}) = \left.\frac{\partial \rho(y_i, \boldsymbol{x}_i, \boldsymbol{\theta})}{\partial \boldsymbol{\theta}}\right|_{\boldsymbol{\theta}=\hat{\boldsymbol{\theta}}}, \tag{13}$$

where $\mathbf{0}$ denotes a column vectors of zeros of the same length as $\boldsymbol{\theta}$. From Huber (1981), the influence function for an M-estimator is given by

$$L_{\hat{\theta}}(y, \boldsymbol{x}) = -\left(\mathbb{E}\left(\frac{\partial \boldsymbol{\psi}(y, \boldsymbol{x}, \boldsymbol{\theta})}{\partial \boldsymbol{\theta}'}\bigg|_{\theta=\hat{\theta}}\right)\right)^{-1} \boldsymbol{\psi}(y, \boldsymbol{x}, \hat{\boldsymbol{\theta}}). \tag{14}$$

This class of M-estimators encompasses many classical estimators, including MLE and ordinary least squares (OLS), two methods commonly employed in the SF literature. The robustness properties of M-estimators have been studied extensively, and several robust M-estimators have been proposed in the literature on robust estimation, though application of these in the SF literature has so far been limited.

4.1 *Corrected Ordinary Least Squares*

Where we have a stochastic frontier model of the form introduced in Sect. 2, the corrected ordinary least squares (COLS) estimation method proceeds by noting that, assuming independence of v_i and u_i, we may re-write Eq. 1 as

$$y_i = \alpha^* + \boldsymbol{x}_i^{*\prime}\boldsymbol{\beta}^* + \varepsilon_i^*, \qquad \varepsilon_i^* = v - su_i^*, \tag{15}$$

$$\alpha^* = \alpha - s\mathbb{E}(u_i), \quad u_i^* = u_i - \mathbb{E}(u_i), \quad \boldsymbol{\beta} = (\alpha, \boldsymbol{\beta}^{*\prime}), \quad \boldsymbol{x}_i = (1, \boldsymbol{x}_i^{*\prime})',$$

where $\mathbb{E}(\varepsilon_i^*) = 0$, and that therefore OLS may be used to obtain unbiased estimates of α^* and $\boldsymbol{\beta}^*$. That is, OLS yields unbiased estimates of all of the frontier parameters apart from the intercept, which is biased downward (upward) in a production (cost) frontier model by $\mathbb{E}(u_i)$. Parameters of the distributions of v_i and u_i are then obtained based on the sample moments of the distribution of the OLS residuals, given a set of distributional assumptions. The moment-based estimator of $\mathbb{E}(u_i)$ is then used to correct our estimated intercept, hence the name *corrected ordinary least squares*.[1] Solutions have been derived for several different models—see Aigner et al. (1977) and Olson et al. (1980) for the N-HN model, Greene (1980) for the N-EXP model, and Greene (1990) for the normal-gamma (N-G) model.

It is straightforward, however, to show that COLS is non-robust. If we partition the parameter vector such that

$$\boldsymbol{\theta} = (\alpha, \boldsymbol{\beta}^{*\prime}, \boldsymbol{\vartheta}')',$$

[1] Not to be confused with the *modified ordinary least squares* (MOLS) approach to estimating a deterministic frontier function. The two terms are often used interchangeably; for an extensive discussion, see Parmeter (2021).

where $\boldsymbol{\vartheta}$ is the vector of parameters of the distribution of the composed error, the COLS estimator $\tilde{\boldsymbol{\theta}}$ may expressed as a function of the OLS estimator

$$\tilde{\boldsymbol{\theta}} = \left(\hat{\alpha}^* + sg\big(\boldsymbol{\mu}(\hat{\boldsymbol{\theta}})\big), \hat{\boldsymbol{\beta}}^{*\prime}, \boldsymbol{h}\big(\boldsymbol{\mu}(\hat{\boldsymbol{\theta}})'\big)\right)', \qquad \hat{\boldsymbol{\theta}} = (\hat{\alpha}^*, \hat{\boldsymbol{\beta}}^{*\prime}), \tag{16}$$

where $\hat{\boldsymbol{\theta}}$ is the OLS estimator, $\boldsymbol{\mu}(\hat{\boldsymbol{\theta}})$ is a vector of moments of the distribution of estimated OLS residuals, and g and $\boldsymbol{h}$ are functions of the latter yielding our moment-based estimators of $\mathbb{E}(u_i)$ and $\boldsymbol{\vartheta}$, respectively. Applying the influence function chain rule gives

$$L_{\tilde{\theta}}(y, \boldsymbol{x}) = L_{\hat{\theta}}(y, \boldsymbol{x})' \frac{\partial \tilde{\boldsymbol{\theta}}}{\partial \hat{\boldsymbol{\theta}}'}, \tag{17}$$

from which it is clear that the robustness properties of COLS follow directly from those of OLS. The non-robustness of OLS is well-known, but one convenient way to demonstrate this is by recognising that OLS is an M-estimator where

$$\rho(y_i, \boldsymbol{x}_i, \boldsymbol{\theta}) = (y_i - \boldsymbol{x}_i'\boldsymbol{\theta})^2,$$

$$\boldsymbol{\psi}(y, \boldsymbol{x}, \hat{\boldsymbol{\theta}}) = -2(y - \boldsymbol{x}'\hat{\boldsymbol{\theta}})\boldsymbol{x}, \qquad \left.\frac{\partial \boldsymbol{\psi}(y, \boldsymbol{x}, \boldsymbol{\theta})}{\partial \boldsymbol{\theta}'}\right|_{\theta=\hat{\theta}} = 2\boldsymbol{x}\boldsymbol{x}',$$

and substituting these expressions in to Eq. 14 leads to a simple formula for influence

$$L_{\hat{\theta}}(y, \boldsymbol{x}) = (y - \boldsymbol{x}'\hat{\boldsymbol{\theta}})\big(\mathbb{E}(\boldsymbol{x}\boldsymbol{x}')\big)^{-1}\boldsymbol{x}, \tag{18}$$

which is clearly unbounded. The gross-error sensitivity of OLS is infinite, and it has the lowest possible breakdown point of $1/n$. Substituting Eq. 18 in to Eq. 17, we then see that COLS inherits these properties and is likewise non-robust.

This is unfortunate for two main reasons. First, COLS is otherwise an attractive estimator in its own right because of its simplicity and ease of implementation, and is possibly second only to MLE in terms of its usage and coverage in the SF literature. In addition, evidence from Monte Carlo experiments suggests that COLS may perform well relative to MLE in small samples (Olson et al., 1980). Second, COLS is often used as a means of obtaining starting values for iterative optimisation algorithms such as those used for MLE. The sensitivity of these starting values to contaminating outliers may be problematic given that in SF modelling, the log-likelihood is not always well-behaved. Similarly, obtaining good starting values can become particularly important in the context of certain robust M-estimators, for which multiple local optima may exist.

Robust Alternatives to COLS

The preceding discussion leads us to consider potential robust alternative to COLS. In principle we could estimate α^* and $\boldsymbol{\beta}^*$ in Eq. 15 using any linear unbiased estimator and, redefining $\hat{\boldsymbol{\theta}}$ as an alternative robust estimator, substitute in to Eq. 16 for the corresponding 'corrected' estimator for the SF model. To give a concrete example, we could use the least absolute deviations (LAD) estimator to obtain robust estimates of α^* and $\boldsymbol{\beta}^*$, and then estimate $\mathbb{E}(u_i)$ and remaining parameters of the marginal distribution of ε_i based on moments of the LAD residuals, and making the necessary correction to the intercept. By analogy with COLS, we might name this the CLAD estimator.

LAD is a well-known estimator with a long history, predating even OLS; for an extensive background and discussion see Dielman (2005). LAD is an M-estimator where

$$\rho(y_i, \boldsymbol{x}_i, \boldsymbol{\theta}) = |y_i - \boldsymbol{x}_i'\boldsymbol{\theta}|, \qquad \boldsymbol{\psi}(y, \boldsymbol{x}, \hat{\boldsymbol{\theta}}) = -\text{sgn}(y - \boldsymbol{x}'\hat{\boldsymbol{\theta}})\boldsymbol{x},$$

however are not able to derive the influence function via Eq. 14 owing to the singularity of the Hessian.[2] Following Koenker (2005), the influence function for the class of quantile regression estimators to which LAD belongs is given by

$$L_{\hat{\theta}}(y, \boldsymbol{x}) = \boldsymbol{Q}^{-1}\text{sgn}(y - \boldsymbol{x}'\hat{\boldsymbol{\theta}})\boldsymbol{x}, \tag{19}$$

$$\boldsymbol{Q} = \int \boldsymbol{x}\boldsymbol{x}' f(\boldsymbol{x}'\hat{\boldsymbol{\theta}})\mathrm{d}G(\boldsymbol{x}), \qquad \mathrm{d}F = \mathrm{d}G(\boldsymbol{x})f(y|\boldsymbol{x}),$$

and by comparing this to the OLS influence function, we can see that LAD gives less weight to outlying observations. Regarding the resistance of the estimator, a distinction needs to be made between the *finite sample breakdown point* or *conditional breakdown point* (Donoho & Huber, 1983) and the ordinary breakdown point. The conditional breakdown point considers contamination in y only, taking $\boldsymbol{x}$ as fixed, while the ordinary breakdown point considers contamination with respect to both y and $\boldsymbol{x}$. It has been shown that the conditional breakdown point of the LAD estimator can be greater than $1/n$ (He et al., 1990; Mizera & Müller, 2001; Giloni & Padberg, 2004). On the other hand, the ordinary breakdown point of LAD is $1/n$. Giloni et al. (2006) propose a weighted LAD estimator with a high breakdown point.

In addition to LAD, there are many other alternatives to OLS notable for their outlier robustness or resistance, which we could consider as the first stage in some corrected regression approach. A comprehensive discussion of robust and resistant regression estimators is beyond the scope of this chapter, but some prominent examples include M-estimation approaches, least median of squares (LMS) (Siegel, 1982), and least trimmed squares (LTS) (Rousseeuw, 1984). Early resistant regression techniques such as LMS and LTS have high breakdown points, but low efficiency. More recently, techniques have been developed with higher efficiency under the assumption of normality, though computational issues can be significant—for a review, see Yu and Yao (2017). Seaver and Triantis (1995) discuss the sensitivity of COLS to

[2] Another consequence of the singularity of the Hessian is that the usual asymptotic results cannot be applied. Bassett and Koenker (1978) show that LAD is asymptotically normal. A simpler proof is given by Pollard (1991).

outliers and compare production function estimates under OLS, LTS, LMS, and weighted least squares (WLS).

Of particular relevance in the context of SF modelling, small sample results from Lind et al. (1992) suggest that LAD significantly outperforms OLS in terms of bias and mean squared error when the error distribution is asymmetric. This suggests that CLAD or and corrected robust or resistant regression estimators could offer significant improvements over COLS.

4.2 Quantile Regression

An alternative estimation approach that has gained some attention in the SF literature in recent years is quantile regression (QR). Introduced by Koenker and Bassett (1978), the QR estimator is an M-estimator where the loss function corresponding to the τth conditional quantile is given by

$$\rho(y_i, \boldsymbol{x}_i, \boldsymbol{\theta}) = \rho_\tau(y_i - \boldsymbol{x}_i'\boldsymbol{\theta}), \qquad \rho_\tau(\varepsilon_i) = \varepsilon_i\big(\tau - \mathbb{I}_{\varepsilon_i<0}\big), \tag{20}$$

where $\mathbb{I}$ denotes the indicator function. Note that when $\tau = 0.5$, we have the conditional median or LAD estimator discussed in Sect. 4.1. Rather than estimating a conditional mean function as in OLS, we are estimating a conditional median or other quantile. The intuitive appeal of QR is clear, since in SF modelling we expect most of the observations to lie below (above) the estimated production (cost) function. Moreover, for an appropriate choice of τ, the estimator will be unbiased, even for the intercept. To see this, note that—assuming that ε_i is independent of $\boldsymbol{x}_i$—the τth conditional quantile of y_i is given by

$$Q(\tau|\boldsymbol{x}_i) = \boldsymbol{x}_i'\boldsymbol{\beta} + F_\varepsilon^{-1}(\tau),$$

where F_ε^{-1} denotes the quantile function of composed error distribution. The QR estimator of the intercept will then be biased by $F_\varepsilon^{-1}(\tau)$, so the problem is to choose some optimal τ^* such that

$$F_\varepsilon^{-1}(\tau^*) = 0,$$

which depends on specific distributional assumptions. Early applications of quantile regression to frontier estimation chose τ arbitrarily, e.g. Bernini et al. (2004), Knox et al. (2007), Liu et al. (2008), and Behr (2010). More recently, Jradi and Ruggiero (2019) noted that a formula for τ^* can be derived by evaluating F_ε, the distribution function for the composed error, at zero, showing that for the N-HN model this gives

$$\tau^* = \frac{1}{2} + s\frac{1}{\pi}\arctan\left(\frac{\sigma_u}{\sigma_v}\right),$$

and Jradi et al. (2021) show that for the N-EXP model

$$\tau^* = \frac{1}{2} + s \exp\left(\frac{1}{2}\left(\frac{\sigma_v}{\sigma_u}\right)^2\right)\Phi\left(-\frac{\sigma_u}{\sigma_v}\right),$$

meaning that we can calculate the appropriate quantile if we know the signal-to-noise ratio σ_u/σ_v. Jradi and Ruggiero (2019) suggest comparing likelihood values for different values of τ, while Jradi et al. (2019) choose the τ that minimises $\tau - \tau^*$. A simpler approach approach based on evaluating F_ε at the expected value of the OLS residuals is proposed by Jradi et al. (2021) and applied to the N-EXP model; Zhao (2021) implement this approach in the N-HN case. Bayesian approaches to QR estimation of the SF model are explored by Tsionas (2020) and Tsionas et al. (2020). For a recent in-depth review and discussion of the application of QR to estimate SF models, see Papadopoulos and Parmeter (2022).

Robustness Properties of Quantile Regression

The perceived robustness of QR estimation relative to MLE has been cited as a key motivation for the its use in a SF context. Papadopoulos and Parmeter (2022) argue that, under usual assumptions about independence of the errors from the regressors, the appeal of QR is diminished and "the one remaining advantage of quantile regression is its robustness to outliers". The robustness and resistance properties of QR have not been discussed in detail in the SF literature, and warrant further consideration. Of particular relevance is the fact that the choice of τ, which has been a focus of the previous discussion, has a direct bearing on the resistance of the estimator to outliers.

In a linear regression model, at least k points will lie exactly on the estimated regression plane under QR, where k is the number of regressors; in other words, $\hat{\varepsilon}_i = 0$ for at least k observations. If we can correctly guess the sign of the residuals, we can run the regression excluding all the observations above and below the estimated regression plane and obtain exactly the same results. This property has been exploited in order to design faster algorithms for QR estimation; see Portnoy and Koenker (1997) and Chernozhukov et al. (2022). But it is also helpful in understanding intuitively the effect of contaminating outliers on the estimator. When $\tau = 0.5$, contaminating outliers are likely to be points above or below the estimated QR plane; yet as we move to more extreme quantiles, it becomes increasingly likely that they will be included in the k observations interpolated by the plane. This presents a potential problem in the context of SF estimation where the quantile τ^* could plausibly be rather extreme, the problem becoming more acute when k is large relative to n.

The influence function and breakdown points of QR have been examined in some detail. As noted previously, the LAD estimator is a special case of QR that arises when $\tau = 0.5$. As such, much of our previous discussion of the properties of the LAD estimator in Sect. 4.1 applies to QR in general. The influence function is as given in Eq. 19, meaning that the influence of an contaminating observation depends only on its sign and $\boldsymbol{x}_i$. In that sense, in line with the previous discussion, QR is robust provided we don't venture too far into the tails of the distribution.

The ordinary breakdown point of QR is $1/n$ and thus no better than that of OLS, while the conditional breakdown point can be larger than $1/n$. This is easily understood in terms of the preceding discussion and the influence function, since we can potentially alter y_i for $n-k$ observations without influencing the estimator, so long as the residuals don't change sign, while even for a single observation, we could choose y_i and $\boldsymbol{x}_i$ such that the QR plane is forced to intersect it. Accordingly He et al. (1990) show that the conditional breakdown point of the QR estimator has an upper bound which decreases with τ.

To summarise, QR is a potentially robust estimation method, and may also have a conditional breakdown point over $1/n$. On the other hand, its resistance to outliers is reduced for more extreme quantiles, potentially lessening its appeal in the context of SF modelling. Further examination of this issue, and of the efficiency of the estimator under various distributional assumptions, is needed to assess the appeal of QR as a robust estimator of the SF model.

4.3 Maximum Likelihood Estimation

Beginning with Aigner et al. (1977) and Meeusen and van Den Broeck (1977), MLE has been the most commonly used estimation method in the SF literature. As mentioned previously, as with COLS and QR estimation, MLE belongs to the class of M-estimators. In the case of MLE

$$\rho(y_i, \boldsymbol{x}_i, \boldsymbol{\theta}) = -\ln f_\varepsilon(y_i - \boldsymbol{x}_i'\boldsymbol{\beta}, \boldsymbol{\theta}),$$
$$\psi(y, \boldsymbol{x}, \hat{\boldsymbol{\theta}}) = -\left.\frac{\partial \ln f_\varepsilon(y - \boldsymbol{x}'\boldsymbol{\beta}, \boldsymbol{\theta})}{\partial \boldsymbol{\theta}}\right|_{\theta=\hat{\theta}},$$

where f_ε denotes the density of the composed error. Substituting these expressions into Eq. 14, the influence function is given by

$$L_{\hat{\theta}}(y, \boldsymbol{x}) = -\left(\mathcal{I}(\hat{\boldsymbol{\theta}})\right)^{-1} \left.\frac{\partial \ln f_\varepsilon(y - \boldsymbol{x}'\boldsymbol{\beta}, \boldsymbol{\theta})}{\partial \boldsymbol{\theta}}\right|_{\theta=\hat{\theta}}, \tag{21}$$

where $\mathcal{I}(\hat{\boldsymbol{\theta}})$ denotes the Fisher information. It is straightforward to calculate the influence for each observation in this way, since both of the terms on the right-hand side of Eq. 15 are ordinarily available as by-products of MLE in most statistical and econometric software packages: the inverse information may be estimated by the covariance matrix which, following the Cramér-Rao Theorem, is typically the inverse of the observed information, and the second term is simply the observation-level score vector.

It is useful at this point to dispel a common misconception that MLE is non-robust in general. In some cases, this misconception arises because the distinction between robustness and resistance is not always followed, and MLE is sometimes said to be 'non-robust' because of its $1/n$ breakdown point (i.e. because it is non-resistant). In fact, this is a property shared by most M-estimators,[3] even those that are robust in the sense of having finite gross-error sensitivity. In other cases, it is perhaps taken for granted that the gross-error sensitivity will always be infinite under MLE, but as we will see, though that is often the case, it is not always so.

From Eq. 21 we can see that the influence function for the maximum likelihood estimator is simply a linear transformation, by a matrix of constants, of the observation-level score vector. In order to determine whether or not MLE is robust, it is therefore sufficient to check whether the score vector is bounded. This is true of M-estimators generally, but in the case of MLE, this depends entirely on the assumed error distribution. It is therefore of interest to consider how the behaviour of the influence function differs under various assumptions considered in the SF literature, and whether or not some distributional assumptions can be considered robust. We will begin by discussing the canonical cases such as the N-HN and N-EXP specifications, which have received the most attention in the literature, and then consider potentially robust alternatives.

Influence Under Canonical Distributional Assumptions

As discussed in Sect. 2, Aigner et al. (1977) and Meeusen and van Den Broeck (1977) originally considered the N-HN and N-EXP cases. An immediate generalisation of the N-HN model is the normal-truncated normal (N-TN) model considered by Stevenson (1980). In each of these cases, f_ε has a convenient closed-form expression, making derivation of the influence function particularly straightforward. Given this, and the large amount of attention given to the N-HN and N-EXP models in the SF literature, it is worth discussing their robustness properties specifically. The 'wrong skew' problem has particular relevance in these cases.

As shown by Waldman (1982), the OLS estimator is a stationary point in the log-likelihood function in the N-HN case, and a sufficient condition for the stability of this stationary point is that the skewness of the OLS residuals has the wrong sign.[4] Horrace and Wright (2020) generalise this result to the N-TN model and the normal-doubly truncated normal (N-DTN) model proposed by Almanidis et al. (2014), and show that the OLS estimator is also a stable stationary point in the N-EXP case. This is well-known in the SF literature as the 'wrong skew' problem. It follows immediately MLE of the SF model is non-robust in these cases: it is straightforward to show via the influence function chain rule that the skewness of the OLS residuals is itself a non-robust functional, implying that even a single contaminating outlier in the 'wrong' direction may result in the OLS estimator being returned. In the event of wrong skew, the influence function therefore becomes that of the OLS estimator.

In other instances, the influence function is more complex. For the N-HN model, where $\boldsymbol{\theta} = (\boldsymbol{\beta}, \sigma_v, \sigma_u)'$,

---

[3] Other M-estimation techniques, e.g. QR, are therefore no more 'robust' than MLE in this sense.

[4] Assuming independence of v_i and u_i, and that the distribution of u_i is positively skewed, $\text{sgn}(\text{Skew}(v_i - su_i)) = \text{sgn}(-s)$, and therefore we expect the OLS residuals to be negatively skewed in the production frontier case, and positively skewed in the cost function case.

$$L_{\hat{\theta}}(y, \boldsymbol{x}) = -\left(\mathcal{I}(\hat{\boldsymbol{\theta}})\right)^{-1} \begin{pmatrix} -\dfrac{\hat{\sigma}_u/\hat{\sigma}_v}{\sqrt{\hat{\sigma}_v^2 + \hat{\sigma}_u^2}}\left(\left(\dfrac{\hat{\sigma}_v}{\hat{\sigma}_u}\right)^2 z + sh(z)\right)\boldsymbol{x} \\ \dfrac{\hat{\sigma}_v}{\hat{\sigma}_v^2 + \hat{\sigma}_u^2}\left(1 - 2\left(\dfrac{\hat{\sigma}_v}{\hat{\sigma}_u}\right)^2 z^2 - s\left(2 + \left(\dfrac{\hat{\sigma}_u}{\hat{\sigma}_v}\right)^2\right) zh(z)\right) \\ \dfrac{\hat{\sigma}_u}{\hat{\sigma}_v^2 + \hat{\sigma}_u^2}\left(1 - 2\left(\dfrac{\hat{\sigma}_v}{\hat{\sigma}_u}\right)^2 z^2 + s\left(\dfrac{\hat{\sigma}_v}{\hat{\sigma}_u}\right)^2 zh(z)\right) \end{pmatrix}, \tag{22}$$

$$h(z) = \frac{\phi(sz)}{1 - \Phi(sz)}, \qquad z = \frac{y - \boldsymbol{x}'\hat{\boldsymbol{\beta}}}{\sqrt{\hat{\sigma}_v^2 + \hat{\sigma}_u^2}} \frac{\hat{\sigma}_u}{\hat{\sigma}_v},$$

where ϕ and Φ are the standard normal probability density and distribution functions, respectively. From this, we can conclude that MLE is non-robust in the N-HN case generally, since each element of the influence function is unbounded, approaching infinity as we increase $\hat{\varepsilon}_i$ in either direction—though the influence increases more slowly for outliers below (above) the production (cost) frontier, there is no robustness to outliers caused by extremely inefficient firms.

The unboundedness of the influence function in the N-TN case is similar and has previously been shown by Song et al. (2017). In the N-EXP case, intuition suggests the model should be better at handling outliers, since the exponential distribution has heavier tails than the half normal distribution. However the influence function for the N-EXP model

$$L_{\hat{\theta}}(y, \boldsymbol{x}) = -\left(\mathcal{I}(\hat{\boldsymbol{\theta}})\right)^{-1} \begin{pmatrix} s\left(\dfrac{1}{\hat{\sigma}_v} h\left(z - \dfrac{\hat{\sigma}_v}{\hat{\sigma}_u}\right) - \dfrac{1}{\hat{\sigma}_u}\right)\boldsymbol{x} \\ \dfrac{\hat{\sigma}_v}{\hat{\sigma}_u^2} - \dfrac{s}{\hat{\sigma}_v}\left(z + \dfrac{\hat{\sigma}_v}{\hat{\sigma}_u}\right) h\left(z - \dfrac{\hat{\sigma}_v}{\hat{\sigma}_u}\right) \\ \dfrac{1}{\hat{\sigma}_u}\left(\dfrac{\hat{\sigma}_v}{\hat{\sigma}_u} h\left(z - \dfrac{\hat{\sigma}_v}{\hat{\sigma}_u}\right) - 1 + z - \dfrac{\hat{\sigma}_v^2}{\hat{\sigma}_u^2}\right) \end{pmatrix}, \tag{23}$$

$$h\left(z - \frac{\hat{\sigma}_v}{\hat{\sigma}_u}\right) = \frac{\phi\left(z - \frac{\hat{\sigma}_v}{\hat{\sigma}_u}\right)}{\Phi\left(z - \frac{\hat{\sigma}_v}{\hat{\sigma}_u}\right)}, \qquad z = -s\frac{y - \boldsymbol{x}'\hat{\boldsymbol{\beta}}}{\hat{\sigma}_u},$$

is likewise unbounded, though it is apparent that influence of outlying observations ought to be less than in the N-EXP case given the contrast between the linear functions of the error term found in Eq. 23 and the quadratic functions that appear in Eq. 22.

Alternative Distributional Assumptions

As mentioned in Sect. 2, many alternative distributional assumptions have been suggested in the SF literature. Following the intuition that outliers should be modelled as instances of large $|v_i|$ rather than contaminating the distribution of u_i, several studies have explored the use of noise distributions with heavier tails than the usual normal distribution, which ought to better accomodate contaminating outliers.

In an early example of this approach, Janssens and van Den Broeck (1993) suggest the use of an 'approximative t' distribution, and derive f_ε for an approximative t-exponential (AT-EXP) SF model. However the authors do not operationalise the model, and to our knowledge its use remains unexplored. More recent proposals for the heavy-tailed noise distributions include the logistic (Stead et al., 2018), Laplace (Nguyen, 2010; Horrace & Parmeter, 2018), Cauchy (Gupta & Nguyen, 2010), scale contaminated normal (Stead et al., Forthcoming), and Student's t (Tancredi, 2002; Wheat et al., 2019) distributions.[5] Note that the contaminated normal and Student's t distributions contain the the normal distribution is nested as a limiting case; Stead et al. (Forthcoming) and Wheat et al. (2019) discuss testing against the N-HN model, which can be interpreted as testing for heavy-tails in the noise distribution.

A common characteristic of these sub-Gaussian SF models is the atypical tail behaviour of the efficiency predictor, which is discussed in Sect. 5. In terms of parameter estimates, results suggest that adopting such distributions for v_i can lead to substantive changes relative to the normal case, even when assumptions about u_i are unaltered. In empirical applications, estimates from the logistic-half normal (LOG-HN), contaminated normal-half normal (CN-HN) and Student's t-half normal (T-HN) models can differ significantly from the N-HN estimates (Stead et al., 2018, Forthcoming; Wheat et al., 2019), especially with respect to σ_u.

An interesting feature of the aforementioned sub-Gaussian noise distributions is that each may be characterised as a scale mixture of normal distributions. That is, in each case

$$f_v(v_i, \boldsymbol{\theta}_v) = \int_0^\infty \frac{1}{\varsigma\sigma_v}\phi\Big(\frac{v_i}{\varsigma\sigma_v}\Big) f_\varsigma(\varsigma)\mathrm{d}\varsigma,$$

where ϕ is the standard normal density and f_ς is the density of the mixing distribution. In the degenerate case, we recover the normal distribution

$$f_\varsigma(\varsigma) = \delta(\varsigma - 1) \Longrightarrow f_v(v_i, \boldsymbol{\theta}_v) = \frac{1}{\sigma_v}\phi\Big(\frac{v_i}{\sigma_v}\Big),$$

while the CN-HN case explored by (Stead et al., Forthcoming) is obtained by assuming that ς follows a generalised multinomial distribution, such that

$$f_\varsigma(\varsigma) = \sum_j [\varsigma = \varsigma_j] p_j \Longrightarrow f_v(v_i, \boldsymbol{\theta}_v) = \sum_j p_j \frac{1}{\varsigma_j\sigma_v}\phi\Big(\frac{v_i}{\varsigma_j\sigma_v}\Big).$$

This discrete approach may be used to approximate any mixing distribution arbitrarily well, though at the cost of increasing the number of parameters to be estimated in more flexible cases. Moving from a finite to an infinite mixture setting, Stefanski (1991) shows that the logistic distribution is obtained when the mixing distribution is Kolmogorov-Smirnov; that is

[5] For packages facilitating estimation some of these models, see https://github.com/AlexStead.

$$f_\varsigma(\varsigma) = 2\varsigma \sum_{m=1}^{\infty} (-1)^{m+1} m^2 e^{-\frac{1}{2}(m\varsigma)^2} \implies f_v(v_i, \boldsymbol{\theta}_v) = \frac{\exp\left(-\frac{v_i}{\sigma_v}\right)}{\sigma_v\left(1+\exp\left(-\frac{v_i}{\sigma_v}\right)\right)^2}.$$

Similarly, Andrews and Mallows (1974) show that the Laplace distribution is obtained when the mixture distribution is exponential with unit variance, such that

$$f_\varsigma(\varsigma) = \frac{1}{2}e^{-\frac{\varsigma}{2}} \implies f_v(v_i, \boldsymbol{\theta}_v) = \frac{1}{2\sigma_v}\exp\left(-\frac{|v_i|}{\sigma_v}\right),$$

and that the Student's t distribution is obtained when ς^2 follows an inverse gamma distribution with scale and shape parameter both equal to $\frac{1}{2\alpha}$, such that

$$f_\varsigma(\varsigma) = \frac{(2\alpha\sqrt{\varsigma})^{-\frac{1}{2\alpha}}}{\Gamma\left(\frac{1}{2\alpha}\right)\varsigma} e^{-\frac{1}{2\alpha\sqrt{\varsigma}}} \implies f_v(v_i, \boldsymbol{\theta}_v) = \sqrt{\frac{2\alpha}{\pi}}\frac{\Gamma\left(\frac{2\alpha+1}{4\alpha}\right)}{\Gamma\left(\frac{1}{2\alpha}\right)\sigma_v}\left(1+2\alpha\left(\frac{v_i}{\sigma_v}\right)^2\right)^{-\frac{2\alpha+1}{4\alpha}}, \tag{24}$$

where α is an inverse degrees of freedom parameter.[6] This characterisation is useful since it provides a way of conceptualising these alternative specifications in terms of departures from normality driven by explicit contamination models. Assuming independence of v_i and u_i, the marginal density of ε_i is given by

$$f_\varepsilon(y_i - \boldsymbol{x}_i'\boldsymbol{\beta}, \boldsymbol{\theta}) = \int_0^\infty \int_0^\infty \frac{1}{\varsigma\sigma_v}\phi\left(\frac{y_i - \boldsymbol{x}_i'\boldsymbol{\beta} + su_i}{\varsigma\sigma_v}\right) f_\varsigma(\varsigma) f_u(u, \boldsymbol{\theta}_u)\mathrm{d}\varsigma\mathrm{d}u_i.$$

Making use of the Fubini-Tonelli theorem, we can interchange the order of integration and then integrate out u_i in order to express the contaminated density as

$$f_\varepsilon(y_i - \boldsymbol{x}_i'\boldsymbol{\beta}, \boldsymbol{\theta}) = \int_0^\infty f_\varepsilon^*\left(y_i - \boldsymbol{x}_i'\boldsymbol{\beta}, (\boldsymbol{\beta}', \varsigma\sigma_v, \boldsymbol{\theta}'_u)'\right) f_\varsigma(\varsigma)\mathrm{d}\varsigma, \tag{25}$$

where f_ε^* denotes the corresponding density that arises when $v_i \sim \mathcal{N}(0, \sigma_v^2)$. For example, in the T-HN case, the expression above becomes

$$f_\varepsilon(y_i - \boldsymbol{x}_i'\boldsymbol{\beta}, \boldsymbol{\theta}) = \int_0^\infty \frac{2}{\sigma}\phi\left(\frac{y_i - \boldsymbol{x}_i'\boldsymbol{\beta}}{\sigma}\right)\Phi\left(\frac{-(y_i - \boldsymbol{x}_i'\boldsymbol{\beta})\lambda}{\sigma}\right)\frac{(2\alpha\sqrt{\varsigma})^{-\frac{1}{2\alpha}}}{\Gamma\left(\frac{1}{2\alpha}\right)\varsigma} e^{-\frac{1}{2\alpha\sqrt{\varsigma}}}\mathrm{d}\varsigma,$$

$$\sigma = \sqrt{\varsigma^2\sigma_v^2 + \sigma_u^2}, \qquad \lambda = \frac{\sigma_u}{\varsigma\sigma_v}.$$

This result is potentially useful in solving for or approximating $f_\varepsilon(\varepsilon_i, \boldsymbol{\theta})$, but also suggests an alternative way of conceptualising MLE under these alternative distri-

[6] Wheat et al. (2019) parameterise the T-HN model in terms of the degrees of freedom $\upsilon = \frac{1}{\alpha}$, but for the purposes of the following discussion, the inverse will be more useful.

butional assumptions; if the true distribution is such that $v_i \sim \mathcal{N}(0, \sigma_v^2)$, we are effectively maximising a pseudolikelihood that reflects an particular contamination model.

Simulation evidence suggests that these models perform well under misspecification—Horrace and Parmeter (2018) show that the Laplace-Exponential (L-EXP) performs better when the true data generating process (DGP) is N-EXP than vice-versa, and likewise Wheat et al. (2019) show that that Student's t-half normal (T-HN) model performs well when the DGP is N-HN, while the N-HN model performs poorly when the DGP is T-HN. In addition, the L-EXP and T-HN models appear to perform better than the N-EXP and N-HN models in the presence of 'wrong skew' (Horrace & Parmeter, 2018; Wheat et al., 2019).

While simulation results offer an encouraging indication that sub-Gaussian SF models may be less sensitive to misspecification in a general sense, none of the aforementioned studies examine the behaviour of the influence function under their models explicitly, leaving the actual robustness of the various specifications to outliers an open question. Examination of the properties of influence function in each case would be tedious, especially in cases where f_ε lacks a convenient analytical expression. Such an approach would also shed relatively little light on how we might choose distributional assumptions with robustness in mind.

Recent results from Stead et al. (2023) are useful in this respect. The authors derive some sufficient conditions for boundedness of the score vector and hence the robustness of MLE of the SF model, and discuss distributional assumptions which satisfy these. An exhaustive discussion of these conditions is beyond the scope of this chapter, however they relate to the boundedness of the logarithmic derivatives of the joint density function $f_{v,u}$. The authors show that these conditions are not satisfied under the canonical distributional assumptions—e.g. N-HN, N-EXP—nor even in most of the sub-Gaussian cases just discussed.

In fact, of the proposed alternative noise distributions discussed, only the Student's t distribution (including its Cauchy special case) is found to satisfy the Stead et al. (2023) conditions for robust MLE. Under the assumption of independence, it may be paired with any one-parameter scale family—e.g. half normal, exponential—for u_i and these conditions will be satisfied. A particularly attractive feature of the Student's t model is that it appears to be robust to contamination in both y and $\boldsymbol{x}$. In other words, it is robust to both outliers and leverage points, and has an ordinary breakdown point greater than $1/n$; its robustness properties in this sense are better than those of the QR estimator. It should be noted that this result may depend on treating the 'degrees of freedom' or shape parameter as a fixed tuning parameter, rather than estimating it via MLE as in Wheat et al. (2019). The authors note that proposed flexible distributions of u_i with two or more parameters, e.g. gamma, do not satisfy their conditions, nor do copulae that have been proposed for modelling dependence between v_i and u_i.

These results demonstrate that simply adopting a sub-Gaussian distribution for v_i, such as the logistic or Laplace distributions, is not enough to ensure robustness of MLE. On the other hand, they indicate the flexibility of the Student's t distribution in robust MLE of the SF model. The T-HN and Student's t-exponential (T-EXP) models may be estimated robustly via MLE, and the results may be used as a guide to identify further robust pairings. Usefully, this offers a path to robust estimation while remaining within the framework of MLE.

4.4 Alternative M-estimators

One particular class of M-estimators generalises MLE by specifying some alternative loss function containing $-\ln f_\varepsilon$ as a limiting case, such that

$$\rho(y_i, \boldsymbol{x}_i, \boldsymbol{\theta}) = \begin{cases} \rho_\alpha(y_i, \boldsymbol{x}_i, \boldsymbol{\theta}), & \alpha > 0 \\ -\ln f_\varepsilon(y_i - \boldsymbol{x}'_{i\beta}, \boldsymbol{\theta}), & \alpha = 0 \end{cases}, \tag{26}$$

$$\boldsymbol{\psi}_\alpha(y, \boldsymbol{x}, \hat{\boldsymbol{\theta}}) = \left.\frac{\partial \rho_\alpha(y, \boldsymbol{x}, \boldsymbol{\theta})}{\partial \boldsymbol{\theta}}\right|_{\boldsymbol{\theta}=\hat{\boldsymbol{\theta}}},$$

where $\alpha \geq 0$ is some tuning parameter and $\boldsymbol{\psi}_\alpha(y, \boldsymbol{x}, \hat{\boldsymbol{\theta}})$ is bounded when $\alpha > 0$. The gross-error sensitivity and efficiency of these estimators both decrease as α increases, creating a trade-off between robustness and efficiency. A loss function $\rho_\alpha(y_i, \boldsymbol{x}_i, \boldsymbol{\theta})$ is chosen such that this trade-off is minimised.

Examples of this approach include minimum density power divergence estimation (MDPDE) (Basu et al., 1998), in which

$$\rho_\alpha(y_i, \boldsymbol{x}_i, \boldsymbol{\theta}) = \int f_\varepsilon^{1+\alpha}(y - \boldsymbol{x}_i'\boldsymbol{\beta}, \boldsymbol{\theta})\mathrm{d}y + \frac{\alpha+1}{\alpha} f_\varepsilon^\alpha(y_i - \boldsymbol{x}_i'\boldsymbol{\beta}, \boldsymbol{\theta}).$$

Comparable methods are maximum L_q-likelihood estimation (ML_qLE) (Ferrari & Yang, 2010) and maximum Ψ-likelihood estimation (MΨLE) (Eguchi & Kano, 2001; Miyamura & Kano, 2006). ML_qLE replaces the logarithm of the density with its Box-Cox transformation

$$\rho_\alpha(y_i, \boldsymbol{x}_i, \boldsymbol{\theta}) = -\frac{\alpha+1}{\alpha}\left(f_\varepsilon^{\frac{\alpha}{\alpha+1}}(y_i - \boldsymbol{x}_i'\boldsymbol{\beta}, \boldsymbol{\theta}) - 1\right),$$

while (MΨLE) makes one of several transformations of the likelihood. Miyamura and Kano (2006) propose

$$\rho_\alpha(y_i, \boldsymbol{x}_i, \boldsymbol{\theta}) = \frac{1}{\alpha+1}\int f_\varepsilon^{\alpha+1}(y - \boldsymbol{x}_i'\boldsymbol{\beta}, \boldsymbol{\theta})\mathrm{d}y - \frac{1}{\alpha} f_\varepsilon^\alpha(y_i - \boldsymbol{x}_i'\boldsymbol{\beta}, \boldsymbol{\theta}),$$

which, up to a constant factor, is identical to the minimum density power divergence estimator. MDPDE, ML_qLE, and MΨLE are equivalent to maximising weighted likelihood functions, where outlying observations are downweighted.[7]

The use of MDPDE as a robust estimator of the SF model is explored by Song et al. (2017), who provide simulation evidence suggesting that the estimator outperforms MLE in the presence of contaminating outliers, as expected, but also that its small-sample performance is comparable to that of MLE. Similar results with respect to MΨLE are shown by Bernstein et al. (2021),[8] who also explore use of ML_qLE, for which the results are by contrast mixed.

We can understand these approaches as maximising some quasi-likelihood function—see White (1982). As discussed in Sect. 4.3, MLE under alternative distributional assumptions can be conceptualised in the same way. In particular, following Eqs. 24 and 25, if $v_i \sim \mathcal{N}(0, \sigma_v^2)$ then the Student's t model can be understood as a Student's t-based robust M-estimator such that,

$$\rho_\alpha(y_i, \boldsymbol{x}_i, \boldsymbol{\theta}) = -\ln \int_0^\infty f_\varepsilon\big(y_i - \boldsymbol{x}_i'\boldsymbol{\beta}, (\boldsymbol{\beta}, \varsigma\sigma_v, \boldsymbol{\theta}_u)'\big) \frac{(2\alpha\varsigma)^{-\frac{1}{2\alpha}}}{\Gamma\big(\frac{1}{2\alpha}\big)\varsigma} e^{-\frac{1}{2\alpha\varsigma}} \mathrm{d}\varsigma,$$

which fits the general framework described by Eq. 26. As with MDPDE, MΨLE, and ML_qLE, we have an tuning parameter α which governs the trade-off between robustness and efficiency, and $-\ln f_\varepsilon$ is recovered when $\alpha = 0$. This provides an alternative motivation for the Student's t model as a robust M-estimator based on a particular model of contamination. This is advantageous, since the contamination model is then incorporated into efficiency prediction in a consistent way, by use of the corresponding Student's t based efficiency predictor. By contrast, other robust M-estimation methods leave robust efficiency prediction as a separate problem.

Further investigation would be useful to compare the performance of these methods in estimating SF models, in terms of robustness to differing contamination models and the trade-off between robustness and efficiency. Intuition suggests that these quasi-likelihood methods may outperform QR and similar methods. The choice of α is crucial in each case. Song et al. (2017) discuss the choice of α under MDPDE, and follow Durio and Isaia (2011) in using an approach to select α based on a measure of the similarity of MDPDE and MLE results. Bernstein et al. (2021) discuss the choice of α under ML_qLE, noting that the estimator is biased except when $\alpha = 0$ and suggest setting α equal to a function of n such that $\alpha \to 0$ in large samples, though in this case we approach MLE. For the Student's t based M-estimator, Wheat et al. (2019) discuss hypothesis testing in the case that α is estimated directly via MLE; if we treat it instead as a fixed tuning parameter, information criteria could be used.

---

[7] In the case of ML_qLE, Ferrari and Yang (2010) use a different formulation for in terms of $q = \frac{1}{1+\alpha}$ and allow for $q > 1$, which effectively upweights outliers.

[8] Note that the authors use the Miyamura and Kano (2006) transformation. As noted previously, this is equivalent to MDPDE, so the similarity of these results is to be expected.

5 Robustness and Efficiency Prediction

As discussed previously, the robustness literature is concerned with ensuring that departures from our model assumptions, such as contaminating observations, cannot push some functional to a boundary of its sample space. In the case of functionals that are defined as measures of technical or cost efficiency, these must belong to the interval (0, 1]. In most cases we might consider a finding that *all* firms are on the frontier, in other words that sample mean efficiency is 1, to be unrealistic. Such a situation can only arise when we are at a boundary of the parameter space, and is an issue of the robustness of the estimation method.

We will be concerned if efficiency predictions start approaching zero. On the other hand, when it comes to predicting firm-specific efficiency scores we may not regard an efficiency prediction of 1 as problematic; identification of firms on the frontier may be of particular interest in some applications. However, when seeking to identify the most efficient firms, we may wish to exclude extreme outliers. This motivates consideration of potentially robust efficiency predictors that are not unduly sensitive to extreme outliers.

As discussed in Sect. 2, firm-specific efficiency prediction is based on the conditional distribution given by Eq. 2. Since the true parameter values are not known in practice, we use some estimator $\hat{\boldsymbol{\theta}}$. By definition, the influence of a contaminating point $(y_i, \boldsymbol{x}_i)$ on some predictor $\hat{u}_j$ of (the natural logarithm of) efficiency evaluated for the jth observation is

$$L_{\hat{u}_j}(y_i, \boldsymbol{x}_i) = \hat{u}_j(y_j - \boldsymbol{x}'_j\hat{\boldsymbol{\beta}}_i, \hat{\boldsymbol{\theta}}_i) - \hat{u}_j(y_j - \boldsymbol{x}'_j\hat{\boldsymbol{\beta}}, \hat{\boldsymbol{\theta}})$$

where $\hat{\boldsymbol{\theta}}_i = \left(\hat{\boldsymbol{\beta}}'_i, \hat{\boldsymbol{\vartheta}}'_i\right)'$ and $\hat{\boldsymbol{\theta}} = \left(\hat{\boldsymbol{\beta}}', \hat{\boldsymbol{\vartheta}}'\right)'$ denote the estimator including and excluding $(y_i, \boldsymbol{x}_i)$, respectively. From this, we can derive the expression

$$\begin{aligned}\hat{u}_l(y_l - \boldsymbol{x}'_l\hat{\boldsymbol{\beta}}_k, \hat{\boldsymbol{\theta}}_k) - \hat{u}_j(y_j - \boldsymbol{x}'_j\hat{\boldsymbol{\beta}}_i, \hat{\boldsymbol{\theta}}_i) =& \hat{u}_l(y_l - \boldsymbol{x}'_l\hat{\boldsymbol{\beta}}, \hat{\boldsymbol{\theta}}) - \hat{u}_j(y_j - \boldsymbol{x}'_j\hat{\boldsymbol{\beta}}, \hat{\boldsymbol{\theta}}) \\ &+ L_{\hat{u}_l}(y_k, \boldsymbol{x}_k) - L_{\hat{u}_j}(y_i, \boldsymbol{x}_i). \quad (27)\end{aligned}$$

When $l = j$, Eq. 27 gives us the influence on the prediction for the jth observation of removing a contaminating point $(y_i, \boldsymbol{x}_i)$ and replacing it with another, $(y_k, \boldsymbol{x}_k)$,

$$\hat{u}_j(y_j - \boldsymbol{x}'_j\hat{\boldsymbol{\beta}}_k, \hat{\boldsymbol{\theta}}_k) - \hat{u}_j(y_j - \boldsymbol{x}'_j\hat{\boldsymbol{\beta}}_i, \hat{\boldsymbol{\theta}}_i) = L_{\hat{u}_j}(y_k, \boldsymbol{x}_k) - L_{\hat{u}_j}(y_i, \boldsymbol{x}_i), \quad (28)$$

while if $j = i, l = k$, we have an expression for the effect of replacing $(y_i, \boldsymbol{x}_i)$ with $(y_k, \boldsymbol{x}_k)$ on the efficiency predictor evaluated at the contaminating point

$$\begin{aligned}\hat{u}_k(y_k - \boldsymbol{x}'_k\hat{\beta}_k, \hat{\boldsymbol{\theta}}_k) - \hat{u}_i(y_i - \boldsymbol{x}'_i\hat{\boldsymbol{\beta}}_i, \hat{\boldsymbol{\theta}}_i) =& \hat{u}_l(y_k - \boldsymbol{x}'_{k\hat{\beta}}, \hat{\boldsymbol{\theta}}) - \hat{u}_i(y_i - \boldsymbol{x}'_{i\hat{\beta}}, \hat{\boldsymbol{\theta}}) \\ &+ L_{\hat{u}_k}(y_k, \boldsymbol{x}_k) - L_{\hat{u}_i}(y_i, \boldsymbol{x}_i). \quad (29)\end{aligned}$$

Equation 28 may be interpreted as the effect of changing the ith observation on the jth efficiency prediction where $i \neq j$, and depends only on the change in influence, whereas Eq. 29 gives the effect of changing the ith observation on its own efficiency prediction, which depends also on the direct effect of evaluating the predictor at a different point. Given that $\hat{u}_i$ is a function of $\hat{\boldsymbol{\theta}}$, the influence of some contaminating observation $(y, \boldsymbol{x})$ on $\hat{u}_i$ can be derived via the influence function chain rule

$$L_{\hat{u}_i}(y, \boldsymbol{x}) = \frac{\partial \hat{u}_i}{\partial \hat{\boldsymbol{\theta}}'} L_{\hat{\theta}}(y, \boldsymbol{x}). \tag{30}$$

From Eq. 30 we can see that one sufficient, though not necessary, condition for the robustness of $\hat{u}_i$ is that robustness of $\hat{\boldsymbol{\theta}}$ and boundedness of $L_{\hat{\theta}}(y, \boldsymbol{x})$. We will now consider various predictors and their robustness properties, in terms of both their influence functions and their tail behaviour.

5.1 Conditional Mean

The most commonly used efficiency predictor is the conditional mean. Following Jondrow et al. (1982), this is given by

$$\mathbb{E}(u_i | y_i - \boldsymbol{x}_i'\boldsymbol{\beta})\Big|_{\theta=\hat{\theta}} = \int_0^\infty u_i f_{u|\varepsilon}(u_i | y_i - \boldsymbol{x}_i'\boldsymbol{\beta}, \boldsymbol{\theta}) \mathrm{d}u_i \Big|_{\theta=\hat{\theta}}, \tag{31}$$

or alternatively following Battese and Coelli (1988), we use

$$\mathbb{E}(e^{-u_i} | y_i - \boldsymbol{x}_i'\boldsymbol{\beta})\Big|_{\theta=\hat{\theta}} = \int_0^\infty e^{-u_i} f_{u|\varepsilon}(u_i | y_i - \boldsymbol{x}_i'\boldsymbol{\beta}, \boldsymbol{\theta}) \mathrm{d}u_i \Big|_{\theta=\hat{\theta}}.$$

It will be convenient to limit discussion to the Jondrow et al. (1982) predictor. From Jensen's inequality, we can see that

$$\exp\big(-\mathbb{E}(u_i | y_i - \boldsymbol{x}_i'\boldsymbol{\beta})\big) \leq \mathbb{E}(e^{-u_i} | y_i - \boldsymbol{x}_i'\boldsymbol{\beta}),$$

and in practice the difference between the two predictors is usually negligible. Let us consider the influence of contaminating observations on the Jondrow et al. (1982) predictor given by Eq. 27. From Eq. 30, the influence function is given by

$$L_{\mathbb{E}(u_i|\varepsilon_i)}(y, \boldsymbol{x}) = \frac{\partial \mathbb{E}(u_i | y_i - \boldsymbol{x}_i'\boldsymbol{\beta})}{\partial \boldsymbol{\theta}}\Big|_{\theta=\hat{\theta}} L_{\hat{\theta}}(y, \boldsymbol{x}).$$

The robustness of the conditional mean efficiency predictor therefore depends not only on the robustness of the estimator $\hat{\boldsymbol{\theta}}$ but also on the derivative of the predictor with respect to the estimated parameter vector, which will depend on the model's distributional assumptions. Again, it would be tedious to examine this derivative under all proposed distributional assumptions, but in the N-HN case

$$\mathbb{E}(u_i|y_i - \boldsymbol{x}_i'\boldsymbol{\beta})\bigg|_{\theta=\hat{\theta}} = \frac{\hat{\sigma}_v\hat{\sigma}_u}{\sqrt{\hat{\sigma}_v^2+\hat{\sigma}_u^2}}\left(\frac{\phi(z_i)}{1-\Phi(sz_i)} - sz_i\right),$$

$$\frac{\partial\mathbb{E}(u_i|y_i - \boldsymbol{x}_i'\boldsymbol{\beta})}{\partial\boldsymbol{\theta}}\bigg|_{\theta=\hat{\theta}} = \begin{pmatrix} -s\dfrac{\hat{\sigma}_u^2}{\hat{\sigma}_v^2+\hat{\sigma}_u^2}\Big(h'(z_i)-1\Big)\boldsymbol{x}_i \\ \left(\dfrac{\hat{\sigma}_u}{\sqrt{\hat{\sigma}_v^2+\hat{\sigma}_u^2}}\right)^3\left(h(z_i) - sz_i - sz_i\hat{\sigma}_v\left(2\dfrac{\hat{\sigma}_v}{\hat{\sigma}_u}+\dfrac{\hat{\sigma}_u}{\hat{\sigma}_v}\right)\big(h'(z_i)-1\big)\right) \\ \left(\dfrac{\hat{\sigma}_v}{\sqrt{\hat{\sigma}_v^2+\hat{\sigma}_u^2}}\right)^3\left(h(z_i) - sz_i + sz_i\dfrac{\hat{\sigma}_u}{\hat{\sigma}_v}\big(h'(z_i)-1\big)\right) \end{pmatrix},$$

$$h'(z_i) = h(z_i)\big(h(z_i) - sz_i\big) \qquad h(z_i) = \frac{\phi(sz_i)}{1-\Phi(sz_i)}, \qquad z = \frac{y_i - \boldsymbol{x}_i'\hat{\boldsymbol{\beta}}}{\sqrt{\hat{\sigma}_v^2+\hat{\sigma}_u^2}}\frac{\hat{\sigma}_u}{\hat{\sigma}_v},$$

which is clearly unbounded. Therefore, even if our estimator $\hat{\boldsymbol{\theta}}$ is robust, contaminating outliers could have an arbitrarily large impact on efficiency predictions for some observations. Additionally, in the N-HN case the tail behaviour of the conditional mean predictor is such that

$$\lim_{s\varepsilon\to\infty}\mathbb{E}(u_i|y_i - \boldsymbol{x}_i'\boldsymbol{\beta}) = \infty, \qquad \lim_{s\varepsilon\to-\infty}\mathbb{E}(u_i|y_i - \boldsymbol{x}_i'\boldsymbol{\beta}) = 0,$$

so that efficiency predictions can approach zero or one as the magnitude of the estimated residual is increased, depending on the sign. Under the assumption of independence of v_i and u_i, Ondrich and Ruggiero (2001) show that, when the distribution of v_i is log-concave, the conditional mean predictor decreases (increases) monotonically as the residual increases (decreases) in a production (cost) frontier setting. Their result implies that this monotonicity is strong when the log-concavity is strong, weak where the log-concavity is weak, and that in the case of log-convex v_i, the direction of the relationship may be reversed. The distribution of v_i is therefore crucial in determining whether or not the efficiency predictions may approach zero or one when the residual is sufficiently large in magnitude. Since the normal distribution is strongly log-concave everywhere, the Ondrich and Ruggiero (2001) result implies that this is the case not only in the N-HN model, but whenever $v_i \sim \mathcal{N}(0, \sigma_v^2)$ and v_i and u_i are independent. It follows from this that the conditional mean cannot be considered robust in these cases.

Under alternative distributional assumptions, the conditional mean may well be bounded or even non-monotonic; Horrace and Parmeter (2018) show that, in the L-TL and L-EXP cases, the predictor is only weakly monotonic, being constant when the estimated residual is positive (negative) for a production (cost) frontier. This is in accordance with the Ondrich and Ruggiero (2001) result, since the Laplace distribution is only weakly log-concave everywhere. The logistic distribution is strongly log-concave everywhere but approaches weak log-concavity at the tails, and accordingly Stead et al. (2018) show that the conditional mean predictor in the Log-HN case appears to approach finite, non-zero limits at the tails. Under the CN-HN specification explored by Stead et al. (Forthcoming), the predictor is non-monotonic, changing direction at the shoulders of the distribution where the scale-contaminated normal distribution is strongly log-convex, but since the tails of the distribution are Gaussian, efficiency predictions nevertheless approach zero or one as the magnitude of the residual becomes large. By contrast, the Student's t distribution being strongly log-convex in its tails, Wheat et al. (2019) show that conditional mean predictor in the T-HN case is also non-monotonic, but that the change in direction is sustained as the residual becomes large in magnitude; in fact, the conditional mean appears to approach the unconditional mean $\mathbb{E}(u_i)$ in both directions as $|\hat{\varepsilon}_i| \to \infty$.

Tancredi (2002) sheds additional light on the latter result, contrasting the limiting behaviour of $f_{u|\varepsilon}$ under the N-HN and Student's t-half t (T-HT) cases; in the former case, $f_{u|\varepsilon}$ becomes increasingly concentrated as $|\varepsilon_i| \to \infty$, while in the latter case $f_{u|\varepsilon}$ becomes increasingly flat. This points to an important qualitative difference in the way the two models handles efficiency prediction for outlying observations—in the N-HN case, prediction uncertainty decreases as $|\hat{\varepsilon}_i| \to \infty$, while in the T-HT case the prediction uncertainty increases. A similar result appears to hold in the T-HN case. This differing tail behaviour of the conditional mean predictor and the conditional distribution generally clearly have important implications in terms of comparing efficiency predictions between firms, and especially in identifying the most and least efficient firms in a sample.

Overall, there appears to be a link between log-convexity of the distribution of v_i and robustness of the conditional mean predictor. This apparent link is interesting, since the results of Stead et al. (2023) indicate a link between log-convexity and robustness in estimation. This suggests that distributional assumptions are key, and that under appropriate distributional assumptions, robust estimation and robust efficiency prediction coincide.

5.2 Conditional Mode

As an alternative to the conditional mean, Jondrow et al. (1982) also suggested using the mode of the conditional distribution. The conditional mode predictor is

$$\mathrm{M}(u_i|y_i - \boldsymbol{x}_i'\boldsymbol{\beta})\Big|_{\boldsymbol{\theta}=\hat{\boldsymbol{\theta}}} = \arg\min_{u_i}\left(-f_{u|\varepsilon}(u_i|y_i - \boldsymbol{x}_i'\boldsymbol{\beta}, \boldsymbol{\theta})\right)\Big|_{\boldsymbol{\theta}=\hat{\boldsymbol{\theta}}}, \tag{32}$$

which, as Jondrow et al. (1982) noted, is analogous to a maximum likelihood estimator. This suggests that the conditional mode predictor will not generally be robust to contaminating outliers. It also suggests possible approaches to robust prediction, drawing on the literature on robust M-estimation. For example, we could apply a Box-Cox transformation to $f_{u|\varepsilon}$ in Eq. 32 for an approach analogous to ML_qLE. The conditional mode has not received as much attention in the SF literature as the conditional mean, but in the N-HN case is given by

$$\mathrm{M}(u_i|\varepsilon_i)\Big|_{\theta=\hat{\theta}} = -s\hat{\sigma}_v\hat{\sigma}_u z_i \mathbb{I}_{sz_i \le 0}.$$

The conditional mode therefore results in an efficiency prediction of 1 whenever $sz_i > 0$, while approaching zero monotonically as $sz_i \to -\infty$. A similar result holds in the N-EXP model—see Jondrow et al. (1982) for a discussion of the behaviour of the conditional mode in both cases. Although, as discussed, we may not regard an efficiency score of 1 as problematic in some cases, the conditional mode is no more robust than the conditional mean in the opposite direction. Likewise, we can see that the derivative of the predictor in the N-HN case

$$\frac{\partial \mathrm{M}(u_i|y_i - \boldsymbol{x}_i'\boldsymbol{\beta})}{\partial \boldsymbol{\theta}}\Bigg|_{\theta=\hat{\theta}} = -s\hat{\sigma}_v \mathbb{I}_{sz_i \le 0} \begin{pmatrix} -\dfrac{\hat{\sigma}_u^3}{\sqrt{\hat{\sigma}_v^2 + \hat{\sigma}_u^2}}\boldsymbol{x}_i \\ \dfrac{\hat{\sigma}_u^2}{\hat{\sigma}_v^2 + \hat{\sigma}_u^2}\dfrac{\hat{\sigma}_u}{\hat{\sigma}_v} z_i \\ \dfrac{3\hat{\sigma}_v^2 + 2\hat{\sigma}_u^2}{\hat{\sigma}_v^2 + \hat{\sigma}_u^2} z_i \end{pmatrix} - \hat{\sigma}_v \delta(sz_i) \begin{pmatrix} \dfrac{\hat{\sigma}_u^3}{\sqrt{\hat{\sigma}_v^2 + \hat{\sigma}_u^2}} z_i \boldsymbol{x}_i \\ \dfrac{\hat{\sigma}_v\hat{\sigma}_u}{\hat{\sigma}_v^2 + \hat{\sigma}_u^2} z_i^2 \\ -\dfrac{2\hat{\sigma}_v^2 + \hat{\sigma}_u^2}{\hat{\sigma}_v^2 + \hat{\sigma}_u^2} z_i^2 \end{pmatrix},$$

is unbounded. Thus a contaminating outlier may have an arbitrarily large influence on not only its own efficiency prediction, but efficiency predictions for other observations using the conditional mode in the N-HN case.

The properties of the conditional mode under alternative distributional assumptions do not seem to have been investigated in detail, though findings on the behaviour of the conditional distribution generally imply that its behaviour is similar to that of the conditional mean—Horrace and Parmeter (2018) find that $f_{u|\varepsilon}$ is constant for $s\varepsilon_i \ge 0$, but varies with ε_i when $s\varepsilon_i < 0$. The behaviour of the conditional mode predictor in other cases, such as the Log-HN or T-HN models, is worth exploring further.

5.3 *Conditional Median*

Yet another potential efficiency predictor is the median of the conditional efficiency distribution, as proposed by Tsukuda and Miyakoshi (2003). This is given by

$$\text{Median}(u_i|y_i - \boldsymbol{x}_i'\boldsymbol{\beta})\bigg|_{\boldsymbol{\theta}=\hat{\boldsymbol{\theta}}} = F_{u|\varepsilon}^{-1}\left(\frac{1}{2}\bigg|y_i - \boldsymbol{x}_i'\boldsymbol{\beta}, \boldsymbol{\theta}\right)\bigg|_{\boldsymbol{\theta}=\hat{\boldsymbol{\theta}}},$$

where $F_{u|\varepsilon}^{-1}$ denotes the quantile function of the conditional distribution. In the N-HN case this has a convenient expression since, as Jondrow et al. (1982) note, the conditional distribution is simply that of a truncated normal random variable. Substituting in the relevant parameters, this is given by

$$\text{Median}(u_i|y_i - \boldsymbol{x}_i'\boldsymbol{\beta})\bigg|_{\boldsymbol{\theta}=\hat{\boldsymbol{\theta}}} = -s\hat{\sigma}_v\hat{\sigma}_u z_i + \frac{\hat{\sigma}_v\hat{\sigma}_u}{\sqrt{\hat{\sigma}_v^2 + \hat{\sigma}_u^2}}\Phi^{-1}\left(\frac{1}{2}\Phi\left(s\sqrt{\hat{\sigma}_v^2 + \hat{\sigma}_u^2}z_i\right)\right), \tag{33}$$

where Φ^{-1} denotes the standard normal quantile function.[9] Since the sample median is robust, in contrast to the sample mean, it may be tempting to intuit that the conditional median predictor ought to be a robust alternative to the conditional mean. However, comparisons shown by Tsukuda and Miyakoshi (2003) for the N-HN case indicate that the two predictors are very similar, and from Eq. 33 we can see that it is monotonic and shares the same limits as the conditional mean. In terms of deriving the influence function, the derivative of the predictor in the N-HN case is given by

$$\begin{aligned}\frac{\partial \text{Median}(u_i|y_i - \boldsymbol{x}_i'\boldsymbol{\beta})}{\partial \boldsymbol{\theta}}\bigg|_{\boldsymbol{\theta}=\hat{\boldsymbol{\theta}}} &= -\frac{1}{2}\frac{s\phi\left(s\sqrt{\hat{\sigma}_v^2+\hat{\sigma}_u^2}z_i\right)}{\phi\left(\Phi^{-1}\left(\frac{1}{2}\Phi\left(s\sqrt{\hat{\sigma}_v^2+\hat{\sigma}_u^2}z_i\right)\right)\right)}\begin{pmatrix}\frac{\hat{\sigma}_u^2}{\sqrt{\hat{\sigma}_v^2+\hat{\sigma}_u^2}}\boldsymbol{x}_i\\ \hat{\sigma}_v z_i\\ -\hat{\sigma}_u z_i\end{pmatrix}\\
&+\Phi^{-1}\left(\frac{1}{2}\Phi\left(s\sqrt{\hat{\sigma}_v^2+\hat{\sigma}_u^2}z_i\right)\right)\begin{pmatrix}0\\ \left(\frac{\hat{\sigma}_u}{\sqrt{\hat{\sigma}_v^2+\hat{\sigma}_u^2}}\right)^3\\ \left(\frac{\hat{\sigma}_v}{\sqrt{\hat{\sigma}_v^2+\hat{\sigma}_u^2}}\right)^3\end{pmatrix} - sz_i\begin{pmatrix}0\\ \hat{\sigma}_u\\ \hat{\sigma}_v\end{pmatrix} + \frac{s\hat{\sigma}_v\hat{\sigma}_u^2}{\hat{\sigma}_v^2+\hat{\sigma}_u^2}\begin{pmatrix}\frac{\sqrt{\hat{\sigma}_v^2+\hat{\sigma}_u^2}}{\hat{\sigma}_v}\boldsymbol{x}_i\\ \frac{2\hat{\sigma}_v^2+\hat{\sigma}_u^2}{\hat{\sigma}_v\hat{\sigma}_u}z_i\\ -\left(\frac{\hat{\sigma}_v}{\hat{\sigma}_u}\right)^2 z_i\end{pmatrix},\end{aligned}$$

which is unbounded. The conditional median therefore appears no more robust than the conditional mean in the N-HN case. Horrace and Parmeter (2018) derive the conditional median predictor for the L-TL and L-EXP models, which again exhibits similar behaviour to the conditional mean. To summarise, the choice of mean, median, or mode of the conditional distribution appears less important than the model's distributional assumptions with respect to robustness. However, further investigation of possible alternative predictors is needed.

[9] An equivalent expression for inefficiency, defined as $1 - \exp(-\text{Median}(u_i|\varepsilon_i))$ is given by Tsukuda and Miyakoshi (2003).

6 Summary and Conclusions

The robustness of stochastic frontier (SF) modelling has been an understudied area, but has been given increased attention in recent years, with the use alternative estimators and distributional assumptions better able to accomodate contaminating outliers explored. Despite this there has been relatively little explicit discussion and comparison of the robustness properties of different estimators and model specifications. In this chapter we have aimed to address this gap, discussing the robustness properties of various approaches in terms of the influence functions, gross error sensitivities, and breakdown points of estimators. The discussion of influence is particularly useful, since the concept is easily extended to efficiency prediction, allowing discussion of the sensitivity of efficiency predictions to contaminating observations.

We show that the influence function for the maximum likelihood estimator is unbounded under standard distributional assumptions, although under alternative distributional assumptions the estimator may be robust. Recent results from Stead et al. (2023) give sufficient conditions for robust maximum likelihood estimation (MLE) of the SF model. Some recent proposals, such as the use of logistic or Laplace noise distributions—see Stead et al. (2018) and Horrace and Parmeter (2018), respectively—do not satisfy these conditions. On the other hand, the Student's t distribution for noise satisfying these conditions when paired with many inefficiency distributions. This offers a route to achieving robust estimation while remaining within the framework of MLE, which is attractive for two main reasons. First, we would like to retain the efficiency of ML estimation. With robust estimation methods, there is generally a trade-off between robustness and efficiency. Second, a key objective of SF modelling is the deconvolution of ε_i into v_i and u_i. Derivation of the Jondrow et al. (1982) and Battese and Coelli (1988) efficiency predictors under alternative distributional assumptions is straightforward. As such, alternative estimation methods can be used to deal with outliers when it comes to estimation, but leave handling outliers in the efficiency prediction stage as a separate problem. Altering distributional assumptions offers a consistent way of dealing with outliers in both stages. Alternative approaches generalising MLE by changing the loss function such that the influence function is unbounded, such as minimum density power divergence estimation, maximum L_q-likelihood estimation, and maximum Ψ-likelihood estimation, have recently been considered by Song et al. (2017) and Bernstein et al. (2021). Under these approaches, the loss function is transformed such that MLE is contained as a limiting case, and a tuning parameter controls the trade-off between robustness and efficiency. We note that MLE under the assumption of Student's t noise can be conceptualised in the same way, where the transformed loss function is derived directly from an explicit model of contamination, which is also reflected in efficiency prediction. We show that the corrected ordinary least squares (COLS) approach to estimation is non-robust, though analogous 'orrected' robust regression methods could be considered. We also consider the application of quantile regression (QR) to SF modelling, which has gained attention recently. QR represents another possible approach to robust estimation of the SF model, though the robustness of the estimator is reduced when we choose extreme quantiles. The appropriate choice of quantile reflects underlying distributional assumptions—see Jradi and Ruggiero (2019) and Jradi et al. (2021), suggesting that as with MLE, the robustness of QR estimation of the SF model depends fundamentally on distributional assumptions. To summarise, recent work on robust estimation of the SF model highlights three main approaches: MLE under appropriate distributional assumptions, alternative robust M-estimation methods, and QR estimation. Each of these belong to the general class of M-estimators, making the derivation and comparison of influence functions straightforward. For large samples, MLE and related approaches may offer a better trade-off between robustness and efficiency than QR when the 'true' model is correctly specified.

With respect to efficiency prediction, we discuss robustness in two related senses: the tail behavior of predictors, and the sensitivity of predictors to contaminating outliers via influence on parameter estimates. The former is relevant when considering the efficiency prediction with respect to gross outliers, and how such outliers may affect the identification of the highest and lowest ranking firms. We discuss derivation of influence functions for efficiency predictors, and their resulting properties. We note that the conditional mean, conditional mode, and conditional median are all non-robust under standard distributional assumptions. Results on the limiting behaviour of the conditional distribution of efficiency again suggest that distributional assumptions are key—for instance, the conditional mean predictor is robust in the Student's t-half normal case.

Our discussion highlights several interesting avenues for future research. With respect to robust estimation, direct comparison of the various robust estimators discussed in the context of SF modelling is needed to establish how they compare in terms of robustness, efficiency, and the trade-off between the two under various settings. In addition, while we identify estimators that are *robust* in the sense of having finite gross error sensitivity, a natural next step would be to identify estimators that are *resistant* in terms of having high breakdown points. The distinction between the ordinary breakdown point and the finite sample breakdown point is crucial here; QR estimation has a finite sample breakdown point greater than $1/n$, but a $1/n$ ordinary breakdown point, while both are greater than $1/n$ for the the Student's t model. This suggests that the latter offers some robustness to leverage points as well as outliers, but the extent of this resistance needs further investigation. In terms of efficiency prediction, further investigation is needed into the link between distributional assumptions and the robustness of efficiency predictors, and also into possible robust approaches to prediction under standard distributional assumptions.

References

Aigner, D., Lovell, C. A. K., & Schmidt, P. (1977). Formulation and estimation of stochastic frontier production function models. *Journal of Econometrics, 6*(1), 21–37.

Almanidis, P., Qian, J., & Sickles, R. C. (2014). Stochastic frontier models with bounded inefficiency. In R. C. Sickles & W. C. Horrace (Eds.), *Festschrift in honor of peter schmidt: Econometric methods and applications* (pp. 47–81). New York, NY: Springer.

Andrews, D. F., & Mallows, C. L. (1974). Scale mixtures of normal distributions. *Journal of the Royal Statistical Society: Series B (Methodological), 36*(1), 99–102.

Bassett, G., Jr., & Koenker, R. (1978). Asymptotic theory of least absolute error regression. *Journal of the American Statistical Association, 73*(363), 618–22.

Basu, A., Harris, I. R., Hjort, N. L., & Jones, M. C. (1998). Robust and efficient estimation by minimising a density power divergence. *Biometrika, 85*(3), 549–59.

Battese, G. E., & Coelli, T. J. (1988). Prediction of firm-level technical efficiencies with a generalized frontier production function and panel data. *Journal of Econometrics, 38*(3), 387–99.

Behr, A. (2010). Quantile regression for robust bank efficiency score estimation. *European Journal of Operational Research, 200*(2), 568–81.

Beran, R. (1977). Minimum Hellinger distance estimates for parametric models. *Annals of Statistics, 5*(3), 445–63.

Bernini, C., Marzia, F., & Gardini, A. (2004). Quantile estimation of frontier production function. *Empirical Economics, 29*(2), 373–81.

Bernstein, D. H., Parmeter, C. F., & Wright, I. A. (2021). Robust estimation of the stochastic frontier model. Working Paper, University of Miami.

Chernozhukov, V., Fernández-Val, I., & Melly, B. (2022). Fast algorithms for the quantile regression process. *Empirical Economics, 62*(1), 7–33.

Coelli, T. J., Rao, D. S. P., & Battese, G. E. (2005). *An introduction to efficiency and productivity analysis*. New York, NY: Springer.

Dielman, T. E. (2005). Least absolute value regression: Recent contributions. *Journal of Statistical Computation and Simulation, 75*(4), 263–86.

Donoho, D. L., & Huber, P. J. (1977). The notion of breakdown point. In P. J. Bickel, K. Doksum, & J. L. Hodges, Jr. (Eds.), *A festschrift for Erich L. Lehmann* (pp. 157–84). Belmont, CA.

Durio, A., & Isaia, E. D. (2011). The minimum density power divergence approach in building robust regression models. *Informatica, 22*(1), 43–56.

Eguchi, S., & Kano, Y. (2001). Robustifing maximum likelihood estimation by psi-divergence. Research Memorandum 802, Institute of Statistical Mathematics.

El Mehdi, R., & Hafner, C. M. (2014) Inference in stochastic frontier analysis with dependent error terms. *Mathematics and Computers in Simulation, 102*(Supplement C), 104–116.

Ferrari, D., & Yang, Y. (2010). Maximum L_q-likelihood estimation. *Annals of Statistics, 38*(2), 753–83.

Giloni, A., Simonoff, J. S., & Sengupta, B. (2006). Robust weighted LAD regression. *Computational Statistics & Data Analysis, 50*(11), 3124–40.

Giloni, A., & Padberg, M. (2004). The finite sample breakdown point of ℓ_1-regression. *SIAM Journal on Optimization, 14*(4), 1028–42.

Greene, W. H. (1980). Maximum likelihood estimation of econometric frontier functions. *Journal of Econometrics, 13*(1), 27–56.

Greene, W. H. (1990). A gamma-distributed stochastic frontier model. *Journal of Econometrics, 46*(1), 141–63.

Greene, W. H. (2008). The econometric approach to efficiency analysis. In H. O. Fried, C. A. K. Lovell, & S. S. Schmidt (Eds.), *The measurement of productive efficiency and productivity growth* (pp. 92–159). Oxford, UK: Oxford University Press.

Gupta, A. K., & Nguyen, N. (2010). Stochastic frontier analysis with fat-tailed error models applied to WHO health data. *International Journal of Innovative Management, Information and Production, 1*(1), 43–48.

Hampel, F. R. (1968). Contributions to the theory of robust estimation. Ph.D. Thesis, University of California, Berkeley.

Hampel, F. R. (1971). A general qualitative definition of robustness. *Annals of Mathematical Statistics, 42*(6), 1887–96.

Hampel, F. R., Ronchetti, E. M., Rousseeuw, P. J., & Stahel, W. A. (1986). *Robust statistics: The approach based on influence functions*. New York, NY: Wiley.

He, X., Jurečková, J., Koenker, R., & Portnoy, S. (1990). Tail behavior of regression estimators and their breakdown points. *Econometrica, 58*(5), 1195–214.

Horrace, W. C., & Parmeter, C. F. (2018). A Laplace stochastic frontier model. *Econometric Reviews, 37*(3), 260–80.

Horrace, W. C., & Wright, I. A. (2020). Stationary points for parametric stochastic frontier models. *Journal of Business & Economic Statistics, 38*(3), 516–26.

Huber, P. J. (1964). Robust estimation of a location parameter. *Annals of Mathematical Statistics, 35*(1), 73–101.

Huber, P. J. (1981). *Robust statistics*. New York, NY: Wiley.

Janssens, G. K., & van den Broeck, J. (1993). Outliers, sample size and robust estimation of stochastic frontier production models. *Journal of Information and Optimization Sciences, 14*(3), 257–74.

Jondrow, J., Lovell, C. A. K., Materov, I. S., & Schmidt, P. (1982). On the estimation of technical inefficiency in the stochastic frontier production function model. *Journal of Econometrics, 19*(2), 233–8.

Jradi, S., & Ruggiero, J. (2019). Stochastic data envelopment analysis: A quantile regression approach to estimate the production frontier. *European Journal of Operational Research, 278*(2), 385–93.

Jradi, S., Parmeter, C. F., & Ruggiero, J. (2019). Quantile estimation of the stochastic frontier model. *Economics Letters, 182*, 15–18.

Jradi, S., Parmeter, C. F., & Ruggiero, J. (2021). Quantile estimation of stochastic frontiers with the normal-exponential specification. *European Journal of Operational Research, 295*(2), 475–83.

Knox, K. J., Blankmeyer, E. C., & Stutzman, J. R. (2007). Technical efficiency in Texas nursing facilities: A stochastic production frontier approach. *Journal of Economics and Finance, 31*(1), 75–86.

Koenker, R. (2005). *Quantile regression*. Cambridge, UK: Cambridge University Press.

Koenker, R., Jr, & Bassett, G. (1978). Regression quantiles. *Econometrica, 46*(1), 33–50.

Kumbhakar, S. C., & Lovell, C. A. K. (2000). *Stochastic frontier analysis*. Cambridge, UK: Cambridge University Press.

Lind, J. C., Mehra, K. L., & Sheahan, J. N. (1977). Asymmetric errors in linear models: Estimation-theory and Monte Carlo. *Statistics: A Journal of Theoretical and Applied Statistics, 23*(4), 305–320.

Liu, C., Laporte, A., & Ferguson, B. S. (2008). The quantile regression approach to efficiency measurement: Insights from Monte Carlo simulations. *Health Economics, 17*(9), 1073–87.

Meeusen, W., & van Den Broeck, J. (1977). Efficiency estimation from Cobb-Douglas production functions with composed error. *International Economic Review, 18*(2), 435–44.

Miyamura, M., & Kano, Y. (2006). Robust Gaussian graphical modeling. *Journal of Multivariate Analysis, 97*(7), 1525–50.

Mizera, I., & Müller, C. H. (2001). The influence of the design on the breakdown points of ℓ_1-type M-estimators. In A. Atkinson, P. Hackl, & C. H. Müller (Eds.), *MODA6-advances in model-oriented design and analysis* (pp. 193–200). Heidelberg, Germany: Physica-Verlag.

Murillo-Zamorano, L. R. (2004). Economic efficiency and frontier techniques. *Journal of Economic Surveys, 18*(1), 33–77.

Nguyen, N. (2010). Estimation of technical efficiency in stochastic frontier analysis. Ph.D. Thesis, Bowling Green State University.

Olson, J. A., Schmidt, P., & Waldman, D. A. (1980). A Monte Carlo study of estimators of stochastic frontier production functions. *Journal of Econometrics, 13*(1), 67–82.

Ondrich, J., & Ruggiero, J. (2001). Efficiency measurement in the stochastic frontier model. *European Journal of Operational Research, 129*(2), 434–442.

Parmeter, C. F. (2001). *Is it MOLS or COLS? Efficiency series paper 04/21*. Departmento de Economía: Universidad de Oviedo.

Papadopoulos, A. & Parmeter, C.F. (2022), Quantile methods for stochastic frontier analysis, *Foundations and Trends in Econometrics, 12*(1), 1–120.

Parmeter, C. F., & Kumbhakar, S. C. (2014). Efficiency analysis: A primer on recent advances. *Foundations and Trends in Econometrics, 7*(3–4), 191–385.

Portnoy, S., & Koenker, R. (1997). The Gaussian hare and the Laplacian tortoise: Computability of squared-error versus absolute-error estimators. *Statistical Science, 12*(4), 279–300.

Pollard, D. (1991). Asymptotics for least absolute deviation regression estimators. *Econometric Theory, 7*(2), 186–199.

Rousseeuw, P. J. (1984). Least median of squares regression. *Journal of the American Statistical Association, 79*(388), 871–880.

Seaver, B. L., & Triantis, K. P. (2006). The impact of outliers and leverage points for technical efficiency measurement using high breakdown procedures. *Management Science, 41*(6), 937–956.

Siegel, A. F. (1982). Robust regression using repeated medians. *Biometrika, 69*(1), 242–244.

Song, J., Oh, D., & Kang, J. (2017). Robust estimation in stochastic frontier models. *Computational Statistics and Data Analysis, 105*, 243–267.

Stead, A. D., Wheat, P., & Greene, W. H. (2018). Estimating efficiency in the presence of extreme outliers: A logistic-half normal stochastic frontier model with application to highway maintenance costs in England. In W. H. Greene, L. Khalaf, P. Makdissi, R. C. Sickles, M. Veall, & M. Voia (Eds.), *Productivity and inequality* (pp. 1–19). Cham, Switzerland: Springer.

Stead, A. D., Wheat, P., & Greene, W. H. (2019). Distributional forms in stochastic frontier analysis In T. Ten Raa, W. H. Greene (Eds.), *Palgrave handbook of economic performance analysis* (pp. 225–274). Cham, Switzerland: Palgrave Macmillan.

Stead, A.D., Wheat, P. & Greene, W.H. (2023), Robust maximum likelihood estimation of stochastic frontier models, *European Journal of Operational Research, 309*(1), 188–201.

Stefanski, L. A. (1991). A normal scale mixture representation of the logistic distribution. *Statistics & Probability Letters, 11*(1), 69–70.

Stevenson, R. E. (1980). Likelihood functions for generalized stochastic frontier estimation. *Journal of Econometrics, 13*(1), 57–66.

Tancredi, A. (2002). Accounting for heavy tails in stochastic frontier models. Working paper 2002.16, University of Padua.

Tsionas, M. G. (2020). Quantile stochastic frontiers. *European Journal of Operational Research, 282*(3), 1177–84.

Tsionas, M. G., Assaf, A. G., & Andrikopoulos, A. (2020). Quantile stochastic frontier models with endogeneity. *Economics Letters, 188*, 108964.

Tsukuda, Y., & Miyakoshi, T. (2003). An alternative method for predicting technical inefficiency in stochastic frontier models. *Applied Economics Letters, 10*, 667–70.

Waldman, D. M. (1982). A stationary point for the stochastic frontier likelihood. *Journal of Econometrics, 18*(2), 275–279.

Stead, A.D., Wheat, P. & Greene, W.H. (Forthcoming), On hypothesis testing in latent class and finite mixture stochastic frontier models, with application to a contaminated normal-half normal model, *Journal of Productivity Analysis*.

Wheat, P., Stead, A. D., & Greene, W. H. (2019). Robust stochastic frontier analysis: A Student's t-half normal model with application to highway maintenance costs in England. *Journal of Productivity Analysis, 51*(1), 21–38.

White, H. (1982). Maximum likelihood estimation of misspecified models. *Econometrica, 50*(1), 1–25.

Yu, C., & Yao, W. (2017). Robust linear regression: A review and comparison. *Communications in Statistics—Simulation and Computation, 46*(8), 6261–6282.

Zhao, S. (2021). Quantile estimation of stochastic frontier models with the normal-half normal specification: A cumulative distribution function approach. *Economics Letters, 206,* 109998.

Zulkarnain, R., & Indahwati, I. (2021). Robust stochastic frontier using Cauchy distribution for noise component to measure efficiency of rice farming in East Java. *Journal of Physics: Conference Series, 1863*, 012031.

Is it MOLS or COLS?

Christopher F. Parmeter

I am a proud member of the PWGQNKX: People Who Have No Idea How Acronyms Work.

modify: make partial or minor changes to (something), typically so as to improve it;

correct: put right (an error or fault).

1 Introduction

Estimation of the structural parameters of the stochastic frontier model through maximum likelihood is undoubtedly the most common approach in applied efficiency analysis. However, a variety of alternative proposals exist, in particular, the two-step practice of first estimating the model via ordinary least squares (OLS) while ignoring the existence of inefficiency, and then shifting the estimated conditional mean upward by some amount. How much to shift this curve up then leads to either the modified OLS estimator (MOLS) or the corrected OLS estimator (COLS). However, depending upon which paper (or textbook) one reads, the use of "C" or "M" may be conflated with the other. I am not immune from this: in Parmeter and Kumbhakar (2014) the MOLS acronym appears while in Kumbhakar et al. (2020) the COLS acronym is used for the same discussion. Both approaches use the same essential

Comments from Chris O'Donnell, Alecos Papadopoulos and Paul Wilson helped improve the draft. I also thank William Greene, Finn Førsund and Knox Lovell for stimulating conversations on the history of this topic. All errors are mine alone.

C. F. Parmeter (✉)
Department of Economics, University of Miami, Coral Gables, USA
e-mail: cparmeter@bus.miami.edu

P. Macedo et al. (eds.), *Advanced Mathematical Methods for Economic Efficiency Analysis*, Lecture Notes in Economics and Mathematical Systems 692,
https://doi.org/10.1007/978-3-031-29583-6_13

ingredient, adjustment of the intercept so that the OLS estimated conditional mean acts as a frontier. However, the amount to shift the conditional mean differs based on the approach.

This paper reviews the historical use of the COLS and MOLS terminology in the field of efficiency analysis. As a point of writing, I will only use the term "correct/corrected" in specific reference to COLS and "modify/modified" in specific reference to MOLS. When the usage is to a generic change I will use "adjust/adjusted" or "shift/shifted" so as to avoid potential confusion. Also, neither "correct" or "modify" is indicative of a specific statistical method so debate over which term to use is purely for historical purposes and the hope that one consistent notation can be used from this point forward.

After this historical tour is completed, we then discuss several of the key econometrics issues that surround these adjustment estimators. In particular, the simplicity of using them when maximum likelihood fails to converge, their ability to highlight if skewness/kurtosis issues are likely to arise, the presence of determinants and the calculation of standard errors.

2 The Methods

To avoid biasing the reader toward my own personal view on which method corresponds to which acronym, I will introduce the two competing methods without distinction.

The two methods have the same end goal, to shift an estimated OLS conditional mean by some amount to produce a "frontier." What is important here is that not only do we care about shifting the estimated conditional mean up, but also by how much. And it is this subtle distinction that lies at the heart of differences in the terminology. Part of my belief as to why there is a potential for the confusion is that there are two distinct entities at play here that coincide to some degree with the development of the field. The first issue is the presence of noise. That is, are we operating under the assumption of a deterministic[1] or a stochastic frontier. The second issue is then how much to shift the estimated conditional mean up. Assuredly, in the context of the stochastic frontier, the shift is such that not all of the data are bounded by the estimated frontier. However, when a deterministic frontier is of interest, the estimated conditional mean can be shifted so that either **all** or **some** of the data are below the estimated frontier. It is this subtle distinction that I believe to be the root source for differences in acronyms.

[1] The use of the word "deterministic" is not meant pejoratively. Rather, the use is in keeping with the argot that arose at the time the field of frontier analysis developed, and is still to some degree used. Both deterministic and stochastic frontier models are fully specified statistical models (Sickles and Zelenyuk, 2019) and each have their own set of assumptions under which estimators for these models perform admirably.

To begin we have the benchmark linear parametric frontier model:

$$y_i = \beta_0 + \boldsymbol{x}_i'\beta + \varepsilon_i, \tag{1}$$

where y_i is the dependent variable (output), β_0 is the intercept, $\boldsymbol{x}_i$ is a $k \times 1$ vector of inputs, β is the $k \times 1$ parameter vector, and ε_i dictates how y_i can deviate from the production frontier.

When $\varepsilon_i = v_i - u_i$, where v_i is stochastic noise and u_i accounts for inefficiency which serves to reduce observed output below the frontier, then the model in Eq. (1) is a stochastic frontier model. Alternatively, when $\varepsilon_i = -u_i$, we have the deterministic frontier model. The classic estimation of the stochastic frontier model is to assume a parametric distribution for v_i and u_i, derive the density of the composite error, $\varepsilon_i = v_i - u_i$, and estimate the model via maximum likelihood. The main approach to estimating the deterministic frontier is the use of programming methods (either linear or quadratic) to enforce the constraint that observed output can be no higher than the estimated production frontier.

The economic meaning of both of these methods is that the presence of inefficiency is a neutral shifter of technology. This is important as it pertains to how the two methods after OLS has been deployed to estimate the conditional mean, to then construct the frontier.

2.1 Method A

Method A is clearly entrenched in the stochastic frontier setting and seeks to use information on the assumed parametric structure of the density of u to recover the parameters of the distribution of u so that $E(u)$ may be estimated. To see this note that the model in Eq. (1) can be written as

$$\begin{aligned} y_i &= \beta_0 + \boldsymbol{x}_i'\beta + v_i - u_i \\ &= \beta_0 + \boldsymbol{x}_i'\beta + v_i - u_i - E(u) + E(u) \\ &= (\beta_0 - E(u)) + \boldsymbol{x}_i'\beta + v_i - (u_i - E(u)) \\ &= \beta_0^* + \boldsymbol{x}_i'\beta + \varepsilon_i^*, \end{aligned} \tag{2}$$

where $\beta_0^* = \beta_0 - E(u)$ is the efficiency biased intercept and $\varepsilon_i^* = v_i - (u_i - E(u))$ is the mean zero error term.

Method A seeks to estimate $E(u)$ from a set of moment conditions based on the parameters of the distribution of u. This estimate of $E(u)$ is then used to adjust the OLS intercept for the downward shift present due to ignorance of the composite error structure. Once this is done $\widehat{\beta} = \widehat{\beta^*} + \widehat{E(u)}$ can be constructed, which will then provide a consistent estimator of the intercept of the stochastic frontier model. Solutions for the unknown parameters based on various distributional assumptions for u and v such that an estimator of $E(u)$ can be constructed abound: the Half Normal solution is found in Aigner et al. (1977) (in passing) and Olson et al. (1980) specifically, the Exponential appears in Greene (1980a), the Gamma in Greene (1990), the Uniform in Li (1996), the Binomial in Carree (2002), the truncated Normal in both Harris (1992) and Goldstein (2003) (in entirely different forms), and the generalized Exponential in Papadopoulos (2021). All of these methods assume that v is Normally distributed and save the truncated Normal, an analytical solution for the unknown parameters of the distribution of u exists. There have also been proposals that have allowed different distributional assumptions on the noise component of ε: Nguyen (2010) provides a solution for the Laplace-Exponential model; Goldstein (2003) discusses the Student's t-Gamma composite error adjustment; and more recently Wheat et al. (2019) provide a closed-form solution when the noise is assumed to stem from a Student's t-distribution and the inefficiency is distributed Half Normal. See Appendix 11 for the solutions to these various settings.

2.2 Method B

Method B seeks to construct a deterministic frontier by various manners which involve shifting the estimated OLS curve up so that some or **all** of the data are below the estimated curve. The simplest approach for this involves direct OLS estimation of the model in Eq. (1) and then adding to $\widehat{\beta_0}$ the largest (positive) residual, $\widehat{\varepsilon}_{(n)} = \max_i \widehat{\varepsilon}_i$. This approach does not specifically require a distributional assumption on u_i.

Alternatively, similar to method A, if a distributional assumption is made on u, then $E(u)$ can be estimated and the intercept adjusted. Methods A and B differ in this regard though because the lack of v in Method B then allows different moments to be used to recover the unknown $E(u)$.[2] Even though v is not present, correction of the OLS intercept in the context of Method B does not insure that the estimated frontier lies everywhere above all of the data.

2.3 What These Methods Aim to Do

Before diving into the specific semantic issues it is clear that both methods involve two distinct stages: first estimate the model ignoring the frontier structure or the explicit presence of u_i via OLS, and second shift/adjust the OLS estimate of β_0 to construct the frontier. The two salient issues of these approaches are if distributional assumptions are imposed on v and u and if v is not degenerate, i.e., a deterministic versus a stochastic frontier. As it turns out the debate over the acronyms depends heavily on which of the two issues one focuses on **first**. As we will hope to confer,

---

[2] The work of Richmond (1974) is salient in this regard.

the main distinction between the two approaches is linked to whether one is using a deterministic or a stochastic frontier, not if distributional assumptions are made on the components of the composite error.

3 History of the Nomenclature

3.1 Prior to 1980

To be clear, both COLS and MOLS involve shifting the estimated OLS conditional mean up by some amount. Perhaps the earliest discussion of just such a shift is Winsten's (1957, p. 283) comment to Farrell (1957): "It would also be interesting to know whether in practice this efficient production function turned out to be parallel to the average production function, and whether it might not be possible to fit a line to the averages, and then **shift** it parallel to itself to estimate the efficient production function" (bolding mine). Here the use of the words "modified" or "correct" are not present, but the idea of shifting the OLS regression line up was clear.

From then it was not until the late 1960s and early 1970s where there was a remarkable increase in interest in the estimation of frontier models. Aigner and Chu (1968) proposed a simple programming approach to estimate a deterministic frontier. Their work ensures that the estimated frontier lies above all of the data and no distributional assumptions are made on "disturbances."[3] It is clear from their baseline formulation that the proposed quadratic program they offer has the flavor of a restricted least squares approach.

The first paper to operate in a deterministic frontier setting, but to explicitly recognize that the estimated frontier may (should) not contain all of the data is Timmer (1971). Specifically, Timmer (1971, p. 779) avers: "The frontier is estimated in a probabilistic fashion by constraining X percent of the observations to fall outside the frontier surface." Another interesting aspect of Timmer (1971) is that his work is clearly not a shift/adjustment as the probabilistic component was enforced on the full estimation of the frontier, which then resulted in different estimates of the slope coefficients **and** the intercept (see Table 1, columns IIIa–IIIc in Timmer (1971)). It is worth mentioning here that Timmer (1970, p. 151) acknowledges that while average and frontier production could look dramatically different in his proposed approach, this was probably not to be expected: "The correlation ... is one further manifestation of the similar nature of frontier and average production functions and the relative neutrality of the shift from average to frontier." Timmer's (1971) approach is similar to Aigner and Chu (1968) except that it explicitly recognizes that without stochastic noise in the model (recall this prior to the creation of the stochastic frontier model),

---

[3] See also Schmidt (1976) who attempts to discuss the Aigner and Chu (1968) setup in the context of OLS as well as the reply by Chu (1978) and Schmidt's (1978) rejoinder.

the programming estimates are likely to be heavily influenced by any extremes in the data. In this case, an arbitrary percentage of firms are allowed to be above the estimated frontier.

While Timmer (1970, 1971) is not concerned with shifting/adjusting an OLS estimated intercept, his work was perhaps the first to recognize that the estimated frontier may not lie above all of the data. Another important paper in this timeline is Afriat (1972) who studies a linear programming problem and introduced the idea of the Beta distribution as a means for introducing multiplicative (hence support on [0, 1]) inefficiency. Afriat (1972, p. 575) discussed this as "This accounts for the 'least-squares' principle of 'estimating' α, β in the Cobb–Douglas production function ...But it is a principle belonging to a general statistical method which does not incorporate an economic meaning for error." A careful reading of Afriat (1972) also reveals that the modify/correct/shift verbiage does not appear and his approach is certainly not centered around least squares estimation of the conditional mean.

Richmond (1974), following the insights of Afriat (1972), proposes an explicit OLS procedure to estimate the parameters of the deterministic frontier model. He assumes that the error term in a multiplicative Cobb–Douglas model is the exponential of a Gamma distribution. With no stochastic error, this then leads to a simple (second order) moment condition of the OLS residuals that allows for estimation of the unknown parameter of the corresponding inefficiency component and the appropriate adjustment for the intercept.[4] The terms modify, correct, shift, or adjust do not appear in Richmond (1974). An exemplar discussion of the literature just discussed here can be found in Sickles and Zelenyuk (2019, Sects. 11.2 and 11.3).

3.2 1980

Why is 1980 an important demarcation? Prior to 1980 the explicit use of either the COLS or MOLS acronym did not appear in any published paper discussing estimation of the frontier model. However, in 1980, the first special issue dealing specifically with the stochastic frontier model appeared in the *Journal of Econometrics* (Aigner and Schmidt, 1980). In this special issue, the COLS terminology was explicitly introduced while the MOLS terminology was implicitly introduced.

From that special issue, we have Olson et al. (1980, p. 16): "The third estimator we consider is a corrected least squares estimator, which we will refer to as COLS, which was discussed briefly in ALS. This estimator is similar in spirit to the estimator suggested by Richmond (1974) in the context of a pure frontier." The corresponding statement in Aigner et al. (1977, p. 28–29) "We note in passing that if estimation of β alone is desired, all but the coefficient in β corresponding to a column of ones in X is estimated unbiasedly and consistently by least squares. Moreover, the components

[4] Note that Richmond (1974) refers to the unknown parameter as n, which may lead some to confuse this with the sample size. Certainly it is not the best notation given the widespread use of n to represent the number of observations.

of σ^2 can be extracted (i.e., consistent estimators for them can be found) based on the least squares results by utilizing Eq. (9) for $V(\varepsilon)$ in terms of σ_u^2 and σ_v^2 and a similar relationship for a higher-order moment of ε, since $V(\varepsilon)$ and higher order mean-corrected moments of ε are themselves consistently estimable from the computed least-squares residuals." In a footnote Aigner et al. (1977, p. 29) then give the exact third moment one would need for the correction to the OLS intercept. Greene (1980a, p. 32) also references this same discussion.

The MOLS terminology can be inferred from Greene (1980a, p. 35): "Of course, the (appropriately modified) OLS estimator is also consistent, and more easily computed." Here though Greene is referring to his earlier discussion on shifting the OLS curve up so that **all** of the observations lie on or below the frontier, i.e., a deterministic frontier. From Greene (1980a, p. 34) "...then the OLS residuals can be used to derive a consistent estimate of α. We need only shift the intercept of the estimated function until all residuals (save for the one support point) have the correct sign."

The issue at this point is that Olson et al. (1980) explicitly use the COLS acronym while Greene (1980a) mentions a modified OLS estimator but never explicitly writes MOLS. Thus, it is clear from these papers what COLS is referring to and what MOLS can be attributed to.

3.3 *After 1980*

Based on the state of the field in 1980, if we were to zoom (no pun intended) through time to today it would be clear that COLS would refer to the adjustment of the OLS estimator by the expected value of the assumed distribution of inefficiency and MOLS would refer to the shifting of the OLS estimator by the largest positive residual. However, time is never so forgiving. Lovell (1993, p. 21), crediting Gabrielsen (1975), baptizes MOLS as the method in which one adjusts the intercept based on a specific set of distributional assumptions (regardless of whether one is estimating a deterministic or stochastic frontier).[5] Prior to Lovell (1993) it is not clear where/when the explicit use of the acronym MOLS was used. While the word modified appeared in both Gabrielsen (1973) (in Norwegian) and Greene (1980a), neither specifically uses the acronym MOLS.

The exact wording in Lovell (1993, p. 21) is "COLS was first proposed by Winsten (1957), although Gabrielsen (1975) is usually credited with its discovery. It makes no assumption concerning the functional form of the nonpositive efficiency component U_j. It estimates the technology parameters of (1.19) by OLS, and corrects the downward bias in the estimated OLS intercept by shifting it up until **all** corrected residuals are nonpositive and at least one is zero" (bolding mine). What is strange[6]

[5] Greene (2008) points the reader to Lovell (1993) as well for this nomenclature.

[6] As Lovell and Schmidt were coauthors on the original 1977 paper which indirectly referenced COLS.

about the Lovell (1993) chapter is that Olson et al. (1980) is not cited, which contains the first published use of the COLS terminology.

Moreover, the line "...although Gabrielsen (1975) is usually credited with its discovery" is odd given that the word "corrected" never appears in Gabrielsen (1975). It also is instructive to consider other substantive papers prior to Lovell (1993) that might have been able to shed light on the COLS/MOLS debate. Between the initial appearance of COLS in Olson et al. (1980) and Lovell's (1993) authoritative review, there is the review of Schmidt (1985), who also provides some glimpses into the evolution of this terminology. Schmidt (1985, p. 302): "This was first noted by Richmond (1974). Furthermore, the estimated intercept can be "corrected" by shifting it upward until no residual is positive, and one is zero. This yields a consistent estimate of A, as shown by Gabrielsen (1975) and Greene (1980a). However, the asymptotic distribution of the "corrected" intercept is unknown ..." This is followed by Schmidt (1985, p. 306): "Also, a simpler correct least squares estimator is possible, which estimates the model by ordinary least squares and then 'corrects' the intercept by adding a consistent estimator of $E(u)$ based on higher (in the half-normal case, second and third) moments of the least squares residuals."

And the use of the word "correct" for both methods is not specific to Schmidt (1985), Førsund et al. (1980, p. 12) also have a similar lack of distinction: "There is also an alternative method of estimation, apparently first noted by Richmond (1974), based on ordinary least squares results; we will call this correct OLS, or COLS. ...estimate by OLS to obtain best linear unbiased estimates of $(\alpha_0 - \mu)$ and of the α_i. If a specific distribution is assumed for u, and if the parameters of this distribution can be derived from its higher-order (second, third, etc.) central moments, then we can estimate these parameters consistently from the moments of the OLS residuals. Since μ is a function of these parameters, it too can be estimated consistently, and this estimate can be used to 'correct' the OLS constant term, which is a consistent estimate of $(\alpha - \mu)$. COLS thus provides consistent estimates of all of the parameters of the frontier." We note that the words "correct" and "modify" do not appear in Richmond (1974).

Continuing, we have Førsund et al. (1980, p. 12): "A difficulty with the COLS technique is that, even after correcting the constant term, some of the residuals may still have the 'wrong' sign so that these observations end up above the estimated production frontier. This makes the COLS frontier a somewhat awkward basis for computing the technical efficiency of individual observations. One response to this problem is provided by the stochastic frontier approach discussed below. Another way of resolving the problem is to estimate (4) by OLS, and then to correct the constant term not as above, but by shifting it up until no residual is positive and one is zero. Gabrielsen (1975) and Greene (1980a) have both shown that this correction provides a consistent estimate of α_0."

All of this discussion on page 12 was in the context of a "deterministic" frontier, so no v was implicit in the model. But the usage of "correct/corrected" was clearly linked to the mean of inefficiency being used to adjust the OLS intercept. Moreover, just a few pages later, when discussing stochastic frontiers, the COLS terminology again appears, Førsund et al. (1980, p. 14): "Direct estimates of the stochastic production frontier model may be obtained by either maximum likelihood or COLS methods. …The model may also be estimated by COLS by adjusting the constant term by $E(u)$, which is derived from the moments of the OLS residuals. …Whether the model is estimated by maximum likelihood or by COLS, the distribution of u must be specified."

The reviews of both Førsund et al. (1980) and Schmidt (1985) are interesting because they use COLS generically to refer to the adjustment of the OLS intercept estimate, whether in a deterministic or stochastic frontier. The MOLS terminology is not present in either review paper.

By 1990 the efficiency community was quite large and the field well developed. Another special issue of the *Journal of Econometrics* (Lewin and Lovell, 1990) contained a survey of the field to date by Bauer (1990, p. 42), who explicitly recognized COLS: "Estimates of this model can be obtained using corrected ordinary least squares (COLS) or by maximizing the likelihood function directly." No citation of Gabrielsen (1975) appears nor does any mention of the MOLS terminology.

Also in this special issue is Greene (1990, p. 152): "…but the OLS constant term is biased …Greene (1980a) obtained estimates for the parameters of the disturbance distribution, and the constant term, in the gamma frontier model by manipulating the OLS residuals." Gabrielsen (1975) is then cited in relation to the work of Greene (1980a). This is then followed by Greene (1990, p. 153): "Only the first two [moments] are actually needed to correct the OLS intercept …"

So it would seem that to resolve this issue Gabrielsen (1975) would need to be consulted. Perhaps this paper had explicitly used the MOLS terminology.

4 The Gabrielsen Dilemma

The Gabrielsen (1975) citation in Lovell (1993) (and several earlier papers) links the unpublished working paper to the Christian Michelsen Institute (CMI), Department of Humanities and Social Sciences, in Bergen, Norway. There is no record of this paper on the current website of the CMI so I contacted Mr. Reidunn Ljones at the Bergen Resource Centre for International Development on July 31, 2017. Note that in 1992, the Department for Natural Science and Technology established the Christian Michelsen Research AS, and the CMR Group. The CMR Group is housed at the University of Bergen.

Upon first contact, Mr. Ljones wrote back to me on August 11, 2017: "I have now tried to find the publication you asked for. I can't find this title within the year 1975, or the number A-85 from another year. Neither his name Arne Gabrielsen. What I have found is that this title was published in Norwegian in 1973 with the number A-85. I think that they have translated the title in a reference/publication list. I can't find any trace suggesting that this publication was translated to English. Your reference must be wrong since 1975, A-85 is by another author with a different title."

After some further correspondence with Mr. Ljones about my desire to receive the paper (even in Norwegian) he wrote to me on August 14, 2017: "I could not find any paper on any of the different references. I have been searching on the author again, and found this reference below also. You will find that this is also the year 1973, but it's DERAP paper; 53 and not A-85. Since all the publications for these different references is missing in our library archive, I have to visit our remote archive. Gabrielsen, Arne Estimering av 'effisiente' produktfunksjoner : eksogene produksjonsfaktorer.—Bergen : CMI, 1973.—33 p. (DERAP paper; 53)."

On September 15, 2017 a scan of Gabrielsen (1973) was delivered to my inbox. I began using Google Translate to initially parse through the paper. From Gabrielsen (1973, p. 2), we have: "First, a modified version of the least squares method is presented." Actual wording is "Først utvikles en modifisert utgave av minstekvadraters metode." The Norwegian "modified" (*modifisert*) does not appear again in the paper nor is the acronym MOLS ever used, but we now have the definitive link to "modified" and the early papers of the 1970s studying estimation of the deterministic stochastic frontier model.

Despite the lack of the exact MOLS, it is clear that Gabrielsen (1973, p. 7) has in mind some form of adjustment to the OLS residual: "We will below develop the least squares estimators for the parameters of the model. We do this because the criterion function from the least-squares method in this model is different from what it would be if the residuals had expectation zero. However, it turns out that the least squares estimators for the limit elasticities in the model will be the usual least squares estimators we would get if the residuals had zero expectation. The difference occurs at the least squares estimates of the efficiency parameter, in our case, the multiplicative constant A and the least squares estimator to the expectation of the residuals."[7] This is the same intent that Richmond (1974) had.

At this point it might seem that Lovell (1993) had correctly attributed MOLS. However, Gabrielsen (1973, Eq. 5) statistical model was "deterministic" in nature; there was only one-sided inefficiency and so the adjustment to the OLS intercept was with the intent of constructing a frontier that lied everywhere above the data, see Gabrielsen (1973, Eqs. 11 and 12), or at least probabilistically lied above the data as in Timmer (1971). Thus it is not clear how the very specific COLS of Olson et al. (1980) for the stochastic frontier model came to be associated with the MOLS of Gabrielsen (1973) for the deterministic frontier model.

[7] Actual text: "Vi skal nedenfor utvikle minstekvadratersestimatorene for parametrene i modellen. Vi gjør dette fordi kriteriefunksjonen ved, minstekvadraters metode i denne modellen er forskjellig fra det den ville være om restleddene hadde forventning null. Det viser seg imidlertid at minstekvadratersestimatorene for grenseelastisitetene i modellen blir de vanlige minstekvadratersestimatorene vi ville fått om restleddet hadde forventning null. Forskjellen inntreffer ved minstekvadratersestimatorene for effisiensparameteren, i vårt tilfelle det multiplikative konstanten A og ved minstekvadratersestimatoren til forventningen av restleddet."

5 Textbook Treatment of the Acronyms

It is also instructive to observe how many of the leading textbooks on efficiency and productivity analysis approach this subject. For example, Kumbhakar and Lovell (2000, p. 70) have a section titled "Corrected Ordinary Least Squares (COLS)" which cites Winsten (1957) and then describes exactly the approach laid out in Greene (1980a) while another section titled "Modified Ordinary Least Squares (MOLS)" (Kumbhakar & Lovell, 2000, p. 71) describes explicitly the approach of Afriat (1972) and Richmond (1974), i.e., the deterministic frontier case. No citation of Gabrielsen (1973) exists. Further on, Kumbhakar and Lovell (2000, p. 91), when describing Olson et al. (1980) refer to the method as MOLS: "This two-part estimation procedure amounts to the application of MOLS to a stochastic production frontier model." As we have seen Olson et al. (1980) never use the MOLS terminology.

Another prominent textbook in this area Coelli et al. (2005, p. 242, Sect. 9.2) also wades into this terminology: "...while Richmond (1974) used a least squares technique, sometimes known as *modified ordinary least squares (MOLS)*." A few pages later, Coelli et al. (2005, p. 245, Sect. 9.3) "One solution to this problem is to correct for the bias in the intercept term using a variant of a method suggest by Winsten (1957)—the resulting estimation is often known as the corrected ordinary least squares (COLS) estimator." There is a footnote in this passage that then states "Winston suggested the COLS estimator in the context of the deterministic frontier ..." Other contemporary textbook treatments are no more specific. Kumbhakar et al. (2015, p. 50) Sect. 3.3.1 entitled "Correct OLS (COLS)" describes the approach of Winsten (1957) and Greene (1980a). See also Sect. 4.3.1 which uses the same terminology. This textbook does not discuss the approach of Olson et al. (1980) whatsoever.

O'Donnell (2018) has separate chapters (7 and 8) dedicated to the estimation of deterministic and stochastic frontiers. Specifically, within each chapter he includes a section on least squares estimation (Sects. 7.3 and 8.2, respectively). In Sect. 7.3 of O'Donnell (2018, p. 268), adjustment of the OLS intercept from the deterministic frontier model is termed COLS: "...is the COLS estimate of the production frontier; by design, it runs parallel to the OLS line of best fit (the dotted line) and envelops **all** the points in the scatterplot" (emphasis mine).

When the least squares estimation of the stochastic frontier is discussed, O'Donnell (2018, p. 302) introduces the MOLS acronym. There is also footnote 4 on the same page: "Elsewhere, these estimators are sometimes referred to as corrected ordinary least squares (COLS) estimators; see, for example, Horrace and Schmidt (1996, p. 260). In this book, the term COLS is reserved for LS estimators for the parameters in deterministic frontier models." An interesting note about Horrace and Schmidt (1996) is that while they deploy the COLS terminology in reference to correction of a least squares estimate of the intercept of a stochastic frontier model, when they introduce (p. 260) a panel data model estimated using generalized least squares (GLS) to account for the presence of random effects, they adjust the corresponding estimate here as well and term the estimator corrected GLS (CGLS), and so again we have use of the "C" for adjustment in a stochastic frontier setting.

Finally, the most recent textbook in the field also discusses both terminologies. Section 11.2 in Sickles and Zelenyuk (2019) is named "Corrected OLS" and refers to what here has been described as MOLS, while Sect. 11.4.1 describes the Olson et al. (1980, p. 372) approach and uses both COLS and MOLS terminology: "...in the usual deterministic COLS method and compared the ALS methodology with their version of COLS, which is often referred to as *modified* OLS or simply MOLS."

As is clear even the various textbooks in our area seem to use both terminologies in different manners.

6 Use of the Acronyms in the *Journal of Productivity Analysis*

To assess how differently the MOLS/COLS usage has been a Google Scholar search[8] for the term COLS in the *Journal of Productivity Analysis* turned up 29 articles. Each of these papers was read to determine if the intent of using COLS was to a stochastic or a deterministic frontier. This reduced the number of articles by five. Two additional relevant papers were found that used corrected OLS (without the COLS acronym). Further, a Google Scholar search for the term MOLS in the *Journal of Productivity Analysis* turned up six articles. Again, each of these papers were read to determine if the intent of using MOLS was to a stochastic or a deterministic frontier. This reduced the number of articles by one. One additional relevant paper was found that used modified OLS (without the MOLS acronym).

Of these 32 articles we have that all six papers using MOLS or modified OLS are always in the context of a stochastic frontier that shifts the OLS intercept based on a presumed distributional assumption for u. However, usage of the COLS acronym is mixed. 15 of the 26 articles use COLS (or corrected OLS) in the context of a deterministic frontier model, while eight articles use COLS (or corrected OLS) in the context of a stochastic frontier while three had no clear distinction between stochastic or deterministic frontier in their use.

Several of the papers invoking the terminology do so in ways that one might wonder how any of these terms came to be. For example, Amsler et al. (2013, p. 294): "[COLS] was first suggested by Winsten (1957)—though it was literally a one-sentence suggestion—and then further developed by Greene (1980a), who proved the consistency of the COLS estimators of α and β." This passage is especially interesting as Peter Schmidt of Olson et al. (1980) was the first to use the COLS acronym (for a stochastic frontier) and here it is being used for a deterministic frontier and the references to both Winsten (1957) and Greene (1980a), as discussed prior, never use this terminology.

[8] Conducted on March 9, 2021.

Similarly, in their influential work Simar et al. (2017, p. 190) proposed nonparametric estimation of the frontier itself coupled with an adjustment of the estimated conditional mean to construct a stochastic frontier. Simar et al. (2017, p. 190): "Our approach can be viewed as a non- or semi-parametric version of the 'modified OLS' (MOLS) method that was introduced as an alternative to MLE method for SFA in parametric setups." Later on page 192 they state "We will extend the idea of the Modified OLS (MOLS), originated in the full parametric, homoskedastic stochastic frontier models (see Olson et al., 1980) for our semi-parametric setup." But as we have discussed above Olson et al. (1980) call their procedure COLS.

The approach of estimating the stochastic frontier model via OLS and then adjusting the estimates to construct a stochastic frontier has also appeared in Wikström (2016) who deployed the MOLS acronym when detailing a panel data stochastic frontier model that involves shifting the estimates of the unobserved heterogeneity to take account of the two part nature. And there is Kumbhakar and Lien (2018, p. 23) who develop the intercept adjustment for residuals estimated from a random-effects panel data model for the generalized panel data stochastic frontier model, without any acronym connection and a simple descriptor of "method of moments estimation."[9] There is the important work of Amsler et al. (2016, p. 281) which discusses COLS estimation in the presence of endogeneity, what they term C2SLS: "This is a straightforward generalization of COLS, which perhaps surprisingly does not appear to have been discussed in the literature." Here it is clear that the intent of C2SLS is with respect to the COLS proposal of Olson et al. (1980).

Lastly, using the acronym OLSE+MME, Huynh et al. (2021, p. 8) reinvent the method without any attribution to the stochastic frontier literature whatsoever: "To the best of our knowledge, the estimation of $\boldsymbol{\beta}$, σ, and λ that we present here is completely new and the estimators have fairly well structured closed forms."

7 Who Cares?

At this point, it might be fair to ask why a discussion of this nature is important. To begin, the COLS estimator of Olson et al. (1980) has many uses. First, for various distributional pairs, the estimator is a crucial diagnostic to determine if wrong skewness (Waldman, 1982; Simar & Wilson, 2010) is likely an impediment for full maximum likelihood. Further, this estimator is quite simple to implement in the case that one does not have access to a nonlinear optimizer or has encountered convergence problems in more traditional maximum likelihood.

One contemporary issue with COLS is that the method as constituted cannot handle determinants of inefficiency. This is of fundamental importance as it essentially limits the use of the method to cases where inefficiency is essentially random. However, in this setting, there is essentially no need for COLS but just straight nonlinear

[9] See also Kumbhakar and Parmeter (2019, Sect. 3.1).

least squares (NLS). Specifically, if we assume that $u \sim \mathcal{N}_+(0, \sigma_u^2(z; \delta))$, we know that $E[u] = \sqrt{2/\pi}\sigma_u(z; \boldsymbol{\delta})$ and so we can rewrite the model in Eq. (2) as

$$\begin{aligned} y_i =& \beta_0 - E(u) + \boldsymbol{x}_i'\beta + v_i - (u_i - E(u)) \\ =& \beta_0 + \boldsymbol{x}_i'\beta - \sqrt{2/\pi}\sigma_u(z; \boldsymbol{\delta}) + \varepsilon_i^*, \end{aligned} \tag{3}$$

where upon parametric specification of $\sigma_u(z; \boldsymbol{\delta})$, such as $\sigma_u(z; \boldsymbol{\delta}) = \sqrt{2/\pi}e^{z_i'\delta}$ the model can be estimated using NLS:

$$\left(\widehat{\beta}_0, \widehat{\boldsymbol{\beta}}, \widehat{\boldsymbol{\delta}}\right) = \min_{\beta_0, \boldsymbol{\beta}, \boldsymbol{\delta}} n^{-1} \sum_{i=1}^{n} \left[y_i - \beta_0 - \boldsymbol{x}_i'\boldsymbol{\beta} + \sqrt{2/\pi}e^{z_i'\delta}\right]^2. \tag{4}$$

In essence the specification of the mean of u automatically corrects the intercept and allows direct estimation of the model in a regression setup through NLS as opposed to MLE. It does not appear that this avenue has been fully explored in the literature. One thing that must be recognized is that the error term, ε_i^*, is heteroskedastic and so robust standard errors should be used following the approach of White using sandwich form of the variance covariance matrix for any inference conducted on the parameters.

Standard errors which are robust to heteroskedasticity can be easily calculated by noting that the nonlinear least squares estimator is simply an M-estimator Huber (1964). Consider an objective function $\Psi(y, \boldsymbol{x}, z, \theta)$, the derivative of which is $\psi(y, \boldsymbol{x}, z, \theta)$. An M-estimator of θ, $\widehat{\theta}$, is defined as

$$\sum_{i=1}^{n} \psi(y_i, \boldsymbol{x}_i, z_i, \widehat{\theta}) = 0. \tag{5}$$

As shown in White (1994, Theorem 6.10), for M-estimators

$$\sqrt{n}\left(\widehat{\theta} - \theta\right) \xrightarrow{d} N(0, S(\theta)), \tag{6}$$

where $S(\theta) = H(\theta)D(\theta)H(\theta)$, with

$$H(\theta) = \left(E[-\psi'(y, \boldsymbol{x}, z, \theta)]\right)^{-1} \text{ and } D(\theta) = Var[\psi(y, \boldsymbol{x}, z, \theta)].$$

$H(\theta)$ is the inverse of the expected value of the second derivative of the objective function, $\Psi(y, \boldsymbol{x}, z, \theta)$, and is most easily viewed as the Hessian matrix. An estimator for $D(\theta)$ is the outer product of the empirical first derivatives of the objective function (or the outer product of gradients estimator):

$$\widehat{D} = n^{-1} \sum_{i=1}^{n} \psi(y_i, \boldsymbol{x}_i, z_i, \widehat{\theta})\psi(y_i, \boldsymbol{x}_i, z_i, \widehat{\theta})'. \tag{7}$$

Note that this approach is heavily parametric. If one is willing to only locally assume the distribution of u then the COLS approach of Simar et al. (2017) is perhaps the most flexible and usable for the field. This framework avoids maximum likelihood, allows nonparametric estimation of the production frontier, and the functional form of the parameter(s) of the distribution of u.

There are likely other issues with COLS that the practitioner should also be aware of. Namely, the use of more flexible distributions (with say more than two parameters) or (spatial) dependence is likely to render COLS infeasible. To see this most clearly, consider the case where u and v are no longer independent. Even though the third moment of v is 0 (assuming a symmetric distribution), the third moment of ε will now depend on the parameters of v. That is, we have $E(\varepsilon^3) = E[v^3] - 3E[v^2(u - E[u])] + 3E[v(u - E[u])^2] - E[(u - E[u])^3]$. Under symmetry and independence the first three terms are all 0 and so as is typical, for a one parameter inefficiency distribution, the parameter can be identified by solving out. However, if an asymmetric distribution is used for v or there is dependence, then that dependence needs to be parameterized (either through a specific bivariate distribution or a specification of the type of copula dependence). Consider the setting of Smith (2008) where $v \sim \mathcal{N}(0, \sigma_v^2)$ and $u \sim \mathcal{N}_+(0, \sigma_u^2)$ and u and v are allowed to be dependent. The complication that this framework introduces is that even though the third moment of v is 0, the dependence now requires that "how" u and v are dependent upon one another will place a crucial role in identifying and estimating the parameters in a COLS-type setup.[10] Extensions in this area would probably mark dramatic improvements in the COLS approach to more interesting and realistic settings.

8 The "Final" Verdict

Given my age and lack of access to a time machine, it is impossible to know for sure the discussions relating to the argot that developed in the frontier literature after Aigner et al. (1977) at the various conferences that arose from the origins of this field. Far be it for me to lay down the gauntlet and suggest which term should refer to which model, especially in light of my own conflation of the COLS/MOLS acronyms. However, it is clear from the literature as reviewed above that the first use of COLS was with respect to a stochastic frontier model that was designed to correct the OLS estimator of the intercept up by the mean of inefficiency and thus not all of the data would be bounded by the corresponding estimated frontier. On the other hand, the use of the MOLS acronym does not directly appear in any of the early literature and the use of the word "modify" **always** appeared in the context of a deterministic frontier.

Given that there exist two different distinctions as to which one might want to adjust an estimated conditional mean, the intent is important. Note that the explicit

[10] Smith (2008) provides a straightforward simulated maximum likelihood estimation routine.

introduction of COLS (Olson et al., 1980) was for a stochastic frontier model, independent of a specific distributional assumption. The introduction of the word modify (Gabrielsen, 1973), while also dependent upon a distributional assumption, was in the context of a deterministic frontier. It is this distinction that I believe to be important when adjudicating between the two acronyms. Should COLS/MOLS be used to refer to how much to shift the frontier up or by the type of frontier one is working with? Given that the type of frontier model being deployed is more important than the amount of adjustment, this should be the dominant force driving the information conveyed to a reader/listener when using either of the terms.

Thus, it is the hope of this article that COLS will be used to refer to those methods which correct the intercept in a "stochastic frontier model" based on some type of distributional assumption on v and u and MOLS will refer to any method that constructs a true frontier where all of the data are bounded by the subsequently estimated frontier.

Or, recognizing the work of Førsund et al. (1980), Schmidt (1985) and Amsler et al. (2013), Peter Schmidt has used COLS in both the deterministic and the stochastic frontier setting. Perhaps it is best to just use COLS and retire MOLS?

9 COLS in the JPA

A Google Scholar search for the term COLS in the *Journal of Productivity Analysis* turned up 29 articles. Each of these papers were read to determine if the intent of using COLS was to a stochastic or a deterministic frontier. This reduced the number of articles by five. Two additional relevant papers were found that used corrected OLS (without the COLS acronym).

Førsund (1992)—clear that intent of COLS is to a deterministic frontier model.

Seaver and Triantis (1992)—cites COLS in relation to Afriat (1972) and Greene (1980b) and it is clear that the intent is for a deterministic frontier model.

Neogi and Ghosh (1994)—cites COLS in relation to Richmond (1974) and clear that the intent is a deterministic frontier model.

Coelli (1995)—clear that intent of COLS is to Olson et al. (1980) and so a stochastic frontier model.

Wilson (1995)—cites COLS in relation to Greene (1980b) and it is clear that the intent is for a deterministic frontier model.

Banker (1996)—cites COLS in relation to Olson et al. (1980) but appears to be concerned with a deterministic frontier model.

Horrace and Schmidt (1996)—clear that intent of COLS is to Olson et al. (1980) and so a stochastic frontier model.

Bardhan et al. (1998)—uses COLS where it is clear that the intent is for a deterministic frontier model.

Gstach (1998)—not clear what the intended usage of COLS is.

Kerkvliet et al. (1998)—cites COLS in relation to Greene (1980b) and it is clear that the intent is for a deterministic frontier model.

Zhang (1999)—cites Olson et al. (1980) and Coelli (1995) and it is clear that the intended usage is to a stochastic frontier model.

Cuesta (2000)—clear that intent of COLS is to a stochastic frontier model.

Førsund and Sarafoglou (2000)—cites COLS in relation to Richmond (1974) but not clear that the intended usage is to a deterministic/stochastic frontier model.

Fuentes et al. (2001)—clear that the intent of COLS is to a deterministic frontier model.

Banker et al. (2002)—cites Richmond (1974), Greene (1980a), and Olson et al. (1980). Not clear what intended usage is for.

Jensen (2005)—cites COLS in relation to Winsten (1957) and it is clear that the intended usage is for a deterministic frontier model.

Smet (2007)—cites COLS in relation to Coelli (1995) and is clear that the intent of COLS is to a stochastic frontier model.

Simar and Wilson (2011)—use corrected OLS where it is clear that the intended usage is for a deterministic frontier model.

Amsler et al. (2013)—clear that the intended usage is for a deterministic frontier model.

Lai (2013)—uses corrected OLS where it is clear that the intended usage is for a stochastic frontier model.

Kuosmanen and Kortelainen (2012)—clear that the intent of COLS is to Greene (1980a) and so is referring to a deterministic frontier model.

Andor and Hesse (2014)—cites COLS in relation to Winsten (1957) and it is clear that the intended usage is for a deterministic frontier model.

Henningsen et al. (2015)—clear that the intent of COLS is to a stochastic frontier model.

Minegishi (2016)—clear that the intent of COLS is to Greene (1980a) and so is referring to a deterministic frontier model.

Wheat et al. (2019)—clear that the intent of COLS is to a deterministic frontier model.

Papadopoulos (2021)—clear that the intent of COLS is to a stochastic frontier model.

10 MOLS in the JPA

A Google Scholar search for the term MOLS in the *Journal of Productivity Analysis* turned up six articles. Each of these papers was read to determine if the intent of using MOLS was to a stochastic or a deterministic frontier. This reduced the number of articles by one. One additional relevant paper was found that used modified OLS (without the MOLS acronym).

Cummins and Zi (1998)—clear that the intended use of MOLS is to Greene (1990) and so is referring to a stochastic frontier model.

Serra and Goodwin (2009)—clear that the intended use of MOLS is to a stochastic frontier model.

Kuosmanen and Kortelainen (2012)—clear that the intent of MOLS is to Olson et al. (1980) and so is referring to a stochastic frontier model.

Minegishi (2016)—uses modified OLS with the intent of distinguishing it from COLS (for a deterministic frontier) and so is clear that this use is for a stochastic frontier model.

Wikström (2016)—cites MOLS in relation to Richmond (1974) and Greene (1980b) but clear that the intent of usage of MOLS is to a stochastic frontier model.

Simar et al. (2017)—clear that the intent of MOLS is to Olson et al. (1980) and so is referring to a stochastic frontier model.

All five of the articles identified used MOLS specifically to refer to the intercept correction in the context of a stochastic frontier model.

11 A Collection of COLS Results

Here we collect various results from papers that have presented COLS estimators. We will try to use a consistent notation throughout this appendix.

To begin, we note that the estimation of the parameters of the distribution of inefficiency is identical across various noise distributions provided they are symmetric (Normal, Laplace, Cauchy, etc.). Let $m_j = n^{-1}\sum_{i=1}^{n}(\hat{\varepsilon}_i - \bar{\hat{\varepsilon}})^j$ denote the jth central moment. We note that since $\bar{\hat{\varepsilon}} = 0$ the central moments of the estimated composite errors are equivalent to the raw moments.

(1) $u \overset{iid}{\sim} \mathcal{E}(\sigma_u)$. We have $\widehat{\sigma}_u = \sqrt[3]{-m_3/2}$;

(2) $u \overset{iid}{\sim} \mathcal{N}_+(0, \sigma_u^2)$. We have $\widehat{\sigma}_u = \sqrt[3]{\left(\frac{\pi}{\pi-4}\right)\sqrt{\pi/2}m_3}$;

(3) $u \overset{iid}{\sim} \mathcal{B}(N, p)$. Define $\xi = (m_4 - 3m_2^2)/m_3$. We have $\widehat{p} = (1/2) + (1/6)\xi + sign(m_3)(1/6)\sqrt{\xi^2+3}$ and $\widehat{N} = -m_3/\left(\widehat{p}(1-\widehat{p})(1-2\widehat{p})\right)$;

(4) $u \overset{iid}{\sim} \mathcal{U}[0, \sigma_u]$. We have $\widehat{\sigma}_u = \sqrt[4]{120(3m_2^2 - m_4)}$;

(5) $u \overset{iid}{\sim} \Gamma(k, \theta)$ where k is the shape and θ is the rate. We have $\widehat{\theta} = -3m_3/(m_4 - 3m_2^2)$ and $\widehat{k} = -\widehat{\theta}^3 m_3/2$;

(6) $u \overset{iid}{\sim} GE(2, \sigma_u, 0)$. We have $\widehat{\sigma}_u = \sqrt[3]{-(4/9)m_3}$.

(7) $u \overset{iid}{\sim} \mathcal{N}_+(\mu, \sigma_u^2)$. In this case, a closed-form solution does not exist. This requires solving a nonlinear system of three equations for three unknowns $(\sigma_v^2, \mu, \sigma_u^2)$. See either Harris (1992) or Goldstein (2003) for precise details.

References

Afriat, S. N. (1972). Efficiency estimation of production functions. *International Economic Review, 13*(3), 568–598.

Aigner, D., & Chu, S. (1968). On estimating the industry production function. *American Economic Review, 58*, 826–839.

Aigner, D. J., Lovell, C. A. K., & Schmidt, P. (1977). Formulation and estimation of stochastic frontier production functions. *Journal of Econometrics, 6*(1), 21–37.

Aigner, D. J., & Schmidt, P. (1980). Editors' introduction. *Journal of Econometrics, 13*(1), 1–3.

Amsler, C., Leonard, M., & Schmidt, P. (2013). Estimation and inference in parametric deterministic frontier models. *Journal of Productivity Analysis, 40*(3), 293–305.

Amsler, C., Prokhorov, A., & Schmidt, P. (2016). Endogeneity in stochastic frontier models. *Journal of Econometrics, 190*, 280–288.

Andor, M., & Hesse, F. (2014). The StoNED Age: The departure into a new era of efficiency analysis? A Monte Carlo comparison of StoNED and the "oldies" (SFA and DEA)', *Journal of Productivity Analysis, 41*(1), 85–109.

Banker, R. D. (1996). Hypothesis tests using data envelopment analysis. *Journal of Productivity Analysis, 7*(1), 139–159.

Banker, R. D., Janakiraman, S., & Natarajan, R. (2002). Evaluating the adequacy of parametric functional forms in estimating monotone and concave production functions. *Journal of Productivity Analysis, 17*(1), 111–132.

Bardhan, I. R., Cooper, W. W., & Kumbhakar, S. C. (1998). A simulation study of joint uses of data envelopment analysis and statistical regressions for production function estimation and efficiency evaluation. *Journal of Productivity Analysis, 9*(2), 249–278.

Bauer, P. W. (1990). Recent developments in the econometric estimation of frontiers. *Journal of Econometrics, 46*(1), 39–56.

Carree, M. A. (2002). Technological inefficiency and the skewness of the error component in stochastic frontier analysis. *Economics Letters, 77*(1), 101–107.

Chu, S.-F. (1978). On the statistical estimation of parametric frontier production functions: A reply and further comments. *The Review of Economics and Statistics, 60*(3), 479–481.

Coelli, T. J. (1995). Estimators and hypothesis tests for a stochastic frontier function: A Monte Carlo analysis. *Journal of Productivity Analysis, 6*(4), 247–268.

Coelli, T. J., Rao, D. S. P., O'Donnell, C. J., & Battese, G. E. (2005). *In Introduction to Efficiency and Productivity Analysis* (2nd ed.). New York, NY: Springer.

Cuesta, R. A. (2000). A production model with firm-specific temporal variation in technical inefficiency: With application to Spanish dairy farms. *Journal of Productivity Analysis, 13*, 139–152.

Cummins, J. D., & Zi, H. (1998). Comparison of frontier efficiency methods: An application to the US. life insurance industry. *Journal of Productivity Analysis, 10*(1), 131–152.

Farrell, M. J. (1957). The measurement of productive efficiency. *Journal of the Royal Statistical Society Series A, General, 120*(3), 253–281.

Førsund, F. (1992). A comparison of parametric and non-parametric efficiency measures: The case of Norwegian ferries. *Journal of Productivity Analysis, 3*(1), 25–43.

Førsund, F. R., Lovell, C. A. K., & Schmidt, P. (1980). A survey of frontier production functions and of their relationship to efficiency measurement. *Journal of Econometrics, 13*(1), 5–25.

Førsund, F. R., & Sarafoglou, N. (2000). On the origins of data envelopment analysis. *Journal of Productivity Analysis, 17*(1), 23–40.

Fuentes, H. J., & E., G.-T. and Perelman, S. (2001). A parametric distance function approach for Malmquist productivity index estimation. *Journal of Productivity Analysis, 15*(1), 79–94.

Gabrielsen, A. (1973), Estimering av "effisiente" produktfunksjoner:eksogene produksjonsfaktorer. DERAP paper; 53. Bergen: CMI.

Gabrielsen, A. (1975). On estimating efficient production functions. Working Paper No. A-85, Chr. Michelsen Institute, Department of Humanities and Social Sciences, Bergen, Norway.

Goldstein, H. (2003). On the COLS and CGMM moment estimation methods for frontier production functions. In B. P. Stigum (Ed.), *Econometrics and the Philosophy of Economics*. Princeton, NJ: Princeton University Press, Chapter 14.

Greene, W. H. (1980a). Maximum likelihood estimation of econometric frontier functions. *Journal of Econometrics, 13*(1), 27–56.

Greene, W. H. (1980b). Maximum likelihood estimation of econometric frontier functions. *Journal of Econometrics, 13*(1), 27–56.

Greene, W. H. (1990). A gamma-distributed stochastic frontier model. *Journal of Econometrics, 46*(1–2), 141–164.

Greene, W. H. (2008). The econometric approach to efficiency analysis. In H. O. Fried, C. A. K. Lovell & S. S. Schmidt (Eds.), *The Measurement of Productive Efficiency and Productivity Change*. Oxford, UK: Oxford University Press, Chapter 2.

Gstach, D. (1998). Another approach to data envelopment analysis in noisy environments: DEA+. *Journal of Productivity Analysis, 9*(1), 161–176.

Harris, C. M. (1992). Technical efficiency in Australia: Phase 1. In R. E. Caves (Ed.), *Industrial efficiency in six nations* (pp. 199–240). Cambridge, MA: The MIT Press, Chapter 5.

Henningsen, G., Henningsen, A., & Jensen, U. (2015). A Monte Carlo study on multiple output stochastic frontiers: A comparison of two approaches. *Journal of Productivity Analysis, 44*(3), 309–320.

Horrace, W. C., & Schmidt, P. (1996). Confidence statements for efficiency estimates from stochastic frontier models. *Journal of Productivity Analysis, 7*, 257–282.

Huber, P. J. (1964). Robust estimation of a location parameter. *The Annals of Mathematical Statistics, 35*(1), 73–101.

Huynh, U., Pal, N., & Nguyen, M. (2021). Regression model under skew-normal error with applications in predicting groundwater arsenic level in the Mekong Delta Region. *Environmental and Ecological Statistics*.

Jensen, U. (2005). Misspecification preferred: The sensitivity of inefficiency rankings. *Journal of Productivity Analysis, 23*(2), 223–244.

Kerkvliet, J. R., Nebesky, W., Tremblay, C. H., & Tremblay, V. J. (1998). Efficiency and technological change in the US brewing industry. *Journal of Productivity Analysis, 10*(2), 271–288.

Kumbhakar, S. C., & Lien, G. (2018). Yardstick regulation of electricity distribution—disentangling short-run and long-run inefficiencies. *The Energy Journal, 38*, 17–37.

Kumbhakar, S. C., & Lovell, C. A. K. (2000). *Stochastic frontier analysis*. Cambridge University Press.

Kumbhakar, S. C., & Parmeter, C. F. (2019). Implementing generalized panel data stochastic frontier estimators. In: M. Tsionas (Ed.), *Panel data econometrics* (pp. 225–249). Academic Press, Chapter 9. https://www.sciencedirect.com/science/article/pii/B9780128143674000095.

Kumbhakar, S. C., Parmeter, C. F., & Zelenyuk, V. (2020). Stochastic frontier analysis: Foundations and advances I. In S. Ray, R. Chambers & S. C. Kumbhakar (Eds.), *Handbook of Production Economics* (Vol. 1). Springer. Forthcoming.

Kumbhakar, S. C., Wang, H.-J., & Horncastle, A. (2015). *A practitioner's guide to stochastic frontier analysis*. Cambridge, England: Cambridge University Press.

Kuosmanen, T., & Kortelainen, M. (2012). Stochastic non-smooth envelopment of data: Semiparametric frontier esitmation subject to shape constraints. *Journal of Productivity Analysis, 38*(1), 11–28.

Lai, H.-P. (2013). Estimation of the threshold stochastic frontier model in the presence of an endogenous sample split variable. *Journal of Productivity Analysis, 40*(2), 227–237.

Lewin, A. Y., & Lovell, C. A. K. (1990). Editor's introduction. *Journal of Econometrics, 46*(1), 3–5.

Li, Q. (1996). Estimating a stochastic production frontier when the adjusted error is symmetric. *Economics Letters, 52*(3), 221–228.

Lovell, C. A. K. (1993). Production frontiers and productive efficiency. In H. O. Fried, C. A. K. Lovell & S. S. Schmidt (Eds.), *The Measurement of Productive Efficiency*. Oxford, UK: Oxford University Press, Chapter 1.

Minegishi, K. (2016). Comparison of production risks in the state-contingent framework: Application to balanced panel data. *Journal of Productivity Analysis, 46*(1), 121–138.

Neogi, C., & Ghosh, B. (1994). Intertemporal efficiency variations in Indian manufacturing industries. *Journal of Productivity Analysis, 5*(3), 301–324.

Nguyen, N. B. (2010). Estimation of technical efficiency in stochastic frontier analysis, Ph.D. thesis, Bowling Green State University.

O'Donnell, C. J. (2018). *Productivity and efficiency analysis*. Singapore: Springer Singapore.

Olson, J. A., Schmidt, P., & Waldman, D. A. (1980). A Monte Carlo study of estimators of stochastic frontier production functions. *Journal of Econometrics, 13*, 67–82.

Papadopoulos, A. (2021). Stochastic frontier models using the generalized exponential distribution. *Journal of Productivity Analysis, 55*(1), 15–29.

Parmeter, C. F., & Kumbhakar, S. C. (2014). Efficiency analysis: A primer on recent advances. *Foundations and Trends in Econometrics, 7*(3–4), 191–385.

Richmond, J. (1974). Estimating the efficiency of production. *International Economic Review, 15*(2), 515–521.

Schmidt, P. (1976). On the statistical estimation of parametric frontier production functions. *The Review of Economics and Statistics, 58*(2), 238–239.

Schmidt, P. (1978). On the statistical estimation of parametric frontier production functions: Rejoinder. *The Review of Economics and Statistics, 60*(3), 481–482.

Schmidt, P. (1985). Frontier production functions. *Econometric Reviews, 4*(2), 289–328.

Seaver, B. L., & Triantis, K. P. (1992). A fuzzy clustering approach used in evaluating technical efficiency measures in manufacturing. *Journal of Productivity Analysis, 3*(3), 337–363.

Serra, T., & Goodwin, B. K. (2009). The efficiency of Spanish arable crop organic farms, a local maximum likelihood approach. *Journal of Productivity Analysis, 31*(1), 113–124.

Sickles, R. C., & Zelenyuk, V. (2019). *Measurement of productivity and efficiency: Theory and practice*. Cambridge, UK: Cambridge Univeristy Press.

Simar, L., Van Keilegom, I., & Zelenyuk, V. (2017). Nonparametric least squares methods for stochastic frontier models. *Journal of Productivity Analysis, 47*(3), 189–204.

Simar, L., & Wilson, P. W. (2010). Inferences from cross-sectional, stochastic frontier models. *Econometric Reviews, 29*(1), 62–98.

Simar, L., & Wilson, P. W. (2011). Two-stage DEA: Caveat emptor. *Journal of Productivity Analysis, 36*(2), 205–218.

Smet, M. (2007). Measuring performance in the presence of stochastic demand for hospital services: An analysis of Belgian general care hospitals. *Journal of Productivity Analysis, 27*(1), 13–29.

Smith, M. (2008). Stochastic frontier models with dependent error components. *The Econometrics Journal, 11*(1), 172–192.

Timmer, C. P. (1970). On measuring technical efficiency. *Food Research Institute Studies in Agricultural Economics, Trade, and Development, 9*(2), 99–171.

Timmer, C. P. (1971). Using a probabilistic frontier production function to measure technical efficiency. *The Journal of Political Economy, 79*(4), 776–794.

Waldman, D. M. (1982). A stationary point for the stochastic frontier likelihood. *Journal of Econometrics, 18*(1), 275–279.

Wheat, P., Stead, A. D., & Greene, W. H. (2019). Robust stochastic frontier analysis: A Student's t-half normal model with application to highway maintenance costs in England. *Journal of Productivity Analysis, 51*(1), 21–38.

White, H. (1994). *Estimation, inference, and specification analysis*. Cambridge, England: Cambridge University Press.

Wikström, D. (2016). Modified fixed effects estimation of technical inefficiency. *Journal of Productivity Analysis, 46*(1), 83–86.

Wilson, P. W. (1995). Detecting influential observations in data envelopment analysis. *Journal of Productivity Analysis, 6*(1), 27–45.

Winsten, C. B. (1957). Discussion on Mr. Farrell's paper. *Journal of the Royal Statistical Society Series A, General, 120*(3), 282–284.

Zhang, X. (1999). A Monte Carlo study on the finite sample properties of the Gibbs sampling method for a stochastic frontier model. *Journal of Productivity Analysis, 14*(1), 71–83.

Stochastic Frontier Analysis with Maximum Entropy Estimation

Pedro Macedo, Mara Madaleno, and Victor Moutinho

1 Introduction

Stochastic frontier analysis (SFA) was first proposed by Aigner et al. (1977), Battese and Corra (1977), and Meeusen and van Den Broeck (1977), and it became one important tool in economic efficiency analysis. The general stochastic frontier model is usually defined as

$$\ln y_n = f(\boldsymbol{x}_n, \boldsymbol{\beta}) + v_n - u_n, \tag{1}$$

where y_n is the scalar output for producer n $(n = 1, 2, \ldots, N)$, $f(\boldsymbol{x}_n, \boldsymbol{\beta})$ represents the production frontier, $\boldsymbol{x}_n$ is a row vector with logarithms of inputs, $\boldsymbol{\beta}$ is a column vector of parameters to be estimated, v is a two-sided statistical noise component, and $u(\geq 0)$ is a one-sided component representing technical inefficiency.

The parameters of the stochastic frontier model in (1) are usually estimated through maximum likelihood (ML), method of moments (MOM), and corrected ordinary least squares (COLS). The random variable v is usually assumed to be normally

P. Macedo (✉)
CIDMA—Center for Research and Development in Mathematics and Applications, Department of Mathematics, University of Aveiro, 3810-193 Aveiro, Portugal
e-mail: pmacedo@ua.pt

M. Madaleno
GOVCOPP—Research Unit in Governance, Competitiveness and Public Policy, Department of Economics, Management, Industrial Engineering and Tourism, University of Aveiro, 3810-193 Aveiro, Portugal
e-mail: maramadaleno@ua.pt

V. Moutinho
NECE—Research Unit in Business Sciences, Department of Management and Economics, University of Beira Interior, 6200-209 Covilhã, Portugal
e-mail: ferreira.moutinho@ubi.pt

P. Macedo et al. (eds.), *Advanced Mathematical Methods for Economic Efficiency Analysis*, Lecture Notes in Economics and Mathematical Systems 692,
https://doi.org/10.1007/978-3-031-29583-6_14

distributed, $N\left(0, \sigma_v^2\right)$, and the u is usually defined by an exponential, a nonnegative half normal, a truncated normal or a gamma distribution (e.g., Kumbhakar & Knox Lovell, 2000; and many influential references therein). Selecting one of these distributions, in particular for the u error component, is usually the main criticism of SFA since different distributional assumptions can lead to different predictions of technical efficiency.

An interesting alternative to ML is maximum entropy (ME) estimation. The ME formalism was first established by Jaynes (1957a, b). Later, Golan et al. (1996) generalized the ME formalism and developed the generalized maximum entropy (GME) and the generalized cross entropy (GCE) estimators, which are stable estimation procedures in ill-posed models, namely in models exhibiting collinearity, in models with small sample sizes (micronumerosity) and non-normal errors, as well as in models where the number of parameters to be estimated exceeds the number of observations available (under-determined models). These estimation techniques were recently embedded into a new area of research entitled info-metrics (Golan, 2018).

Since frontier models are usually ill-posed, an increasing interest in ME estimation in SFA has emerged in the literature; e.g., Campbell et al. (2008), Rezek et al. (2011), Macedo and Scotto (2014), Macedo et al. (2014), Robaina-Alves et al. (2015), Moutinho et al., (2018a, 2018b), Galindro et al. (2018) and Silva et al. (2019). The main interest comes from the advantages of ME estimation mentioned previously. Moreover, and as will be discussed later, in SFA with ME estimation, the composed error structure is used without distributional assumptions, and thus ME estimators appear to be a promising approach in efficiency analysis.

The SFA finds now extensive use in several areas of efficiency measurement and has been extended in several ways (Löthgren, 1997). However, as discussed in a previous chapter, these models are limited provided they are specified for scalar (single-output) dependent variables, thus limiting the applicability of the frontier models when dual approaches cannot be used (Löthgren, 1997).

When data is scarce, noisy, and limited, maximum entropy could play an important role in information extraction. Silva et al. (2019) proposed a stochastic frontier approach using generalized maximum entropy estimation and data envelopment analysis over cross-section data on thirteen European electricity distribution companies. Results point out that the same companies are found to be in the highest and lowest efficiency groups revealing a weak sensitivity to the prior information used with ME estimation. Previously, Moutinho et al. (2018b) estimated the agriculture economic-environmental efficiency for European countries. The agriculture gross value added over greenhouse gas emissions is the measure of eco-efficiency. Disparities in levels of eco-efficiency through the years have been pointed out. Robaina-Alves et al. (2015) specified a stochastic frontier model and used maximum entropy estimation to assess technical efficiency combining information from the data envelopment analysis to evaluate the resource and environmental efficiency problem of European countries. Technical efficiency in a wine region of Portugal is also investigated through this methodology by Macedo and Scotto (2014). For the energy sector, Antunes et al. (2022) explore whether energy performance-related inputs and outputs share mutual information to allow producers to learn simultaneously from each other's distributional behaviors.

In this work, the GME and the GCE estimators are reviewed and their implementation in the SFA context is discussed in Sect. 2. Some parts of this section represent an extension of the discussion started in Macedo (2019). Section 3 follows the empirical illustration and some concluding remarks are provided in Sect. 4.

2 Maximum Entropy Estimation

2.1 Generalized Maximum (Cross) Entropy Estimator

To briefly present the GME and GCE estimators, the linear regression model should be first introduced. As usually it is defined as

$$y = X\beta + e, \tag{2}$$

where $\boldsymbol{y}$ denotes a $(N \times 1)$ vector of noisy observations, $\boldsymbol{\beta}$ is a $(K \times 1)$ vector of unknown parameters to be estimated, $\boldsymbol{X}$ is a known $(N \times K)$ matrix of explanatory variables, and $\boldsymbol{e}$ is the $(N \times 1)$ vector of random errors.

Golan et al. (1996) presented the reformulation of the model in (2) as

$$y = XZp + Vw, \tag{3}$$

where

$$\beta = Zp = \begin{bmatrix} z_1' & 0 & \dots & 0 \\ 0 & z_2' & \dots & 0 \\ \vdots & \vdots & \ddots & \vdots \\ 0 & 0 & \dots & z_K' \end{bmatrix} \begin{bmatrix} p_1 \\ p_2 \\ \vdots \\ p_K \end{bmatrix} \tag{4}$$

and

$$e = Vw = \begin{bmatrix} v_1' & 0 & \dots & 0 \\ 0 & v_2' & \dots & 0 \\ \vdots & \vdots & \ddots & \vdots \\ 0 & 0 & \dots & v_N' \end{bmatrix} \begin{bmatrix} w_1 \\ w_2 \\ \vdots \\ w_N \end{bmatrix}, \tag{5}$$

with $\boldsymbol{Z}$ and $\boldsymbol{V}$ being a $(K \times KM)$ and $(N \times NJ)$ matrices of support spaces, respectively; $\boldsymbol{p}$ and $\boldsymbol{w}$ being a $(KM \times 1)$ and a $(NJ \times 1)$ vectors of unknown probabilities to be estimated; $M(\geq 2)$ and $J(\geq 2)$ being the number of points in the support spaces. Thus, subject to the model constraints,

$$\boldsymbol{y} = \boldsymbol{X}\boldsymbol{Z}\boldsymbol{p} + \boldsymbol{V}\boldsymbol{w}, \tag{6}$$

the additivity constraints for $\boldsymbol{p}$,

$$\mathbf{1}_K = \left(\boldsymbol{I}_K \otimes \mathbf{1}_M^{'}\right)\boldsymbol{p}, \tag{7}$$

and the additivity constraints for $\boldsymbol{w}$,

$$\mathbf{1}_N = \left(\boldsymbol{I}_N \otimes \mathbf{1}_J^{'}\right)\boldsymbol{w}, \tag{8}$$

the GME estimator can be expressed as

$$\text{argmax}_{\boldsymbol{p},\boldsymbol{w}}\left\{-\boldsymbol{p}'\ln\boldsymbol{p} - \boldsymbol{w}'\ln\boldsymbol{w}\right\}, \tag{9}$$

and the GCE estimator can be expressed as

$$\text{argmin}_{\boldsymbol{p},\boldsymbol{w}}\left\{\boldsymbol{p}'\ln\frac{\boldsymbol{p}}{\boldsymbol{q}_1} + \boldsymbol{w}'\ln\frac{\boldsymbol{w}}{\boldsymbol{q}_2}\right\}, \tag{10}$$

where $\otimes$ represents the Kronecker product, $\boldsymbol{q}_1$ and $\boldsymbol{q}_2$ are the vectors with prior information concerning the parameters and the errors of the model, respectively.

The GME and GCE estimators have no closed-form solutions, which means that numerical optimization techniques are required to solve these statistical problems, and then the estimated probability vectors can be used to obtain point estimates of the unknown parameters and the unknown errors. It is important to highlight that the GME estimator is a particular case of the GCE estimator when both vectors of prior information are non-informative (i.e., expressed through uniform distributions). A detailed matrix illustration of the GME estimator in a simple linear regression model is available in Macedo (2022).

Wu (2009) proposed a weighted generalized maximum entropy (W-GME) estimator with a data-driven weight. The objective function in (9) is then expressed as

$$\text{argmax}_{\boldsymbol{p},\boldsymbol{w}}\left\{-(1-\gamma)\boldsymbol{p}'\ln\boldsymbol{p} - \gamma\boldsymbol{w}'\ln\boldsymbol{w}\right\}, \tag{11}$$

where $\gamma \in (0, 1)$ defines the weights assigned to each entropy component, and it is selected through the minimization of the sum of the squared prediction errors with a least squares cross-validation scheme. The author compares the W-GME with the conventional GME and shows that the proposed W-GME estimator provides superior performance under different simulated scenarios. The same idea will be used next in the SFA context.

Additional details on maximum entropy estimation can be found, for example, in Golan et al. (1996), Jaynes (2003), and Golan (2018), and the references therein.

2.2 ME Estimation in SFA

Considering the general stochastic frontier model in (1) defined in matrix form as

$$\ln \boldsymbol{y} = f(\boldsymbol{X}, \boldsymbol{\beta}) + \boldsymbol{v} - \boldsymbol{u}, \tag{12}$$

the reparameterization of the $(K \times 1)$ vector $\boldsymbol{\beta}$ and the $(N \times 1)$ vector $\boldsymbol{v}$ follows the same procedures as in the traditional regression model with GME and GCE estimators discussed in Sect. 2.1, i.e., each parameter and each error are treated as discrete random variables with compacts supports, considering $M \geq 2$ and $J \geq 2$ possible outcomes, respectively. Thus, the reparameterizations are given by $\boldsymbol{\beta} = \boldsymbol{Z}\boldsymbol{p}$, where $\boldsymbol{Z}$ is a $(K \times KM)$ matrix of support points and $\boldsymbol{p}$ is a $(KM \times 1)$ vector of unknown probabilities to be estimated, and $\boldsymbol{v} = \boldsymbol{A}\boldsymbol{w}$, where $\boldsymbol{A}$ is a $(N \times NJ)$ matrix of support points and $\boldsymbol{w}$ is a $(NJ \times 1)$ vector of unknown probabilities to be estimated.

Now, and most importantly, the reparameterization of the $(N \times 1)$ vector $\boldsymbol{u}$ could be similar to the one conducted for the random variable representing noise, $\boldsymbol{v}$, although considering that $\boldsymbol{u}$ is a one-sided random variable, which implies that the lower bound for the supports (with $L \geq 2$ points) is zero for all error values. The reparameterization of $\boldsymbol{u}$ can then be defined by $\boldsymbol{u} = \boldsymbol{B}\boldsymbol{\rho}$, where $\boldsymbol{B}$ is a $(N \times NL)$ matrix of support points and $\boldsymbol{\rho}$ is a $(NL \times 1)$ vector of unknown probabilities to be estimated.

Thus, following a similar weighted structure of the W-GME proposed by Wu (2009) and the discussion started with Macedo (2019), the GME estimator can be expressed as

$$\operatorname{argmax}_{\boldsymbol{p},\boldsymbol{w},\boldsymbol{\rho}} \left\{ -(1-\theta)\boldsymbol{p}' \ln \boldsymbol{p} - \frac{\theta}{2}\boldsymbol{w}' \ln \boldsymbol{w} - \frac{\theta}{2}\boldsymbol{\rho}' \ln \boldsymbol{\rho} \right\}, \tag{13}$$

and the GCE estimator can be expressed as

$$\operatorname{argmin}_{\boldsymbol{p},\boldsymbol{w},\boldsymbol{\rho}} \left\{ (1-\theta)\boldsymbol{p}' \ln \frac{\boldsymbol{p}}{\boldsymbol{q}_1} + \frac{\theta}{2}\boldsymbol{w}' \ln \frac{\boldsymbol{w}}{\boldsymbol{q}_2} + \frac{\theta}{2}\boldsymbol{\rho}' \ln \frac{\boldsymbol{\rho}}{\boldsymbol{q}_3} \right\}, \tag{14}$$

both of them are subject to the same model and additivity constraints, namely

$$\begin{aligned} \ln \boldsymbol{y} &= \boldsymbol{X}\boldsymbol{Z}\boldsymbol{p} + \boldsymbol{A}\boldsymbol{w} - \boldsymbol{B}\boldsymbol{\rho}, \\ \boldsymbol{1}_K &= \left(\boldsymbol{I}_K \otimes \boldsymbol{1}_M'\right)\boldsymbol{p}, \\ \boldsymbol{1}_N &= \left(\boldsymbol{I}_N \otimes \boldsymbol{1}_J'\right)\boldsymbol{w}, \\ \boldsymbol{1}_N &= \left(\boldsymbol{I}_N \otimes \boldsymbol{1}_L'\right)\boldsymbol{\rho}, \end{aligned} \tag{15}$$

where $\otimes$ represents the Kronecker product, the vectors $\boldsymbol{q}_i$ $(i = 1, 2, 3)$ represent prior information and $\theta \in (0, 1)$ assigns different weights to the components of the objective function. By default, $\theta = 2/3$, although the value can be obtained by some kind of cross-validation through the minimization of a given loss function, as proposed by Wu (2009) with the W-GME.

The support matrices $\boldsymbol{Z}$ and $\boldsymbol{A}$ are defined by the researcher based on prior information as in the classical regression modeling. Some discussions and guidelines for the choice of these supports are provided, for example, in Golan et al. (1996), pp. 137–142, Preckel (2001), Caputo and Paris (2008), Henderson et al. (2015) and Golan (2018), pp. 262–264 and pp. 380–382. Macedo (2022) presents a summary of general principles, mainly motivated by empirical and simulation works: (1) the bounds and the center of the parameters' supports are problem specific and should be chosen with care. The center is usually defined at zero if no prior information exists. The bounds are usually defined based on theoretical constraints, based on information from previous works, based on confidence intervals with high confidence levels from other robust and stable estimation techniques; (2) the number of support points is usually between three and seven; (3) for the error component are usually considered three points in supports and the bounds are usually defined by the three-sigma rule, considering the standard deviation of the noisy observations; (4) the supports should be symmetric about zero and should be defined with equally spaced points between the lower and upper bounds.

In the SFA context, the definition of matrix $\boldsymbol{B}$ is the most concern. It is known that the traditional distributional assumptions concerning the error inefficiency component are chosen in empirical studies due to the expected behavior in the distribution of technical inefficiency predictions. For example, in the discussion of the normal—half normal model, a popular one in empirical work, Kumbhakar and Knox Lovell (2000, p. 74) argued that the assumption of a nonnegative half normal distribution "is based on the plausible proposition that the modal value of technical inefficiency is zero, with increasing values of technical inefficiency becoming increasingly less likely." Thus, following this reasoning, using GME or GCE estimation, the same beliefs can be expressed in the models through the error supports (in GME) or through the vectors with prior information (in GCE).

Campbell et al. (2008) suggested the use of the mean of the data envelopment analysis (DEA) and SFA efficiency predictions to define the supports with a specific upper bound (ub) defined as

$$\boldsymbol{b}_n^{'} = [0, 0.005, 0.01, 0.015, \text{ub}], \tag{16}$$

and Macedo et al. (2014) suggested supports defined as

$$\boldsymbol{b}_n^{'} = [0, 0.01, 0.02, 0.03, -\ln(\text{DEA}_n)], \tag{17}$$

where DEA_n can represent the lower prediction of technical efficiency obtained by DEA in the N observations of the sample, considering the analysis from different DEA specifications (constant returns to scale, variable returns to scale, etc.). The definition of this prior information undoubtedly deserves future research, but some preliminary results in Moutinho et al. (2018b) suggest, for a specific empirical scenario, that the rankings of efficiency predictions are not very sensitive to the definition of these supports.

Regarding the GCE estimator, the vector $\boldsymbol{q}_3$ can follow a general structure such as

$$\boldsymbol{q}_3 = [0.70, 0.20, 0.05, 0.03, 0.02]', \tag{18}$$

for each observation, where the cross entropy objective shrinks the *posterior* distribution to have more mass near zero. The elements of $\boldsymbol{q}_3$ must be defined in descending order and their sum must be one. The supports in matrix $\boldsymbol{B}$ can be defined with five equally spaced points in the interval $[0, -\ln(\text{DEA}_n)]$; see Macedo and Scotto (2014) for further details.

The choice of the three central values in (16) or (17), that define the prior mean and the skewness within the GME estimator, or the vector $\boldsymbol{q}_3$ within the GCE estimator in (18), represents the main concerns with ME estimation in SFA, even if there are some possibilities to obtain prior information (normalized entropy; information from the residuals of the OLS (GME or GCE) estimation; moment generating function of the truncated normal distribution (Meesters, 2014), or of the skew normal distribution (Chen et al., 2014)). Despite these concerns, as mentioned by Rezek et al. (2011), p. 364, the selection "of these vectors sets a prior expectation of mean efficiency; however, it does not preordain that result.", and since incorrect prior information does not constrain the solution if it is not consistent with the data (Golan et al., 1996), it is expected that the GCE estimator remains stable when the vector $\boldsymbol{q}_3$ comprises possible incorrect prior information. Finally, it is important to note that with ME estimation the DEA is used only to define an upper bound for the supports, which means that the main criticism of DEA (all deviations from the production frontier are considered as technical inefficiency) is used as an advantage, and the main criticism on SFA is partially overwhelmed because the composed error structure is used without specific statistical distributional assumptions.

3 An Application to Eco-Efficiency Analysis of European Countries

An upgrade of the empirical study performed by Robaina-Alves et al. (2015) is now used to illustrate SFA with GCE estimation. Gross Domestic Product (GDP) is considered the desirable output and Greenhouse Gas (GHG) emissions as the undesirable output. Fossil fuel consumption, Renewable Energy Consumption, Capital, and Labor are regarded as the inputs. The GDP/GHG ratio is the output considered in the stochastic frontier model assuming a log-linear Cobb–Douglas production function, thus being our measure of eco-efficiency. We should bear in mind that although economic efficiency does not imply environmental efficiency (Robaina-Alves et al., 2015), in the presence of economic inefficiency we might expect environmental inefficiency.

Nowadays, economic growth should rely on the decoupling concept, namely, increasing economic growth and simultaneously decreasing the use of resources and GHG emissions (Haberl et al., 2020). Inefficient use of energy resources will increase GHG emissions (Robaina-Alves et al., 2015; Sueyoshi et al., 2017), turning mandatory to reach energy efficiency to increase the odds of having economic efficiency, although not implying environmental efficiency mandatorily. The concept of economic efficiency was aroused by Farrell (1957), influenced by Koopmans's (1951) formal definition and Debreu's (1951) measure of technical efficiency. Overall efficiency may be decomposed into its technical and allocative components (Murillo-Zamorano, 2004). Farrell characterized the different ways in which a productive unit can be inefficient either by obtaining less than the maximum output available from a determined group of inputs (technically inefficient; Knox Lovell, 1993) or by not purchasing the best package of inputs given their prices and marginal productivities (allocatively inefficient).

To allow possible comparisons with the results from Robaina-Alves et al. (2015), the same general procedures are followed here for the GCE estimator, namely: the vector $\boldsymbol{q}_3$ is defined as in (18); the supports in matrix $\boldsymbol{Z}$ are defined with five equally spaced points in the interval $[-15, 15]$; the supports in matrix $\boldsymbol{A}$ are defined symmetrically and centered on zero using the three-sigma rule with the empirical standard deviation of the noisy observations; and the supports in matrix $\boldsymbol{B}$ are defined with five equally spaced points in the interval $[0, -\ln(\text{DEA}_n)]$.

Two years before the COVID-19 pandemic (2018 and 2019) and the first year of the pandemic (2020) are considered in the analysis. This division will allow us to evaluate the difference between the levels of eco-efficiency before and at the beginning of the pandemic. Eco-efficiency predictions through SFA with GCE estimation are presented in Table 1 (and Fig. 1) and the corresponding rankings of eco-efficiency predictions (ordered from highest to smallest) are presented in Table 2. Finally, and for comparison purposes, Table 3 compares the rankings and the eco-efficiency predictions (ordered from highest to smallest) through SFA with GCE and ML (normal—truncated normal model) estimation for 2020. The SFA with GCE estimator is computed in MATLAB (code adapted from Macedo (2017) that is available on-line) and SFA with ML is obtained through the "frontier" package in R (Coelli & Henningsen, 2013).

It is interesting to note that although the eco-efficiency predictions are different (as expected), the rankings established by GCE and ML estimation are very similar (Table 3). To help reading results, the closer the value of eco-efficiency is to one, the more efficient the country is, which indicates that it uses in the best way the available resources (our considered inputs) to produce the maximum possible output (GDP) while minimizing environmental impacts (reducing GHG emissions). Table 1, Fig. 1, and Table 2 indicate that the best-positioned country in terms of eco-efficiency is Sweden in all years (2018, 2019, and 2020), whereas the worst in terms of eco-efficiency has been Estonia in 2018, 2019, and 2020. Therefore, the one which was

Table 1 Eco-efficiency predictions through SFA with GCE estimation

	2018	2019	2020
Belgium	0.7035	0.7121	0.6941
Bulgaria	0.7146	0.6648	0.7026
Czech Republic	0.6672	0.6787	0.6740
Denmark	0.7905	0.7963	0.7978
Germany	0.7212	0.7332	0.7302
Estonia	0.5218	0.6041	0.5630
Ireland	0.6864	0.6186	0.6504
Greece	0.8273	0.8030	0.8066
Spain	0.7305	0.7325	0.7411
France	0.7783	0.8047	0.8073
Italy	0.7667	0.7802	0.7790
Cyprus	0.7719	0.7875	0.7772
Latvia	0.8246	0.8080	0.8043
Lithuania	0.8184	0.7968	0.7837
Luxembourg	0.8299	0.8346	0.8461
Hungary	0.7765	0.7656	0.7525
Netherlands	0.7286	0.7430	0.7275
Austria	0.7740	0.7754	0.7798
Poland	0.7337	0.7308	0.7224
Portugal	0.7074	0.6817	0.7137
Romania	0.7879	0.7815	0.7656
Slovenia	0.8240	0.8417	0.8431
Slovakia	0.8129	0.8261	0.8312
Finland	0.6934	0.6877	0.7007
Sweden	0.8765	0.8767	0.8714

best positioned before the COVID-19 pandemic was still in the highest position in 2020. The worst was still the worst score in all years as well, even if we noticed a slight improvement in eco-efficiency in 2019 (meaning it was recovering), to decrease again at the beginning of the emergence of the COVID-19 pandemic.

In all years rankings and eco-efficiency predictions (ordered from highest to smallest) through SFA with GCE and ML estimation evidenced higher values in terms of eco-efficiency score using the ML estimation (normal—truncated normal model). As previously noticed, the highest score and the lowest scored country do not change, but positions in the overall ranking and for all years changed, although with very small differences in positions (see Table 3 for an example in the year 2020). Regarding the rankings and answering to the question "Do Distributional Assumptions Matter?", Kumbhakar and Knox Lovell (2000, p. 90) noted that "Sample mean

Fig. 1 Eco-efficiency predictions through SFA with GCE estimation

efficiencies are no doubt apt to be sensitive to the distribution assigned to the one-sided error component (…) What is not so clear is whether a ranking of producers by their individual efficiency scores, or the composition of the top and bottom efficiency score deciles, is sensitive to distributional assumptions. Indeed there is some evidence that neither rankings nor decile compositions are particularly sensitive." Although further research is still needed on this topic, some results (e.g., Moutinho et al., 2018b) suggest a weak sensitivity of the rankings to different supports that could be defined under maximum entropy estimation.

Additionally, we should bear in mind that the eco-efficiency indicator provides an overall outcome for the economic and environmental efficiency of the joint use of production factors, the inputs considered in the present study. It was not our goal here to verify which of these have contributed to the improvement or not in the overall ranking of eco-efficiency as in Robaina-Alves et al. (2015).

Table 2 Rankings of eco-efficiency predictions (ordered from highest to smallest) through SFA with GCE estimation

2018	2019	2020
Sweden	Sweden	Sweden
Luxembourg	Slovenia	Luxembourg
Greece	Luxembourg	Slovenia
Latvia	Slovakia	Slovakia
Slovenia	Latvia	France
Lithuania	France	Greece
Slovakia	Greece	Latvia
Denmark	Lithuania	Denmark
Romania	Denmark	Lithuania
France	Cyprus	Austria
Hungary	Romania	Italy
Austria	Italy	Cyprus
Cyprus	Austria	Romania
Italy	Hungary	Hungary
Poland	Netherlands	Spain
Spain	Germany	Germany
Netherlands	Spain	Netherlands
Germany	Poland	Poland
Bulgaria	Belgium	Portugal
Portugal	Finland	Bulgaria
Belgium	Portugal	Finland
Finland	Czech Republic	Belgium
Ireland	Bulgaria	Czech Republic
Czech Republic	Ireland	Ireland
Estonia	Estonia	Estonia

4 Concluding Remarks

Although there are some open questions, namely the choice of supports for the inefficiency error component, given the usual characteristics of stochastic frontier models and the advantages of maximum entropy estimation, the use of these information-theoretical methods seems to be promising in efficiency analysis. The illustration conducted in this work to obtain eco-efficiency predictions of European countries strengthens this idea, namely when the results from GCE estimation are compared with the ones from the well-known ML estimator, because although the eco-efficiency predictions are different, the rankings established by GCE and ML estimation are very similar. From the empirical application performed in this chapter, it was possible to understand that before and during the COVID-19 pandemic (2018–2020) the highest eco-efficient country was Sweden and the lowest eco-efficient country was Estonia,

Table 3 Rankings and eco-efficiency predictions (ordered from highest to smallest) through SFA with GCE and ML estimation in 2020

ML		GCE	
Sweden	0.9353	Sweden	0.8714
Luxembourg	0.9018	Luxembourg	0.8461
Slovenia	0.8957	Slovenia	0.8431
Slovakia	0.8922	Slovakia	0.8312
Greece	0.8905	France	0.8073
France	0.8856	Greece	0.8066
Denmark	0.8808	Latvia	0.8043
Italy	0.8636	Denmark	0.7978
Latvia	0.8612	Lithuania	0.7837
Austria	0.8484	Austria	0.7798
Lithuania	0.8447	Italy	0.7790
Romania	0.8279	Cyprus	0.7772
Spain	0.8269	Romania	0.7656
Germany	0.8195	Hungary	0.7525
Cyprus	0.8187	Spain	0.7411
Hungary	0.8053	Germany	0.7302
Netherlands	0.7929	Netherlands	0.7275
Poland	0.7846	Poland	0.7224
Portugal	0.7832	Portugal	0.7137
Finland	0.7621	Bulgaria	0.7026
Bulgaria	0.7460	Finland	0.7007
Belgium	0.7428	Belgium	0.6941
Czech Republic	0.6931	Czech Republic	0.6740
Ireland	0.6544	Ireland	0.6504
Estonia	0.5555	Estonia	0.5630

and this happened for all years, with only slight improvements and decreases noticed for the rest of the countries.

Acknowledgements This work is supported by the Center for Research and Development in Mathematics and Applications (CIDMA), the Research Unit on Governance, Competitiveness and Public Policies (GOVCOPP), and the Research Unit in Business Science and Economics (NECE-UBI) through the Portuguese Foundation for Science and Technology (FCT—Fundação para a Ciência e a Tecnologia), references UIDB/04106/2020, UIDB/04058/2020 and UID/GES/04630/2021, respectively.

References

Aigner, D., Knox Lovell, C., & Schmidt, P. (1977). Formulation and estimation of stochastic frontier production function models. *Journal of Econometrics, 6*(1), 21–37.

Antunes, J., Gupta, R., Mukherjee, Z., & Wanke, P. (2022). Information entropy, continuous improvement, and US energy performance: A novel stochastic-entropic analysis for ideal solutions (SEA-IS). *Annals of Operations Research, 313*, 289–318.

Battese, G., & Corra, G. (1977). Estimation of a production frontier model: With application to the pastoral zone of Eastern Australia. *The Australian Journal of Agricultural and Resource Economics, 21*(3), 169–179.

Campbell, R., Rogers, K., & Rezek, J. (2008). Efficient frontier estimation: A maximum entropy approach. *Journal of Productivity Analysis, 30*, 213–221.

Caputo, M. R., & Paris, Q. (2008). Comparative statics of the generalized maximum entropy estimator of the general linear model. *European Journal of Operational Research, 185*, 195–203.

Chen, Y.-Y., Schmidt, P., & Wang, H.-J. (2014). Consistent estimation of the fixed effects stochastic frontier model. *Journal of Econometrics, 181*(2), 65–76.

Coelli, T., & Henningsen, A. (2013). Frontier: Stochastic frontier analysis. R package version 1.1. http://CRAN.R-Project.org/package=frontier

Debreu, G. (1951). The Coefficient of Resource Utilization. *Econometrica, 19*(3), 273–292.

Farrell, M. J. (1957). The measurement of productive efficiency. *Journal of the Royal Statistical Society* (A, general), *120*, 253–281.

Galindro, A., Santos, M., Santos, C., Marta-Costa, A., Matias, J., & Cerveira, A. (2018). Wine productivity per farm size: A maximum entropy application. *Wine Economics and Policy, 7*(1), 77–84.

Golan, A. (2018). *Foundations of info-metrics: Modeling, inference, and imperfect information.* Oxford University Press.

Golan, A., Judge, G., & Miller, D. (1996). *Maximum entropy econometrics: Robust estimation with limited data.* Chichester: Wiley.

Haberl, H., Wiedenhofer, D., Virág, D., Kalt, G., Plank, B., Brockway, P., Fishman, T., Hausknost, D., Krausmann, F., Leon-Gruchalski, B., Mayer, A., Pichler, M., Schaffartzik, A., Sousa, T., Streeck, J., & Creutzig, F. (2020). A systematic review of the evidence on decoupling of GDP, resource use and GHG emissions, part II: Synthesizing the insights. *Environmental Research Letters, 15*(6), 065003.

Henderson, H., Golan, A., & Seabold, S. (2015). Incorporating prior information when true priors are unknown: An Information-Theoretic approach for increasing efficiency in estimation. *Economics Letters, 127*, 1–5.

Jaynes, E. T. (1957a). Information theory and statistical mechanics. *Physical Review, 106*(4), 620-630.

Jaynes, E. T. (1957b). Information theory and statistical mechanics. II. *Physical Review, 108*(2), 171-190.

Jaynes, E. T. (2003). *Probability theory - the logic of science.* Cambridge University Press.

Knox Lovell, C. (1993). Production frontiers and productive efficiency. In Fried, H. O., Knox Lovell, C., Schmidt, S. S. (Eds.), *The measurement of productive efficiency: Techniques and applications* (pp. 3–67). Oxford: Oxford University Press.

Koopmans, T. C. (1951). An analysis of production as efficient combination of activities. In Koopmans, T. C. (Eds,), *Activity analysis of production and allocation, cowles commission for research in economics.* Monograph no. 13. New York.

Kumbhakar, S. C., & Knox Lovell, C. (2000). *Stochastic frontier analysis.* Cambridge University Press.

Löthgren, M. (1997). Generalized stochastic frontier production models. *Economics Letters, 57*(3), 255–259.

Macedo, P. (2017). Ridge regression and generalized maximum entropy: An improved version of the ridge–GME parameter estimator. *Communications in Statistics - Simulation and Computation, 46*(5), 3527–3539.

Macedo, P. (2022). A two-stage maximum entropy approach for time series regression. *Communications in Statistics - Simulation and Computation*. https://doi.org/10.1080/03610918.2022.2057540

Macedo, P., & Scotto, M. (2014). Cross-entropy estimation in technical efficiency analysis. *Journal of Mathematical Economics, 54*, 124–130.

Macedo, P., Silva, E., & Scotto, M. (2014). Technical efficiency with state-contingent production frontiers using maximum entropy estimators. *Journal of Productivity Analysis, 41*(1), 131–140.

Macedo, P. (2019). A note on the estimation of stochastic and deterministic production frontiers with maximum entropy. In *Proceedings of the 26th APDR congress, July 4–5*. Portugal: University of Aveiro.

Meesters, A. (2014). A note on the assumed distributions in stochastic frontier models. *Journal of Productivity Analysis, 42*(2), 171–173.

Meeusen, W., & van Den Broeck, J. (1977). Efficiency estimation from Cobb-Douglas production functions with composed error. *International Economic Review, 18*(2), 435–444.

Moutinho, V., Madaleno, M., Macedo, P., Robaina, M., & Marques, C. (2018a). Efficiency in the European agricultural sector: Environment and resources. *Environmental Science and Pollution Research, 25*(18), 17927–17941.

Moutinho, V., Robaina, M., & Macedo, P. (2018b). Economic-environmental efficiency of European agriculture – a generalized maximum entropy approach. *Agricultural Economics – Czech, 64*(10), 423–435.

Murillo-Zamorano, L. R. (2004). Economic efficiency and frontier techniques. *Journal of Economic Surveys, 18*(1), 33–77.

Preckel, P. V. (2001). Least squares and entropy: A penalty function perspective. *American Journal of Agricultural Economics, 83*(2), 366–377.

Rezek, J., Campbell, R., & Rogers, K. (2011). Assessing total factor productivity growth in Sub-Saharan African agriculture. *Journal of Agricultural Economics, 62*(2), 357–374.

Robaina-Alves, M., Moutinho, V., & Macedo, P. (2015). A new frontier approach to model the eco-efficiency in European countries. *Journal of Cleaner Production, 103*, 562–573.

Silva, E., Macedo, P., & Soares, I. (2019). Maximum entropy: A stochastic frontier approach for electricity distribution regulation. *Journal of Regulatory Economics, 55*(3), 237–257.

Sueyoshi, T., Yuan, Y., & Goto, M. (2017). A literature study for DEA applied to energy and environment. *Energy Economics, 62*, 104–124.

Wu, X. (2009). A weighted generalized maximum entropy estimator with a data-driven weight. *Entropy, 11*, 917–930.

www.ingramcontent.com/pod-product-compliance
Lightning Source LLC
Chambersburg PA
CBHW071335240425
25656CB00006B/193
9783031295829